Oceans of
Kansas

LIFE OF THE PAST
James O. Farlow, Editor

INDIANA UNIVERSITY PRESS
Bloomington and Indianapolis

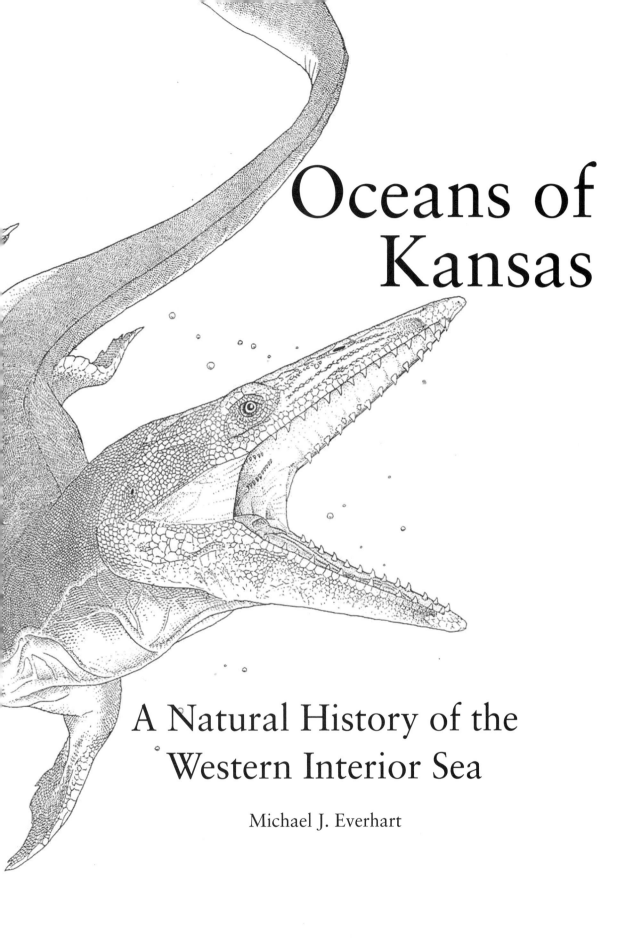

Oceans of
Kansas

A Natural History of the
Western Interior Sea

Michael J. Everhart

This book is a publication of
Indiana University Press
601 North Morton Street
Bloomington, IN 47404-3797 USA

http://iupress.indiana.edu

Telephone orders 800-842-6796
Fax orders 812-855-7931
Orders by e-mail iuporder@indiana.edu

The paper used in this publication meets the
minimum requirements of American National
Standard for Information Sciences—Permanence
of Paper for Printed Library Materials, ANSI
Z39.48-1984.

Manufactured in the United States of America
All photographs by Michael J. Everhart
Library of Congress Cataloging-in-Publication Data

Everhart, Michael J.
  Oceans of Kansas : a natural history of the
western interior sea / Michael J. Everhart.
    p. cm. — (Life of the past)
  Includes bibliographical references and index.
  ISBN 0–253–34547–2 (cloth : alk. paper)
  1. Marine animals, Fossil—Kansas. 2.
Paleontology—Cretaceous. I. Title. II. Series.
  QE766.E89 2005
  560'.457'09781—dc22
                                        2004025824

1 2 3 4 5 10 09 08 07 06 05

Dedicated to my parents, Jack McKay Everhart (1922–2000) and Betty Lou Everhart (1923–1994)

"But the reader inquires, What is the nature of these creatures thus left stranded a thousand miles from either ocean? How came they in the limestones of Kansas, and were they denizens of land or sea?"

Cope, E. D., 1872

CONTENTS

Color plates follow page 128.

Although I am almost a native Kansan and proud of my state, I have to admit that the drive across Kansas on Interstate 70 is not a major scenic experience if you are expecting mountains or other dramatic landscapes. While there are low hills and river valleys to be crossed along the way, the most visible change from east to west is going from a moderate number of trees to almost no trees. That being said, Kansas has many charms that are well hidden from those who are traveling as fast as they can to get across the state, and even from those who have lived here their entire lives. To me, as a paleontologist, that means a wealth of rock exposures that faithfully reveal a fossil record compiled over millions of years when Kansas was covered by a succession of Paleozoic and Mesozoic oceans. Note that I use the word "oceans" here in the broadest sense, since these bodies of water covered portions of a submerged continent (North America) and are more properly called "seas."

I have been interested in fossils for about as long as I can remember. In grade school I had the usual curiosity about fossil shells and crinoids that were found in the limestone rocks to the east of where I lived. Growing up south of Wichita, Kansas, I spent quite a bit of time exploring along the banks of the Arkansas River, other streams, and the spoil piles of local sandpits. Occasionally I would find the teeth or bones of Pleistocene mammals. I think that was when I realized that they represented the remains of extinct animals that lived long ago in a very different Kansas. Somewhere along the way, I saw my first shark teeth and fish bones from the chalk of western Kansas. A field trip during a vertebrate paleontology course in college provided me with my first experience collecting fossils in the Smoky Hill Chalk, and I have been hooked on that particular time span ever since.

Kansas has a wide variety of fossils, from very old Mississippian rocks (more than 340 million years old) in the extreme southeastern corner of the state to the Late Cretaceous rocks (about 75 million years old) of northwestern Kansas, that were deposited as bottom muds in the series of Paleozoic and Mesozoic oceans that covered Kansas. More recently deposited Tertiary sediments (non-marine) containing the remains of extinct terrestrial animals are found along and in streams and rivers statewide. That means you can find fossils just about anywhere in the state. If you are interesting in collecting fossils in Kansas, the first thing to decide is what kind of fossils you want to collect. The best place to start, in that regard, is a library or bookstore. I would recommend two books as "must have" references for amateur fossil hunters in Kansas: *Kansas Geology*, edited by Rex Buchanan (1984), and *Roadside Kansas*, by R. C. Buchanan and J. R. McCauley (1987). Once you have an idea of what kind of fossils you want to find and where to look for them, you're ready to get serious about it. There are many places where rocks are exposed and accessible to collectors. The most important thing to remember, however, is that most fossils in Kansas are on private property. You must have permission from the owner to go on the land and to collect. Always respect the property of others, take proper safety precautions, never leave your trash behind, and, most of all, have fun.

The intent of this book is to provide information about many of the animals that lived during the Late Cretaceous and, to some extent, the people who discovered their fossil remains and described them. There are a number of excellent sources of information available in print and on the Internet regarding other kinds of fossils that can be found in Kansas, and the reader is encouraged to spend some time learning about paleontology in general.

Acknowledgments

There are so many people who, over the years, have helped me learn and understand the paleontology of Kansas during the Late Cretaceous that it is difficult to know where to begin in expressing my appreciation. You've all heard the adage that it takes a village to raise a child. I can certainly testify that it also takes one to write a book.

First, I thank my wife, Pamela Everhart, for her support and companionship in the field (not to mention her superior ability to find interesting things for me to dig up). She was also properly impressed when I brought a mammoth tooth to class while we were in high school. This book and a lot of other projects in paleontology would not have been possible for me without her. The rest of the list follows in no particular order. My fifth grade teacher, Vivian Louthan, encouraged my interest in "rocks" and took me to a gem and mineral show that left a lasting impression. John Ransom, Harry Rounds, and Don Distler, among many other teachers, guided my interest in the study of living things and the remains of things that lived millions of years ago. Paul Tasch introduced me to the Smoky Hill Chalk on a vertebrate paleontology field trip in 1968, and I immediately was hooked on it. David Parris, Barbara Grandstaff, and J. D. Stewart were all supportive as well as being excellent teachers and resources of otherwise unknown and mysterious information when we were getting started in our serious study of the Smoky Hill Chalk in the late 1980s. Although he may not remember it, David Parris sponsored my membership in the Society of Vertebrate Paleontology (SVP) many years ago. I owe a major debt of thanks to J. D. Stewart and Donald Hattin for our continuing discussion of the stratigraphy of the Smoky Hill Chalk. Pete Bussen has been a source of valuable information on history, paleontology, weather, and a variety of other useful subjects, which he gained in pursuit of his "Doctor of Disagree-ology" degree. I have found that Pete has a learned opinion on just about everything, whether I asked for it or not. A number of landowners in western Kansas, including the Albins, Birds, Babcocks, Bentleys, Bodeckers, Bonners, Cheneys, Collinses, Millers, and Surratts have generously allowed us access to their property over the years and obviously made many of our discoveries possible. Richard Zakrzewski, Larry Martin, and Ken Carpenter made the collections in their charge available for study and provided guidance and answers to even more questions. Their friendship has always been appreciated. Richard Zakrzewski assisted in getting my appointment as an Adjunct Curator of Paleontology at the Sternberg Museum. Dale Russell, Gorden Bell, Jim Martin, Bruce Schumacher, Kenshu Shimada, David Schwimmer, Glenn Storrs, Takehito Ikejiri and David Cicimurri indulged me in discussions of a variety of subjects, but mostly mosasaurs, plesiosaurs, and sharks. Earl Manning, in particular, has been a wonderful resource of useful information on identification of specimens, reference material, historical issues in paleontology, and general all-around common sense. At one time or another, Bob Purdy, Michael Brett-Surman, Earle Spamer, Ted Daeschler, Charles Schaff, Desui Maio, Michael Morales, George Corner, Greg Liggett, and many others have helped me track down fossils that originated in Kansas. Dan Varner, Doug Henderson, Russell Hawley, and other artists have provided me with their enlightened reconstructions of a lost world I can only dream about. Steve Johnson's questions regarding a few mosasaur vertebrae led us both to a once-in-a-lifetime discovery. The time spent in the field with Tom Caggiano, Steve Balliett, and Andy Abdul has been as educational for me as I hope it has been for them. I will always "owe" Tom Caggiano for his discovery of the missing dinosaur foot. Shawn Hamm and Keith Ewell dragged me "kicking and screaming" into the study of fossil sharks from other formations in Kansas. A number of anonymous reviewers and some who aren't anonymous have made me aware (tactfully, of course) of my shortcomings as a writer and paleontologist—their assistance has been invaluable to me and I hope they see that I learned something in the process. Access to the Internet has allowed me to meet and converse with co-workers in distant places that I will probably never meet in person. Nevertheless, I count them as good friends and thank them for their help over the years. A number of library people from all over the world, including Angie from Kansas and Steve from New Zealand, have helped me to accumulate necessary paleontology reference materials, some of which are extremely rare or difficult to find.

And in closing, I certainly appreciate all those people from around the world who have communicated with me in English because I know I would have never been able to return the favor in their language. While this turned out to be a long list, I'm sure I've unintentionally left out several people, so please forgive the oversight. As I said earlier, it takes a village . . .

AMNH	American Museum of Natural History, New York, New York
ANSP	Academy of Natural Sciences of Philadelphia, Philadelphia, Pennsylvania
CMC	Cincinnati Museum Center, Cincinnati, Ohio
CMNH	Carnegie Museum of Natural History, Pittsburgh, Pennsylvania
DMNH/DMNS	Denver Museum of Nature and Science, Denver, Colorado
ESU	Emporia State University Geology Museum, Emporia, Kansas
FFHM	Fick Fossil and History Museum, Oakley, Kansas
FHSM	Sternberg Museum of Natural History, Fort Hays State University, Hays, Kansas
FMNH	The Field Museum, Chicago, Illinois
KU	University of Kansas Invertebrate Paleontology Collection, Lawrence, Kansas
KUVP	University of Kansas Vertebrate Paleontology Collection, Lawrence, Kansas
LACMNH	Los Angeles County Museum of Natural History, Los Angeles, California
MCZ	Museum of Comparative Zoology, Harvard, Cambridge, Massachusetts
NAMAL	North American Museum of Ancient Life, Lehi, Utah
NJSM	New Jersey State Museum, Trenton, New Jersey

ROM	Royal Ontario Museum, Toronto, Ontario, Canada
SDSMT	South Dakota School of Mines and Technology, Rapid City, South Dakota
UCM	University of Colorado Museum, Boulder, Colorado
UNO	University of New Orleans, New Orleans, Louisiana
UPI	Uppsala Paleontological Institute, Uppsala University, Sweden
UNSM	University of Nebraska State Museum, Lincoln, Nebraska
USNM	United States National Museum, Washington, D.C.
YPM	Yale Peabody Museum, New Haven, Connecticut

Oceans of
Kansas

One

Introduction:
An Ocean in Kansas?

One Day in the Life of a Mosasaur

The bright midday sun glinted off the calm waters of the Inland Sea and silhouetted the long, sinuous form of a huge mosasaur lying motionless amid the floating tangle of yellow-green seaweed. Twenty years old and more than thirty feet in length, the adult mosasaur was almost full-grown and was much larger than any of the fish or sharks that lived in the shallow seaway. A swift and powerful swimmer over short distances, the mosasaur used surprise and the thrust of his muscular tail to outrun his prey with a short burst of speed. His jaws were more than four feet long and were lined with sharp, conical teeth that he used to seize and hold his prey. Several unusual adaptations in his lower jaws allowed them to flex in the middle and enabled him to easily swallow the large fish and other animals he caught. This ability was essential to the mosasaur because he had to hold on to his prey with his teeth or risk losing it. If he let go of his prey in the middle of the ocean, there was a good chance it would sink to the bottom and be lost, or be grabbed by a hungry shark.

He was floating at the surface with his eyes and nostrils just above the water. His dark upper body absorbed hot rays of the Late Cretaceous sun as dozens of tiny fish emerged from hiding in the seaweed and darted cautiously around his submerged bulk. They were feeding on parasites and other small invertebrates that had attached themselves to his scaly hide. He breathed slowly and quietly through his nostrils as his ears and other senses remained on the alert for the telltale sounds made by approaching prey. He was a patient hunter, preferring to let his victims come to him instead of wasting energy swimming around the vast seaway in search of food.

Overhead, winged reptiles of various sizes floated lazily through the cloudless sky, riding the wind currents above the warm water while looking for schools of small fish feeding near the surface. Occasionally, one would skim the surface of the water and grab an unwary fish with its narrow beak. The mosasaur had recently tried to eat the floating carcass of a dead pteranodon, but found the thin wings difficult to get into his mouth. After tearing off the small body, he had let the rest of the animal sink to the bottom. The living ones overhead could see him clearly from above and avoided feeding near him.

Amid an ever-changing mixture of background noises made by a variety of creatures in the ocean, he noticed that a faint buzz of clicking sounds was getting louder, alerting him to a group of hard-shelled ammonites that was feeding nearby. Though not his favorite prey, they were all that had approached him since he had taken a large, solitary fish early in the morning. Even with that recent meal, his appetite was still unsatisfied and hunger was beginning to gnaw at him. The adaptations that made it possible for mosasaurs to return to the sea included an increased metabolism and a requirement for larger amounts of food to support a more active lifestyle.

Exhaling most of the air from his lungs, he slowly submerged his head, leaving behind only the faintest of ripples. His large eyes immediately found the brightly colored, coiled shells of the ammonites as they approached, bobbing and darting below him. Propelled by water forced through their internal siphons, they moved generally backwards through the water, with their short tentacles trailing behind them. Instinctively, he knew that their large shells would hide him from their view until they had moved well past him. He would make his attack from above, long before they had a chance to sense the danger.

Using his four large paddles, the mosasaur carefully maneuvered his long, snakelike form into an attack position, watching intently for any indication that the ammonites had detected the danger from above. Singling out a slightly larger ammonite at the edge of the group, he dived downward with a powerful slash of his long, broad tail. The ammonites reacted quickly and instinctively to the disturbance, scattering in all directions below him, but not before his heavy jaws closed across the soft forebody of his victim. His sharp teeth shattered the front edge of the ammonite's shell, destroying the buoyancy of the shell and rendering the ammonite helpless.

Without his captive pocket of air, the ammonite would sink swiftly to the bottom of the seaway. With practiced ease, the mosasaur flexed his body upward and raised the ammonite toward the surface. Then he released it and grabbed the tentacles of the immobilized creature with his teeth as it began to sink. Far too late, the ammonite released a cloud of jet-black ink into the water. The mosasaur ignored the bitter taste of the ammonite's last defense as he gave a quick jerk of his head to pull the ammonite's soft body from its shell. The heavy shell and several fragments slipped sideways through the water and quickly disappeared into the murky depths. Opening and closing his jaws rapidly, he swallowed the fleshy morsel in a single gulp.

Looking around for more prey, he saw another ammonite swimming in confused circles nearby. A swift lunge and his sharp teeth crunched through the ammonite's hard shell. Moments later, the soft body of the second ammonite followed the first into the mosasaur's stomach. The rest of the ammonites had jetted away as fast as they could and were no longer in view. His hunger briefly satisfied, he rose slowly to the surface to breathe and resume his ambush position rather than chasing after the fleeing shellfish.

He had hardly settled into waiting when he sensed the noises made by the approach of another mosasaur. Female mosasaurs tended to band together in pods for the protection of their young, while the males were solitary and very territorial. The approaching mosasaur was probably a young male searching for his own place in the expanse of the Inland Sea. In one swift, fluid motion, the older mosasaur turned and began to swim rapidly toward the sound of the approaching intruder. With his flippers held tightly against his body, the mosasaur moved quickly through the water just beneath the surface. His tail broke through the water's surface repeatedly as he intentionally made as much noise as possible. He wanted to sound as threatening as possible to the other mosasaur. Although he was prepared to fight for his territory, he would first try to frighten other males away with his size and ferocity. Long-healed scars on his body showed that even the winners in such fights could be badly hurt. He had been lucky several times earlier in his life and had survived injuries that easily could have been fatal. As he had gotten older, he had learned to avoid such battles whenever possible.

His course intercepted the other mosasaur broadside in a patch of open water. Turning quickly to face the threat, the smaller animal displayed a mouth filled with sharp teeth. Despite being nearly ten feet shorter and much less massive, the invader refused to turn and flee. The big mosasaur circled warily around his now stationary foe, watching intently as the other animal almost doubled back upon itself as it continued to show its open jaws. Trying to appear as threatening as possible, the younger animal still refused to turn and run.

The larger mosasaur was in no mood for such tactics. Making a large splash with his tail to distract the intruder, he surged forward and seized the smaller animal across the throat and back of the head. For a moment, the smaller mosasaur struggled helplessly as the powerful grip of the larger animal threatened to crush his skull. Then the larger mosasaur moved his head quickly, snapping the other mosasaur's neck. The smaller mosasaur gave a brief shudder, then went limp. Angrily, the big mosasaur shook the slender body again, making certain that his foe was no longer a threat.

Realizing that his victim was too large for him to swallow, the mosasaur released his grip and moved away. The body of the dead mosasaur rose slowly toward the surface and floated there until most of the remaining air had escaped from its lungs. Then it began to sink headfirst toward the bottom. Still enraged by the invasion of his territory, the big mosasaur searched about for any other interlopers as he swam in a large circle back to his ambush site. The commotion caused by the brief battle had frightened any other prey away and would certainly draw sharks to the area to feed on the remains of the dead mosasaur. Sharks also seemed to be attracted to the movement of a mosasaur's tail. Although he was too large for them to be much of a threat to him, any shark bite could cause a wound that could become seriously infected. He already had several healed scars from past bites on his tail and flippers.

Later in the afternoon, he sensed the noisy approach of a group of swimming birds. Large and wingless, these birds migrated through the seaway on the way to and from their nesting grounds to the north. They were fast swimmers and fed on the abundance of small fish and squid that lived in the sea, catching them in their toothed beaks.

He submerged quietly until he was well below the surface, then swam slowly toward the birds. From the sounds he heard, he could tell they were feeding. In the past, he had been able to ambush careless stragglers from below as they rested between dives for food. Nearing the flock, he could see the darker bodies of the birds silhouetted against the sunlit surface as they dived and fed on a school of small, silvery fish they had trapped. Slashing his powerful tail from side to side, he surged upward toward the body of the nearest bird. His mouth opened just before he reached the surface and quickly closed on the bird as his momentum carried his upper body several feet out of the water. Crushed by his powerful jaws, the bird struggled briefly and died.

When he was certain his prey would not escape, he moved the limp body around in his mouth until it was pointed headfirst into his throat. Then he lifted his head out of the water and allowed gravity to help him swallow the bird. The noise made by the rest of the retreating flock was already fading in the distance.

The hours passed by and dark clouds of an approaching storm covered the sun as it sank toward the horizon. Driven by the changing weather, the waves became larger and larger. The mosasaur found it increasingly difficult to maintain his stationary position and nearly impossible to sense the approach of possible prey against the increasing background noise caused by the wind and rain. Instinctively, he knew it was time to move to open water. Moving forward with rhythmic undulations of his tail, he headed toward the edge of the seaweed mat.

Imagine if you will the middle of North America covered by a vast inland sea. Most of Texas, New Mexico, Oklahoma, Colorado, Kansas, Nebraska, South Dakota, North Dakota, Wyoming and Montana, parts of Missouri, Iowa, and Minnesota, as well as central Canada are underneath a shallow ocean. Not just any ocean, but one that stretched for hundreds of miles from Utah to Minnesota, and from the Gulf of Mexico past the Arctic Circle. At times, this ancient ocean was as large as the present-day Mediterranean Sea and was the home of many kinds of strange creatures that have been extinct for more than 65 million years. This sea covered Kansas and the rest of the Midwest during most of the last 70 million years of the Age of Dinosaurs, and almost until the very end of the Cretaceous period (Fig. 1.1).

The Cretaceous period lasted from about 144 million years ago until 65 million years ago. For much of that time, most of the land we now call the Midwest was completely covered by a shallow, saltwater ocean. Drainage from the older North American continent to the east and the mountains rising on the new land to the west carried vast amounts of soil, sand, and gravel into this seaway, creating intermixed layers of sandstone, shale, and mudstones along the shorelines. In the clear waters at the center of the seaway, calcium carbonate shells of untold billions of microscopic creatures produced thick layers of limestone and chalk.

In Kansas, the geological record of the Cretaceous begins with marine and near-shore deposits of the Cheyenne Sandstone and Kiowa Shale formations that lie on top of the Morrison Formation (Jurassic) in the west, and the Wellington Formation (Permian) in the central part of the state. The last Cretaceous rocks are the dark gray beds of the Pierre Shale in the northwest corner. Almost 70 million years of geologic history is preserved in between. One of the deposits near the top layer of the Cretaceous rocks in Kansas is referred to as the Niobrara Formation. It contains a unique upper member called the Smoky Hill Chalk. This chalk is composed mostly of calcium carbonate, much like that found in the white cliffs of Dover, England. The Smoky Hill Chalk was deposited in Kansas during a five-million-year time span, roughly between 87 and 82 million years ago. During that time, the Western Interior Sea was gradually retreating from its greatest expansion. The deposition of these chalky marine sediments occurred during the last half of the Cretaceous period, and ended about 17 million years before the end of the Age of Dinosaurs.

In Kansas, the Smoky Hill Chalk is about 600 feet thick and lies above the Fort Hays Limestone and below the Pierre Shale. For the most part, the chalk is composed of compacted shells (cocco-

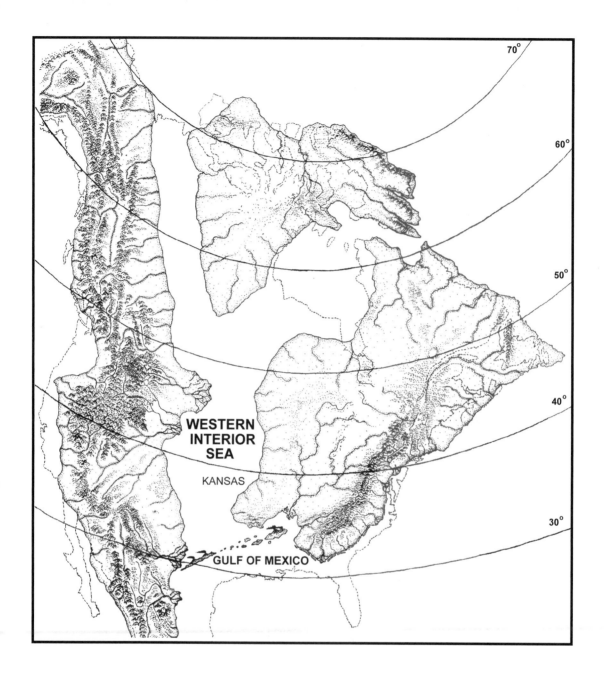

*Figure 1.1. The above map shows the approximate boundaries of the Western Interior Sea during the deposition of the Smoky Hill Chalk. Present-day exposures of the chalk are located just above the "K" in Kansas. Adapted from Schwimmer, 2002 (*King of the Crocodylians, *Indiana University Press); base map by Ron Hirzel.*

liths) of microscopic, golden-brown algae (Chrysophyceae) that lived and died by the untold billions in the warm, shallow sea. Besides making up the chalk, these microscopic plants were the basis for the complex food chain that supported vast numbers of fish and many large predators, including sharks, mosasaurs, pteranodons, and birds.

The Western Interior Sea, sometimes just called the Inland Sea, was formed by the flooding of low-lying areas of the North American continent during a period of the earth's history when there were no polar ice caps and sea level was at its highest. Near the center of the sea, the water was probably less than 600 feet deep (Hattin, 1982) and the limey mud bottom was relatively flat and featureless. In the area where Kansas is now located, the sediments were deposited at a rate which would ultimately produce about one inch of hardened chalk for every 700 years of time (ibid.). The chalk also contains more than a hundred thin layers of bentonite clay, most of which are rusty red in color, that are the remains of ash deposited from periodic, major volcanic eruptions in what is now Nevada, Utah, Idaho, and Montana. These ash deposits can be traced for miles across the chalk beds and are currently used as chronological markers when describing the stratigraphy of the formation. In addition, several species of vertebrate and invertebrate marine life that lived in the Western Interior Sea at different times during the deposition of the chalk are useful in determining the age and biostratigraphy of widely separated exposures (Chapter 13).

This shallow ocean was home to a variety of marine animals that are now extinct. These included giant clams, rudists, crinoids, squid, ammonites, numerous sharks and bony fish, turtles, plesiosaurs, mosasaurs, pteranodons, and even several species of primitive marine birds with teeth. Although it seems unlikely that you would find dinosaur fossils in the middle of the Western Interior Sea, a number of them (a hadrosaur and several nodosaurs) have been collected from the Smoky Hill Chalk, and their remains have been well documented (Carpenter et al., 1995). The bodies of these dinosaurs must have somehow floated hundreds of miles out to sea before sinking to the bottom. It is possible that they died during catastrophic floods and were carried out to sea in large, tangled mats of trees and other vegetation.

Over a period of about five million years, the remains of many of these animals were preserved as fossils in the soft, chalky mud of the sea bottom. When this mud was compressed under the weight of thousands of feet of overlying shale, it became a deposit of chalk that is about 600 feet thick in western Kansas. Most of the massive chalk formation that once covered Kansas, however, has been eroded away over the last 60 million years and is now exposed only in a relatively small area in the northwest quarter of the state. The eastern edge of this part of Kansas is also known as the Smoky Hills, which provided the name for the Smoky Hill River and ultimately for the geological formation known as the Smoky Hill Chalk.

During the last 130 years or so, the Smoky Hill Chalk has been the source of thousands of fossil specimens, many of which are on exhibit today in museums around the world. A large number of these were collected by or for such famous paleontologists as E. D. Cope, O. C. Marsh, S. W. Williston, and the Sternberg family. These specimens include a large portion of the Yale Peabody Museum collection that resulted from the Yale College scientific expeditions of the 1870s. Much of the early work on the Cretaceous fossils from Kansas was published in volumes 2, 4, and 6 of the

University Geological Survey of Kansas (1897, 1898 and 1900). Descriptions of these strange, "prehistoric" animals from their often fragmentary remains were sometimes bizarre by today's standards and often resulted in inaccurate reconstructions drawn under the direction of the various paleontologists. E. D. Cope was one of the most imaginative in his descriptions of not only how the animals looked, but how they lived and interacted. Cope (1872) noted in regard to mosasaurs that "their heads were large, flat, and conic, with eyes directed partly upward; that they were furnished with two pairs of paddles like the flippers of a whale, but with short or no portion representing the arm. With these flippers and the eel-like strokes of their flattened tail they swam, some with less others with greater speed. They were furnished, like snakes, with four rows of formidable teeth on the roof of the mouth. Though these were not designed for mastication, and, without paws for grasping, could have been little used for cutting, as weapons for seizing their prey they were very formidable."

Another good example of this sort of fanciful (and in this case highly inaccurate) prose was provided by the "Father of American Paleontology," Joseph Leidy (1870, p. 10), in his description of *Elasmosaurus:* "We may imagine this extraordinary creature, with its turtle-like body, paddling about, at one moment darting its head a distance of upwards of twenty feet into the depths of the sea after its fish prey, at another into the air after some feathered or other winged reptile, or perhaps when near shore, even reaching so far as to seize by the throat some biped dinosaur."

Many fossil specimens from the chalk have been found by amateur collectors, and many of these have been significant additions to paleontology. Both the Sternberg Museum of Natural History at Fort Hays State University in Hays, Kansas, and the Museum of Natural History at the University of Kansas in Lawrence, Kansas, have excellent collections and exhibits of fossils from the Smoky Hill Chalk. The Denver Museum of Nature and Science in Denver, Colorado; the Sam Noble Museum of Natural History in Norman, Oklahoma; the Field Museum in Chicago, Illinois; the Philadelphia Academy of Natural Sciences; and the American Museum of Natural History also have many Kansas fossils. Many Kansas fossils were also sold to major museums in Europe and elsewhere around the world by the Sternberg family and others. Unfortunately, we don't have a good record of where all these fossils have gone, and even worse, we realize that some of them were destroyed during two world wars.

Kansas during the Cretaceous: A Timeline

For the most part, this book will discuss discoveries regarding the natural history of the Western Interior Sea during the deposition of the Smoky Hill Chalk during a period roughly between 87 and 82 million years ago (mya). However, in order to better understand that time interval, it is useful to look at the Kansas oceans during that portion of the Cretaceous for which we have a geological record in the state.

The Mesozoic (Age of Reptiles) is divided into three major periods: the Triassic, the Jurassic, and the Cretaceous. Based on the 1999 geologic time scale published by the Geological Society of America, the Mesozoic lasted roughly 180 million years. The Cretaceous period is the last of the three, unequal divisions, roughly from 144 million mya to about 65 mya, or a time interval just short of 80 million years. In Kansas, the geological record as shown in the surface rocks is missing for all of the Triassic, almost all of the Jurassic, and most of the Early Cretaceous. Rocks of the latter part of the Early Cretaceous lie non-conformably upon shales of the Permian period in the central part of the state, and on the edge of the Morrison Formation in the far west. In this case, "non-conformably" means there is a gap of about 140 million years in the geological record between the top of the Permian rocks and the bottom of the Cretaceous rocks (middle Albian). The gap is less in the western part of the state, where there is "only" 40 million years of "time" missing between the Morrison Formation (Late Jurassic) and the latter part of the Early Cretaceous. In other words, rocks from most of the Mesozoic (140 of 180 million years) are missing in most of Kansas. While we presume that Kansas was above sea level and eroding away for a least part of that time interval, whatever was going on in Kansas during the Triassic and Jurassic will probably never be known for certain.

We can, however, say quite a lot about the fossil record in the Cretaceous rocks that are preserved in Kansas. These layers of Cretaceous rocks, representing a fairly continuous progression of time from the oldest to the youngest, are stacked upon one another in an orderly fashion. One of the advantages to understanding the geology of Kansas is that it is relatively simple to visualize compared to places like Colorado or Arizona. There are no mountains to contend with and everything is pretty flat, at least in the large-scale view. The youngest Cretaceous rocks are in the northwest corner of the state, and the oldest rocks (Mississippian) occur in the southeast corner. Put another way, as you travel roughly 430 miles from St. Francis (Cheyenne County) in the northwest corner to Baxter Springs (Cherokee County) in the southeast, you are descending 250 million years through time (geologically speaking) at an average of about 580,000 years to the mile. It is almost like being in a time machine.

In this book, however, I am going to be describing the discovery of animals that lived during a much shorter period of time. The book will concentrate on the relatively brief geological period when the Smoky Hill Chalk was deposited near the middle of the Western Interior Sea, but I will take occasional "side trips" into other parts of the Cretaceous in Kansas where the rocks were deposited as a series of near-shore sandstones (including a river delta) and off-shore shales and deeper water limestones and chalks from about 112 mya to 75 mya. I hope that when I'm done, you'll better understand how the geological record of that period was preserved at the bottom of the oceans of Kansas. Merriam's (1963) *The Geologic History of Kansas* is one of the better references in regard to the surface and subsurface rocks occurring in the state. Buchanan (1984) and Buchanan and McCauley (1987) provide introductory guides to the rocks, fossils, and roadside geology of Kansas. Additional information on the Cretaceous fossils and geology of Kansas is available on the Internet through the Oceans of Kansas Paleontology website (http://www.oceansofkansas.com).

The oldest Cretaceous formation in Kansas is the Cheyenne Sandstone. This is relatively pure (beach?) sand that was deposited in south central Kansas (Clark and Kiowa counties) along the northern or northeastern shoreline of the sea as it advanced from the south. This flooding occurred near the end of the Early Cretaceous during an age called the Albian. As the sea advanced, covering the much older and heavily eroded Permian Shales, the sediments that were deposited changed from sand to sandy shales and then dense gray shales (Scott, 1970). These shales represent the remnants of rocks that were being eroded from nearby land masses and being carried into the ocean by rivers flowing from the land. The gray shales that make up the overlying Kiowa Shale preserve evidence of abundant life in the shallow sea and the nearby shoreline (many invertebrates, teeth of sharks, bones of fish, turtles, plesiosaurs, and crocodiles). The sea continued to expand north and eastward across Kansas throughout the Albian age. As it did, it buried the Cheyenne Sandstone and Kiowa Shale under blankets of mud and other sediments that rained down on the sea floor (Fig. 1.2).

About 99 mya, the Albian stage (Early Cretaceous) ended and the Cenomanian stage (Late Cretaceous) began. In north-central Kansas, this is evidenced by sand and other sediments that were deposited in a huge delta by a major river or rivers flowing into the east edge of the sea from the northeast. Iron-rich rocks from as far away as Wisconsin and Michigan were being eroded away bit by bit and carried to the edge of the sea in central Kansas, where they were laid down layer after layer, forming the Dakota Sandstone. This formation is visible today as buttes and other multi-colored (off-white to dark reddish-purple) erosional features to the northwest of McPherson and around Kanopolis Lake in the central part of the state (McPherson, Saline, Ellsworth, and Russell Counties). Historically, major vertebrate fossils have been limited to a crocodile (*Dakotasuchus kingi*: Mehl, 1941; Vaughn, 1956) and a few general reports that mention the teeth of sharks and bony fish. Recent collections of the Upper Dakota, however, indicate a rich marine fauna (Everhart et al., 2004) of sharks, rays, and bony fish that existed during the transition from a non-marine to a near-shore marine environment (Hattin and Siemers, 1978). Many thousands of leaf impressions were collected by G. M. Sternberg, B. F. Mudge, C. H. Sternberg, and others from the Dakota (Lesquereux, 1868) beginning in the mid-1860s.

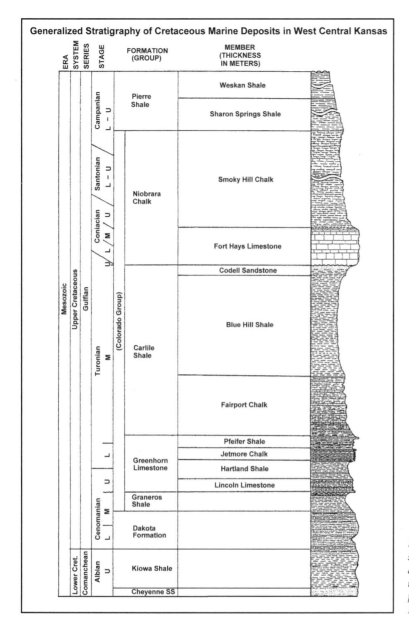

Generalized Stratigraphy of Cretaceous Marine Deposits in West Central Kansas

ERA	SYSTEM	SERIES	STAGE	FORMATION (GROUP)	MEMBER (THICKNESS IN METERS)

Figure 1.2. A generalized stratigraphic column of the Early and Late Cretaceous formations in west-central Kansas. Adapted from Shimada, 1996. (L = Lower; M = Middle; U = Upper).

Near the middle of the Cenomanian, sea levels rose again, and a dark gray shale called the Graneros covered the Dakota Sandstone, effectively burying the river delta under some ten meters of mud (Hattin and Siemers, 1978). The sea continued to deepen and the Graneros Shale was replaced by the Lincoln Limestone Member of the Greenhorn Limestone during the Upper Cenomanian. The deposition of limestone continued thorough the Lower Turonian. At the high-water mark of this expansion (transgression) of the sea, a 30-cm (10–12-in.) layer of resistant limestone was laid down, forming the Fencepost Limestone bed at the top of the Greenhorn Limestone Formation.

Figure 1.3. A map of western Kansas, circa 1868–1869, just prior to the completion of the Kansas (Union) Pacific Railroad. Many of the discoveries described by Cope and Marsh from Kansas occurred south of the railroad, along the Smoky Hill River in Wallace, Gove, and Trego counties. Note that present-day Logan County is the eastern half of what was Wallace County in 1870.

At the beginning of the middle Turonian, the sea began to recede (regress) from the middle of the continent. A chalky limestone called the Fairport Chalk Member of the Carlile Shale was deposited for a time and then was replaced by the dark gray Blue Hill Shale Member. Near the end of the middle Turonian, the coastline was approaching rapidly and the shale was replaced by the near-shore Codell Sandstone.

By the beginning of the Upper Turonian, the ocean was gone again from Kansas, or at least no record of deposition remains from that time. Major erosion occurred in the central part of the state, where almost all the Codell Sandstone has been removed. During the following transgression, when the seas returned at the beginning of the Coniacian, they deepened rapidly and the clear-water Fort Hays Limestone (Merriam, 1963) was deposited on top

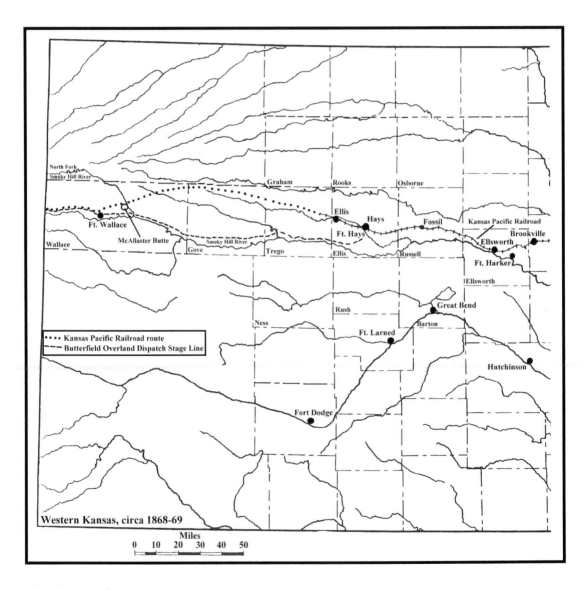

Western Kansas, circa 1868-69

• • • • Kansas Pacific Railroad route
– – – – Butterfield Overland Dispatch Stage Line

Miles
0   10   20   30   40   50

of the remaining Codell Sandstone. The Fort Hays Limestone is the lower member of the Niobrara Formation, formed during a period when the sea was at its widest and the water near the center (Kansas) was the deepest. By the middle of the Coniacian, the sea was again regressing and the limestone was replaced by the Smoky Hill Chalk.

As mentioned earlier, the chalk was deposited over a period of about five million years, between 87 and 82 mya. This period of time includes the late Coniacian, all of the Santonian and the early Campanian. By the early Campanian, sea levels were still dropping and the chalk was replaced by the Sharon Springs Member of the Pierre Shale. In Kansas, there are indications that the western portion of the chalk may have been raised above sea level for a brief time. When the sea returned, a dark gray shale was deposited through most of the rest of the Campanian. The geological forces lifting the Rocky Mountains, however, were also lifting much of the Great Plains to the east. At some point before the end of the Cretaceous, Kansas rose above the sea for the last time and the sea bottom was exposed to the forces of nature. Over the last 70 million years or so, surface erosion has removed much of the upper Pierre Shale in Kansas, and the geological record of the Western Interior Sea ends in the state well before the end of the Cretaceous.

What remains, however, is one of the best records that we know of concerning the marine life that flourished in the oceans of Kansas. In the following pages, I will discuss the fossil discoveries which have occurred during the last 130 years in the rocks covering the western third of Kansas (Fig. 1.3). Kansas was very different during the 1870s, and paleontology was as much an adventure as it was a scientific pursuit. The fossils that were found there were new to science and generated an intense interest in paleontology that continues today. To me, the discovery of these fossils provides a fascinating look at the strange creatures that lived and died in the seas of the Late Cretaceous. The stories of their discovery and identification also provide an interesting view of the early years and the growth of paleontology in the United States.

Two

Our Discovery of the Western Interior Sea

When Lewis and Clark set out in 1804 on their westward trek to explore the Louisiana Purchase, they had no idea they would also be crossing the expanse of an ancient ocean that once covered the middle of North America. It was early in the trip when they found the only fossil from the collections made by the expedition that survives today (Chapter 5). Along the Missouri River, near the northwest corner of what is now Iowa, they came across a fossil that Meriwether Lewis described in a note that is curated along with the specimen as "the petrified jaw bone of a fish" (see Spamer et al., 2000, for a more detailed account). The chalk that is exposed along the river in this area is not far from the Niobrara Formation further to the west in South Dakota. The "fish jaw" of Meriwether Lewis was eventually presented to the American Philosophical Society, where it was studied and then misidentified some years later by Dr. Richard Harlan (1824) as the jaw of a new species of marine reptile, *Saurocephalus lanciformis*. He believed it to be most closely related to the marine reptiles called ichthyosaurs. Joseph Leidy (1856, p. 302), however, noted that the specimen (ANSP 5516) was a fragment of a "maxillary bone with teeth, of a peculiar genus of sphyrænoid fishes, from the cretaceous formation of the Upper Missouri."

The journals of the Lewis and Clark expedition also tell of another mysterious fossil. In 1818, Dr. Samuel Mitchell (p. 406) wrote, "What shall we think of the genus and species of that petrified skeleton of a very large fish, seen in the Sioux county, up the Missouri by Patrick Gass? In his Journal to the Pacific ocean with Messrs. Lewis and Clark in 1804–06, he relates that it was forty-five feet long and lay on top of a high cliff." As noted on a copy of Clark's original map made years later for the Maximilian-Bodmer western journey (Moulton, 1983–1997, Clark-Maximilian Sheet 9, vol. 1), the remains were discovered along a stretch of the Missouri River in what is now northwest Gregory County, south-central South Dakota, and probably came from the Late Cretaceous Pierre Shale. At least four members of the Lewis and Clark expedition noted the discovery of the large skeleton in their journals on Monday, September 10, 1804 (see Moulton, 1983–1997). Clark's description (ibid., vol. 3, p. 61) is perhaps the most complete: "[B]elow the Island on the top of a ridge we found a back bone with the most of the entire [length] laying Connected for 45 feet. [T]hose bones are petrified, some teeth & ribs also connected." John Ordway (ibid., vol. 9, p. 57) described the remains simply as "the rack of Bones of a verry [sic] large fish" while Joseph White-house (ibid., vol. 11, p. 72) wrote that they "saw lying on the banks on the South side of the River, the Bones of a monstrous large Fish, the back bone of which measured forty-five feet long." Gass (ibid., vol. 10, p. 38) also noted that "part of these bones were sent to the City of Washington." While the bones they collected and sent back to Washington were apparently lost, the description appears most likely to be that of a large mosasaur. Moulton (vol. 3, p. 63) indicated it may have been a plesiosaur (elasmosaur?) but provided no further evidence in that regard.

Ten years after naming of *Saurocephalus,* Dr. Harlan misidentified fragments of another fossil from the Western Interior Sea. In this instance, Harlan (1834, p. 405) noted that the remains had been found by "a trader from the Rocky mountains . . . [who] observed, in a rock, the skeleton of an alligator-animal, about seventy [?] feet in length; he broke off the point of the jaw as it projected, and gave it to me. He said that the head part appeared to be about three or four feet long." Ignoring the field observations of the fur trader, just as he had those of Meriwether Lewis, Harlan decided that the remains were those of an ichthyosaur and gave it the name *Ichthyosaurus Missouriensis.* An examination of the accurately drawn figure published with his paper clearly shows the fragment to be the anterior end of the premaxillary from a mosasaur skull. The mistake was noted relatively quickly by his contemporaries, but that is only the beginning of the story.

From this point onward, the tale becomes more complicated. Several years later, an articulated skull, lower jaws, and vertebrae of a strange beast (a mosasaur) were recovered from the same Big Bend of the Missouri area in South Dakota. In this case, "articulated" means that the bones of the skull of the mosasaur were still

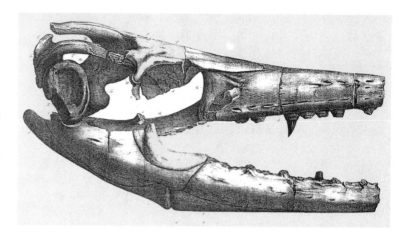

Figure 2.1. The skull of Mosasaurus Maximiliana *as published by Goldfuss (1845). This skull was the first articulated skull ever found of a mosasaur. Note that the anterior ends of the premaxilla and both dentaries are missing from this specimen and were described earlier by Dr. Richard Harlan (1834) as the remains of his mistaken* Ichthyosaurus missouriensis.

arranged in their original or natural positions. The remains came into the possession of a government Indian Agent named Major Benjamin O'Fallon and were displayed in the formal garden of his home in St. Louis (Goldfuss, 1845, p. 3). The specimen eventually attracted the attention of Prince Maximilian zu Wied (1782–1867) during his travels through the American West from 1832 to 1834, and he acquired the specimen. The prince shipped the specimen back to Germany, where a well-known naturalist, Dr. August Goldfuss, spent several years preparing and describing it. Although it was encased in a hard limestone concretion, the skull was preserved fully articulated and uncrushed. In many respects, it was the best example of a mosasaur skull found up until that time in terms of understanding the construction of the mosasaur skull, much better than the disarticulated specimen of *Mosasaurus Hoffmanni* from the Netherlands. It was, however, missing the tips of the lower jaws and the anterior end of the premaxilla (Fig. 2.1). In what was an excellent paper that was subsequently ignored by many other early workers on mosasaurs, Goldfuss (1845) described the specimen completely and gave it the name *Mosasaurus Maximiliana* in honor of his benefactor.

Russell (1967) noted that soon after the Goldfuss paper was in printed in 1845, a letter from Hermann von Meyer to "Professor Bronn," published in the German journal *Neues Jahrbuch für Mineralogie, Geognosie, Geologie und Petrefaktenkunde,* provided the first mention that Harlan's "ichthyosaur" fragment was probably the missing premaxilla of the Goldfuss mosasaur. Although the Goldfuss skull is still in the collection of the Institut für Geologie and Paläeontologie in Bonn, Germany, Harlan's fragments were thought to be lost (Russell, 1967). In 2004, however, the missing premaxilla was rediscovered quite unexpectedly by Gordon Bell and Mike Caldwell in Paris, France, where it had been safely stored for more than 150 years (see Chapter 9).

Harlan's legacy remains, however, because the name "*Mosasaurus Maximiliana* Goldfuss 1845" is still the junior synonym of *Mosasaurus Missouriensis* (Harlan, 1824). While it is unlikely that Goldfuss will ever receive the credit he deserves for his meticulous work on the first articulated skull of a mosasaur ever found, Baur (1892) did note that "if this important paper had been studied more carefully by subsequent writers [e.g., Cope, Marsh, and others], much confusion could have been spared. Williston (1895) elaborated further on the subject when he said, "As Baur has said, had later authorities studied this paper more attentively they would not have claimed as new a number of discoveries made and published long before, among which may be mentioned the position of the quadrate bone, the presence of the quadrato-parietal and malar arches, and the sclerotic plates."

Following the American Civil War, the pace of westward expansion in the United States increased significantly. Gold had been discovered in Colorado, and Denver was growing rapidly. Communication between Kansas City and Denver was largely by a stagecoach and wagon freight line along the Butterfield Trail that ran westward across the prairie through Kansas and eastern Colorado. At about the same time, the Union Pacific portion of the transcontinental railroad was being completed across Nebraska, and a southern route, the Kansas Pacific, was being built from Kansas City to Denver. Along with the settlers in this westward expansion came survey crews and construction workers for the railroads. The resulting encroachment on Indian lands in the West resulted in conflicts and made it necessary for the government to establish a military presence between Kansas City and Denver. Several forts were built along the Butterfield Trail and elsewhere in the western half of Kansas. Along with the troops and guns came military doctors who were arguably among the best-educated men in Kansas at the time. For the period between 1866 and 1872, three of these doctors were among the first fossil collectors and paleontologists in Kansas.

Fort Harker was established originally in 1866 as Fort Ellsworth in central Kansas, near present-day Kanopolis, on the trail to Denver and the eventual route of the Kansas Pacific railroad. Dr. George M. Sternberg (1838–1915), older brother of the Charles H. Sternberg who would later become famous as a fossil hunter, was the military surgeon assigned to the fort. While Dr. Sternberg would later be known for his work in bacteriology and as the Surgeon General of the Army during the Spanish-American War, he also began routinely collecting fossils in western Kansas before anyone else. According to Rogers (1991, p. 130), Dr. Sternberg's younger brother Charles credited him with alerting "O. C. Marsh, Joseph Leidy and other paleontologists to the existence of Kansas's vast fossil beds, worthy of exploration. It was his brother who made possible the first placement of Sternberg fossils in the halls of the Smithsonian." Dr. Sternberg began by collecting fossil leaves from the Dakota Sandstone (early Late Cretaceous) near Fort Harker and then made significant collections of vertebrate fossils from the Smoky Hill Chalk and the Pierre Shale of western

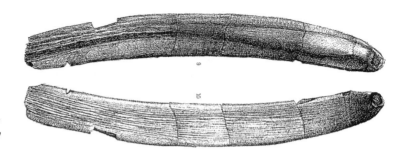

Figure 2.2. Leidy's (1873, pl. 17) figure of the large fin ray (USNM 52) discovered by Dr. George M. Sternberg in western Kansas. It was from this specimen that Leidy (1870) named the giant teleost fish Xiphactinus audax. *The fin ray is shown in dorsal and ventral view and is approximately 40 cm (16 in.) in length.*

Kansas while serving with General Sheridan's campaign against the Indians from 1868 to 1870. Almost all his specimens were donated to the U. S. Army Medical Museum in Washington, D.C., and from there were transferred to the Smithsonian (United States National Museum—USNM). By my rough count during a visit in 2001, Dr. Sternberg is attributed as the collector of more than thirty mosasaur specimens in the USNM collection. He actually signed his name to each bone in most cases! He also discovered the large fish fin (Fig. 2.2) that Leidy (1870) described as the type specimen for *Xiphactinus audax*. Cope (1872b) later more fully described and named the same fish from more complete specimens, but his "*Portheus molossus* Cope 1872" name will always be the junior synonym of Dr. Sternberg's discovery of *Xiphactinus audax* Leidy 1870 (Chapter 5).

Dr. Sternberg was also one of the first collectors of fossils from the Pierre Shale in far western Kansas in 1869 and 1870. However, the first vertebrate fossil found in the Pierre Shale and the first major Cretaceous vertebrate to be described from Kansas was the type specimen of *Elasmosaurus platyurus* Cope 1868 which was found by another Army surgeon in the spring of 1867. Dr. Theophilus H. Turner (1841–1869), the assistant surgeon at Fort Wallace in western Kansas, discovered the remains of a very large marine reptile eroding from a ravine in the Pierre Shale about twelve miles northeast of the fort (Almy, 1987). Later that summer, he gave three of the vertebrae to John LeConte, a member of a party that was in the process of surveying the route for the Union Pacific railroad (LeConte, 1868). After the survey was completed, LeConte delivered the vertebrae to E. D. Cope at the Academy of Natural Sciences of Philadelphia (ANSP) in November 1867. Cope immediately recognized them as belonging to a large plesiosaur and wrote to Turner, asking him to procure the remainder of the specimen and send it to Philadelphia at the expense of the ANSP (Almy, 1987). With the help of other soldiers at Fort Wallace, Turner returned to the site in late December 1867 and secured some 900 pounds of bones and matrix. Near the end of February, 1868, at the urging of Cope, Turner arranged to transport the specimen by military wagon train some ninety miles east to where the approaching Kansas Pacific railroad was being built. From there, the remains were shipped by rail to Philadelphia. Cope received the crates containing the specimen in March, examined the remains, and, as would soon become the custom in his rivalry with O. C. Marsh, hurriedly described and named the specimen. At the March 24, 1868 meeting of the Academy of Natural Sciences of Philadelphia (ANSP), Cope (1868a) reported the discovery "of an animal related to the *Plesiosaurus*" which he called *Elasmosaurus platyurus*. At about the same time, a short note from Cope (1868b), also including the new name, was published in LeConte's (1868) railroad survey report. The controversy that followed regarding the restoration of *Elasmosaurus* with the head on the wrong end (Cope, 1869b) completely overshadowed the fact that Dr. Turner had discovered and successfully collected one of the largest vertebrate fossils known at the time, under primitive conditions, and with no prior experience (Fig. 2.3). Turner certainly deserves more recognition for this feat than he has received so far (Chapter 7).

The next major fossil to be reported from Kansas was a mosasaur discovered in the Smoky Hill Chalk in western Gove County. Williston (1898a) reported that the partial skull of the type specimen of "*Tylosaurus proriger* Cope [1869a] was collected by

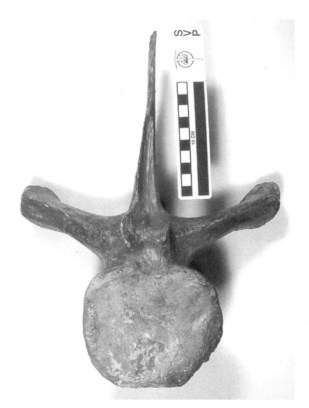

Figure 2.3. A recent photograph of the dorsal vertebra of the giant plesiosaur Elasmosaurus platyurus (ANSP 10081), discovered and collected in western Kansas by Dr. Theophilus Turner in 1867 and figured by Cope (1869, pl. II). The centrum of this vertebra is about 12 cm (5 in.) across.

Colonel Cunningham and Mr. Minor in the vicinity of Monument station [Gove County], and sent by them to Prof. Louis Agassiz. The locality is probably Monument Station of the overland route, in the vicinity of Monument Rocks, in the valley of the Smoky Hill River."

After becoming a state in 1861, Kansas established the Kansas State Agricultural College (now Kansas State University) at Manhattan in 1863. It was there that Professor Benjamin F. Mudge (1817–1879) began the first systematic collection of fossils from the Western Interior Sea. Professor Mudge (1866a) reported on fossil footprints he had collected in 1865 from the Dakota Sandstone (early Late Cretaceous) fifty miles north of Junction City, Kansas. Mudge was also the first to note and publish (1866b) the presence of fossil leaves in the Dakota Sandstone, although at the time he was unsure of their age or geological provenance. In a footnote, LeConte (1868, p. 7) indicates that "Professor Mudge, of the Kansas Geological Survey, has procured specimens from the same locality." Lesquereux (1868) credited B. F. Mudge for a collection of leaf fossils from the Dakota Formation that he examined in the Smithsonian for his publication "On Some Cretaceous Fossil Plants From Nebraska." However, the fossil leaves were actually collected by Mudge from just north and east of Fort Harker in Kansas, not from near "Fort Ellsworth in Nebraska" as indicated by Lesquereux.

According to Williston (1898a), Mudge's "first expedition, as I remember, was up the Republican and Solomon rivers into the wholly uninhabited region, the home then of the bison and roving bands of marauding Indians. It was made shortly after the close of the college year in 1870." Mudge sent many of his fossils to E. D. Cope in Philadelphia, giving Cope the opportunity to name most of the mosasaur species found in the Smoky Hill Chalk even though O. C. Marsh was actually in Kansas first. After being fired from his teaching job in 1873 over a dispute with the administration, Mudge began collecting fossils for O. C. Marsh and Yale College. His discoveries include a toothed bird (*Ichthyornis dispar*), several species of mosasaurs, and other important specimens that are now in major museums in the United States, including at least 303 specimens in the Yale Peabody Museum collection.

Peterson (1987, p. 228) indicates that B. F. Mudge was the "most active" of the early fossil collectors in Kansas. In 1866, he went further west to the area around Ellsworth (near Fort Harker) and near the fork of the Solomon River to the north to collect invertebrates and plant fossils. Peterson (ibid., pp. 228–229) said that "[o]nly Mudge is known to have done much fossil collecting in 1869, and he did not venture very far into western Kansas. During the summer he went up the Republican River as far as the northern state line where he found many fossil plants and other specimens to add to the KSAC collection. In October, he accompanied Kansas Senator Edmund Ross and two others on an expedition, with a military escort, up the Solomon River to visit troops stationed in the area and to determine the valley's potential for agricultural and railroad purposes. Mudge concentrated on soils and geology, including fossils, but also noted evidences of ancient Indian occupation. In what is now Phillips County, Mudge found a number of fossils in Cretaceous formations including the vertebrae and other portions of an eight-foot saurian. Returning down the South Solomon River, Mudge noted much exposed magnesium limestone but found fewer fossils. Although the results of this trip were not spectacular, it was important for introducing Mudge to the large, well preserved fossils of new species awaiting discovery in western Kansas."

In 1870, Mudge collected fossils from the Cretaceous formations around Fort Wallace; the vertebrates he forwarded to Cope in Philadelphia and the invertebrates he sent to F. B. Meek at the Smithsonian (Peterson, 1987, pp. 229–230). E. D. Cope's relationship with Mudge was further solidified in 1871 when Cope visited him at the state agricultural college and examined Mudge's collection of mosasaur and fish specimens (Cope, 1872a). Cope (1871, p. 405; later figured in Cope, 1875, pl. 26, fig. 3) also named a new species of mosasaur (*Liodon mudgei*) in Mudge's honor "in recognition of the valuable results of his investigations as State geologist of Kansas." In his review of American mosasaurs, however, Russell (1967, pp. 181–182) considered *L. mudgei* to be *Platecarpus nomen vanum*, either *P. ictericus* or *P. coryphaeus*.

Mudge continued to collect in 1871. According to Peterson (1987, p. 230), Mudge "went to western Kansas where he found many more vertebrates. Although he kept specimens for the KSAC cabinet [collection], he sent the best items to experts in the East for identification and publication. The molluscs were sent to Meek, most of the plant material to Lesquereux, and most of the vertebrates to Cope. Louis Agassiz at Harvard and James D. Dana at Yale also received items. Many of Mudge's finds were published by Cope and Lesquereux in the yearly report of the U. S. Geological Survey of the Territories. His fossil plants of 1871 included seven new species, including one species of oak that Lesquereux named *Quercus mudgeii*." (See also Preliminary Report of the United States Geological Survey of Montana and Portions of Adjacent Territories Washington, 1872, pp. 301–304).

"The collectors had another very productive year in 1872. Professor Mudge again ventured north of the Smoky Hill River. After stopping near Hays where he found fossil shells and fish, he went north to Smith County where he met the rest of his party—Prof. G. C. Merrill of Washburn [University in Topeka, Kansas]; Prof. P. H. Felker of Michigan Agricultural College; R. Warder of the Indiana Geological Survey; and seven KSAC students. They explored the geology of the valleys of Prairie Dog Creek and several branches of the Solomon River and found many vertebrate fossils. Later in the year, Mudge spent two weeks examining the geology of the Arkansas River valley that became a subject for a paper he presented before the Kansas Academy of Science. In the Fall, he had visits from both Marsh and Cope. It appears that he gave most of his saurian fossils to Cope, who found fourteen new species among them. The rarer birdlike fossils he passed to Marsh. In July a third visitor was Leo Lesquereux, who spotted a number of new species in Mudge's plant material. Lesquereux spent the summer examining plant fossils and sites in the West. On his way to the Rocky Mountains, he stopped to collect fossil leaves at Fort Harker where he met Charles Sternberg who thereafter sent all his plant material to Lesquereux" (Peterson, 1987, p. 231).

Leidy (1868) made a brief report to the Academy of Natural Sciences of Philadelphia regarding the discovery and photographs of a huge Kansas mosasaur. The recorded remarks (ibid., p. 316) are brief but enigmatic: "Dr. Leidy exhibited some photographs of fossil bones, received from Mr. W. E. Webb, Sec. [Secretary] of the National Land Co., at Topeka, Kansas. They represent vertebrae, and fragments of the jaws with teeth of a skeleton of *Mosasaurus,* reported by Mr. Webb to be about 70 feet in length, recently discovered on the great plains of Kansas, near Fort Wallace." Stories regarding a huge but unverified mosasaur from western Kansas are found in various other publications, including *Buffalo Land,* Webb's (1872) semi-fictional account of early explorations of the plains, and by Cope (1872c, p. 333 and 1872d, p. 279). It is likely that the specimen is actually the type specimen of *Tylosaurus proriger* first described by Cope (1869a) (and figured in Cope, 1870) from a specimen in the Museum of Comparative Zoology. The remains were obtained by Prof. Agassiz in his trip to western Kansas in 1868 (see Almy, 1987, p. 193). Cope (1869) stated that the type specimen "belonged to Prof. Agassiz" and then noted (1875) that the original description "was based on material in the Museum of Comparative Zoology, Cambridge, Mass., brought by Prof. Agassiz from the Cretaceous beds in the neighborhood of Monument, Kans. and near the line of the Kansas Pacific Railroad." Note here that the present-day location of "Monument, Kansas" is about twenty-five miles north-northwest of the original Monument station, which was located on the Butterfield stage line near the Smoky Hill River. When the railroad replaced the stage line in the early 1870s, the inhabitants of the station moved to a town site on the railroad line along with some of the buildings. Surrounded by miles of relatively flat prairie that is now used for raising wheat and corn, there are no Late Cretaceous marine fossils to found in or near present-day Monument, Kansas. A newspaper description of the town of Monument in the Topeka *Weekly Leader* (Anonymous, 1868) does not paint a pretty picture of the living conditions at the time: "There are twenty cloth houses at Monument, besides the quarters tents [U.S. Army] . . . [f]our saloons, two fruit stands, one boarding house and news depot and post office. At Monument is a level plain as far as the eye can reach on either side. Twelve miles westward is Sheridan, and yet seventeen further and you reach Fort Wallace, where Harrison Nichols keeps hotel. We'd soon be in mid ocean in a wash-tub as in the centre of these plains on foot."

While Kansas has often been suggested as the place where the "Bone Wars" between E. D. Cope and O. C. Marsh began, it is more likely that their rivalry started over Cretaceous fossils being found in New Jersey in the late 1860s. The competition and other confrontations certainly intensified in Kansas during the early 1870s and then reached their peak further north and west with the discovery of dinosaurs and giant mammals. Marsh was the first of the two men to collect fossils in Kansas in 1870, although by then Cope had been the recipient of many specimens, including the giant plesiosaur *Elasmosaurus platyurus* (Chapter 7) and the type specimen of *Tylosaurus proriger,* from various collectors in Kansas. Marsh arrived in November 1870 and collected for two weeks near Fort Wallace with his first Yale College scientific expedition. He returned to Kansas with another expedition composed of Yale students in the summer of 1871 and collected many fossils from the chalk. Cope's only visit to the state was in the fall of 1871. Although Marsh returned in 1872, both men would hire others to collect for them on more or less a full-time basis. Marsh hired Professor B. F. Mudge to collect for him in Kansas in 1873. Mudge then recruited several of his students, including Samuel W. Williston, to help him in the field. Williston would go on to work for Marsh at Yale for several years and would then return to the University of Kansas. Cope, on the other hand, hired a young Charles H. Sternberg as his man in Kansas. While initially at a disadvantage due to the superior knowledge and experience of Marsh's collectors, Sternberg would become a well-recognized supplier of fossils for museums all over the world. His sons, George F., Levi, and Charles M. Sternberg would later go on to careers in paleontology of their own, with George eventually returning to Kansas to continue the exploration of the Cretaceous and Tertiary deposits in the state. George settled in Hays and began collecting fossils for what would later become the Sternberg Museum of Natural History at Fort Hays State University (originally Kansas State Teachers College).

Stratigraphy

One of the important issues that will surface again and again in this book is the lack of stratigraphic information on the occurrence of most of the fossils that have been collected from the Smoky Hill Chalk. In this usage, "stratigraphic occurrence" refers to what approximate chronological level the fossil was found within the 200-m (600-ft.) chalk unit. As noted earlier, this chalk was deposited in the Western Interior Sea over a period of about five million years. Fossils found near the bottom of the chalk are millions of years older than those found near the top. Knowing the ages of specimens is useful in understanding when species appeared and when they became extinct, and what ecological relationships may have existed. Relatively few of the fossil remains collected since 1870 have even good locality data, let alone stratigraphic information. While this may not seem too important at first, it means that we cannot establish when the animal lived in relation to other remains that we find in the chalk. This certainly limits the usefulness of any fossil in the study of the ecosystem of the Western Interior Sea. Even though an accurate frame of reference for locating fossils within the five-million-year depositional period of the chalk has been in place since Hattin (1982), few people are aware of it and fewer use it in the field.

The geology of western Kansas was not well understood when most of the early collecting was done. According to Zakrzewski (1996), geological studies of the western interior of the United States had begun as early as the 1850s, but had proceeded slowly and sporadically. Meek and Hayden (1861) first referred to the chalk and limestone strata as the Niobrara Division in their description of exposures along the Missouri River near the mouth of the Niobrara River in Nebraska. In Kansas, these Upper Cretaceous strata were referred to simply as the "Niobrara" by the geologists and paleontologists of the day (Hattin, 1982). E. D. Cope (1872a) wrote the earliest substantial account of vertebrate fossils from the "Niobrara Beds." Though he described the geology of the chalk and each of the various species in some detail, he made no attempt to delineate their stratigraphic occurrence within the formation.

In 1889, the University Geological Survey of Kansas, which eventually became the Kansas Geological Survey, was established by the state legislature. Samuel W. Williston was appointed to the faculty of the University of Kansas the following year. These two events provided the basis for much of the early progress that was to be made in the study of geology and paleontology in Kansas. At the time, even reaching an agreement on what to name the formations that cropped out in the western part of the state was no easy matter. Most of the early terminology used to describe the Smoky Hill Chalk was based on the occurrence of the predominant fossils and added little stratigraphic information to individual specimens. The "Niobrara Division" of Meek and Hayden (1861) was made up of two distinct units, a lower limestone member and an upper chalk member. Logan (1897) was the first to call the upper chalk member "The Pteranodon Beds" in apparent recognition of the abundance of well-preserved *Pteranodon* material that had been discovered there (Chapter 10). That same year, demonstrating his support for Logan's descriptive terminology, Williston (1897) further divided the Pteranodon Beds into the lower Rudistes Beds and the upper Hesperornis Beds, providing essentially the first biostratigraphic subdivisions of what was to become the Smoky Hill Chalk.

Williston (1897) briefly discussed the stratigraphic occurrence of mosasaurs in the Pteranodon Beds for the first time. He also made the observation that *Clidastes* does not occur in the lower Rudistes Beds, indicating that other genera (*Platecarpus* and *Tylosaurus*) probably occurred within 100 feet of the contact of the chalk with the underlying Fort Hays Limestone. Williston (ibid., p. 245) was certainly aware of the lack of good stratigraphic data for Niobrara vertebrate fossils when he wrote, "I need not call the attention of future collectors to the importance of locating the horizon of specimens more accurately than has been done heretofore."

Williston (1898b) also published the first comprehensive description of the systematics and comparative anatomy of mosasaurs from the Smoky Hill Chalk, and he discussed their range and distribution in comparison with specimens discovered earlier in New Zealand and Europe. He commented that *Tylosaurus,* "so far as was known, begins near the lower part of the Niobrara [Smoky Hill Chalk] and terminates at its close or in the beginning of the Fort Pierre [Pierre Shale]." Of *Platecarpus,* he stated that the species on which the genus is based are "known nowhere outside of Kansas and Colorado, and are here restricted exclusively to the Niobrara." He again concluded that the lowest horizon of *Clidastes* "is the upper part of the Niobrara in Kansas."

It was not until after the turn of the twentieth century that the exploration for oil and gas in western Kansas enabled rapid advances in understanding the geology of the entire Niobrara Formation. According to Hattin (1982), Moore and Hays (1917) were the first to regard the Kansas Niobrara as a formation and the first to give member status to the currently recognized divisions, the Fort Hays Limestone and the Smoky Hill Chalk.

Russell (1967) reviewed mosasaur specimens in the Yale Peabody collection and suggested that the Smoky Hill Chalk could be divided into a lower, *Clidastes liodontus–Platecarpus coryphaeus–Tylosaurus nepaeolicus* zone and an upper, *Clidastes propython–Platecarpus ictericus–Tylosaurus proriger* zone. He also suggested that the increased abundance of *Clidastes* specimens in the upper portion of the chalk was an indication of a gradual change from a mid-ocean to a near-shore environment. Russell (1970) noted significant differences between the distribution of mosasaur species in the Smoky Hill Chalk compared to the Gulf Coast species occurring in the Selma Formation of Alabama. In his initial paper concerning the biostratigraphy of the Smoky Hill Chalk, Stewart (1988) stated that he was aware of several exceptions to Russell's stratigraphic distribution of mosasaurs in the Smoky Hill Chalk that caused him to regard it with "a degree of skepticism."

It was only after Hattin (1982) published his composite measured section of the Smoky Hill Chalk that significant progress could be made in understanding the vertebrate biostratigraphy of this formation. Hattin used bentonites and other geological features to delineate his twenty-three lithologic marker units, and he divided the chalk into five biostratigraphic zones based on the occurrence of invertebrate species. In doing so, he provided field workers with the first dependable method of determining their stratigraphic location in the section.

Stewart (1988) used the distribution of the fish genus *Protosphyraena* to further delineate biostratigraphic zones in the Smoky Hill Chalk. Stewart (1990) then incorporated Hattin's marker units as upper and lower boundaries for his six proposed biostratigraphic zones (Table 2.1). This report provided the first comprehensive description of the distribution of known invertebrate and vertebrate species in the Niobrara Formation and was the first at-

TABLE 2.1.

Biostratigraphy of the Smoky Hill Chalk (adapted from Hattin, 1982; Stewart, 1990; Everhart, 2001). Time (MYA = Millions of Years Ago) and boundaries as indicated by Hattin's (1982) marker units (MU) are approximate. The Smoky Hill Chalk Member of the Niobrara Chalk was deposited between 87 and 82 million years ago. Note that the Late Coniacian is unequally divided into two zones: a lower zone (*P. perniciosa*) from the base of the chalk to about MU 5 and a brief upper zone (*Spinaptychus* n. sp).

Age	Marker Units	MYA	Biostratigraphic zone of:
Early Campanian	MU 16 to MU 23	83–82	*Hesperornis*
Late Santonian	MU 11 to MU 16	84–83	*Spinaptychus sternbergi*
Middle Santonian	MU 8 to MU 11	85–84	*Clioscaphites vermiformis* and *C. choteauensis*
Early Santonian	MU 6 to MU 8	86–85	*Cladoceramus undulatoplicatus*
Late Coniacian	Base to MU 6	87–86	*Protosphyraena perniciosa / Spinaptychus* n. sp.

tempt to assign specific stratigraphic ranges for mosasaur species within the Smoky Hill Chalk. Even with the substantial improvements over previous attempts, Stewart believed that his biostratigraphy was flawed by the lack of reliable stratigraphic data for even those specimens collected during the previous twenty years. He stated that his framework was "submitted in the hopes that other researchers will test it and improve upon it." More recently, publications on the stratigraphic occurrence of mosasaurs in the Smoky Hill Chalk (Schumacher, 1993; Sheldon, 1996; Everhart 2001) have benefited from the framework provided by Hattin (1982) and Stewart (1990). Most other papers regarding specimens from the chalk published since 1990 have generally included accurate stratigraphic information.

Even though fossils have been collected from the Smoky Hill Chalk for more than 130 years, there is still additional collecting and describing to be done. A number of new species have been discovered in the past fifteen years and there are probably others that will be found in the future. In addition, many species that were discovered and/or named by early workers need better specimens and stratigraphic information if we are to fully understand their occurrence and significance in the Western Interior Sea.

Recommended Reading about Early Paleontologists and Kansas Fossils

Edward D. Cope (1840–1897)

Davidson, J. P. 1997. *The Bone Sharp*. Philadelphia: The Academy of Natural Sciences of Philadelphia. 237 pp.

Osborn, H. F. 1931. *Cope: Master Naturalist*. Princeton, N.J.: Princeton University Press. 740 pp. (Reprint, New York: Arno Press, 1978.)

Joseph Leidy (1823–1891)

Warren, L. 1998. *The Last Man Who Knew Everything.* New Haven, Conn.: Yale University Press. 320 pp.

Othniel C. Marsh (1831–1899)

Schuchert, C., and C. M. LeVene. 1940. *O. C. Marsh—Pioneer in Paleontology.* New Haven, Conn.: Yale University Press. 541 pp.

Benjamin F. Mudge (1817–1879)

Peterson, J. M. 1987. "Science in Kansas: The Early Years, 1804–1875." *Kansas History Magazine* 10(3): 201–240.

Charles H. Sternberg (1850–1943) and George F. Sternberg (1883–1969)

Liggett, G. A. 2001. Dinosaurs to Dung Beetles: Expeditions Through Time. Guide to the Sternberg Museum of Natural History. Sternberg Museum of Natural History, Hays, Kansas, 127 pp.

Rogers, K. 1991. *A Dinosaur Dynasty: The Sternberg Fossil Hunters.* Missoula, Mont.: Mountain Press Publishing Company. 288 pp.

Sternberg, C. H. 1909. *The Life of a Fossil Hunter.* New York: Henry Holt and Company. (Reprint, Bloomington: Indiana University Press, 1990.)

George Miller Sternberg (1838–1915)

Sternberg, M. L. *George Miller Sternberg: A Biography.* Chicago: American Medical Association. 331 pp., 10 pls.

Samuel W. Williston (1851–1918)

Shor, E. N. 1971. *Fossils and Flies: The Life of a Compleat Scientist—Samuel Wendell Williston, 1851–1918.* Norman: University of Oklahoma Press. 285 pp.

Three

Invertebrates, Plants, and Trace Fossils

Little of the bright sunlight above ever reached the bottom of the Western Interior Sea. However, the steady rain of organic detritus from the plankton and other organisms in the water column supported an unusual abundance of life on the soft, muddy sea floor. Huge, flat clams called inoceramids literally covered the bottom, nearly edge to edge in many places, and if the scene had been visible to the eyes of a human visitor, it would have stretched outward in all directions for mile after monotonous mile. Life had been good for these strange animals for many thousands of years now, because the sluggish circulation pattern of the shallow sea overhead brought just enough food and oxygen to the bottom to supply their meager needs. It wasn't always so, and the oxygen levels would decrease time and again, killing most of the strange forms living in the nearly total darkness that enveloped the limey mud surface. Layer after layer of the flattened shells of previous generations were hidden below the surface. The surrounding mud was made up mostly of untold billions of dead shells from the microorganisms that formed the base of the food chain that prospered several hundred feet overhead in the sunlit waters near the surface. Living in a precarious balance, invertebrate life on the bottom came and went in irregular pulses measuring hundreds of thousands of years.

The animals living here on the bottom weren't picky about their food. Most were filter feeders, continuously moving large quantities of seawater through their various feeding mechanisms to remove small bits of whatever happened to be suspended near them. Over many millions of years, inoceramid clams had evolved into larger and larger forms, eventually reaching a maximum diameter of nearly 1.5 m (5 ft.). Whether this increase in size provided them with more surface area for their huge gills in a low-oxygen environment or a more efficient system for filtering bits of food out of the water is unknown. What is known is that their large, flat shells literally covered the bottom at times, they provided the only available hard substrate for many other animals to colonize, and in some cases they provided a shelter in which to hide.

Most of the inoceramids were covered with a layer of oysters packed edge to edge; frequently, several generations of oysters lived together, attached to the empty shells of the previous occupants of that space. There was little diversity among the major invertebrates on the sea bottom; the giant clams and their attached oysters made up most of the biomass. Smaller epizoans lived wherever they could find an attachment point. A few, mostly solitary rudists lived in heavy, funnel-shaped shells set deep in the mud. They were, however, at the extreme north end of their preferred range and were never found here in large numbers. In warmer waters to the south, they were colonial reef builders.

An observer standing on the sea floor would have initially been able to see little in the surrounding darkness. As the watcher's eyes adjusted to the darkness, tiny motes of colored light may have become visible, and appeared to float or dart around unattached to a recognizable life form. As in modern oceans, it is likely that many forms had evolved bioluminescence as a means of attracting prey or a mate. The squid and their shelled cousins the ammonites that lived in these dark waters probably had such faint markings.

Schools of small fish sheltered in and around the open shells of the larger bivalves covering the sea floor. Rarely, there were small, deep-bodied fish called pycnodonts (Chapter 5) that nibbled at the variety of smaller invertebrates growing on the oyster-encrusted larger shells. An occasional mud-grubbing ptychodontid shark would emerge from the gloom as it searched for the smaller, thinner-shelled clams among the larger ones.

Other than coccolithophores (a single-celled kind of golden-brown algae), coccoliths (the disk-shaped, calcite "scales" that surround each algal cell), and the fecal pellets (compacted waste products of animals that fed on the coccolithophores) that compose the bulk of the chalk itself, the most common biological remains preserved in the chalk are the shells of larger invertebrates. In some areas, especially in the lower chalk, it is impossible to walk across an exposure without stepping on numerous fragments of giant inoceramids and oysters. Loose bits and pieces of shell occur in layers that accumulate and blanket the surface in some areas because they are more resistant to erosion than the chalk that surrounds them. In that regard, they can easily confuse both the novice and the expert alike, appearing in odd shapes and textures which mimic the bones, and especially the jaws, of vertebrates. In the years I have collected in the chalk, I am sure I've picked up thousands of shell fragments just to be sure I wasn't missing something more important.

The first Cretaceous fossils reported from western Kansas were actually not those of invertebrates or vertebrates. They were instead the imprints of the leaves of deciduous trees found in the Dakota Sandstone north and west of Salina. While the Dakota is now recognized as being deposited in the early part of the Late Cretaceous, its age was confusing to the early geologists who were exploring in wilds of Kansas and Nebraska. Hawn (1858) had initially indicated that the darkly stained Dakota sandstones were of Triassic age, but Meek and Hayden (1859) countered that suggestion with the report of the discovery of fossil leaves which came from trees unknown earlier than the Cretaceous. Their source (ibid.), Dr. J. S. Newberry, an "authority on fossil botany," noted that "they include so many highly organized plants, that were there not among them several genera exclusively Cretaceous, I should be disposed to refer them to a more recent era. . . . A single glance is sufficient to satisfy any one they are not Triassic."

B. F. Mudge (1866) also noted the presence of fossil leaves in the Dakota Sandstone (Late Cretaceous) found in north central Kansas. At the time, however, he was unsure of their age. John LeConte (1868) described the geography/geology of Kansas in his survey report for the Union Pacific railroad and noted a bed of rocks [the Dakota Sandstone] that "continue to Fort Harker, where the sandstone becomes less ferruginous, of a reddish and pale yellow color, and contains leaves of trees of exogenous [trees that grow annual rings] growth." He also indicated that a collection had been made there in November of 1866 and sent to Leo Lesquereux (1806–1899) for examination.

It is quite likely that some of these leaf imprints were also collected and sent to the U.S. Army Medical Museum in Washington, D.C., by Dr. George M. Sternberg, the surgeon at Fort Harker. Sternberg collected many fossils while serving with the U.S. Army in Kansas, but he kept few field notes. However, he did write that "the bluffs north and east of the fort (Fort Harker) are composed of a recent red sandstone [the Dakota Sandstone] which contains the impressions of the leaves of trees of existing species (oak, ash, willow, etc.)" (Sternberg, 1920, p. 12).

In a footnote, LeConte (1868, p. 7) indicated that "Professor Mudge, of the Kansas Geological Survey, has procured specimens from the same locality." That same year, Lesquereux credited B. F. Mudge for a collection of leaf fossils from the Dakota Formation that he examined in the Smithsonian for his 1868 publication, even though he had the locality confused. The title of the paper, "On Some Cretaceous Fossil Plants From Nebraska," was misleading since the fossil leaves were actually collected by Mudge from a few miles north and east of Fort Harker (present-day Kanopolis) in Kansas, and not from near "Fort Ellsworth in Nebraska." The collection is mentioned again by Hayden (1873, p. 266): "Species of sweet-gum, poplar, willow, birch, beech, oak, sassafras, tulip-tree, magnolia, maple and others have been described from the fossils."

Thousands of fossil leaves were collected, particularly by Charles Sternberg, and sent to museums around the world. Sternberg's collecting localities near Fort Harker, however, are still something of a mystery and in any case are certainly located on private land. Fossil leaves are found occasionally in road cuts and other exposures. Years ago I did visit one locality west of Salina in a rocky pasture where just about every piece of sandstone I picked up had an impression of at least one leaf.

LeConte (1868) was certainly one of the first to mention invertebrates from western Kansas, listing fourteen "new" species from the Kiowa Shale (?) just west of Salina as he started his survey for the Union Pacific railroad. Then from the "clay and limestone" (probably the Greenhorn Formation) near present-day Bunker Hill in Russell County, he found "a *Belemnite* not described and *Inoceramus problematicus*," along with shark teeth and vertebrae. Further west, he noted a thin bed of oysters (*Pseudoperna congesta*) and "many teeth of fish and sharks." Several days later, still traveling west, he found fragments of "gigantic *Inocerami*" in the lower Smoky Hill Chalk.

*Figure 3.1. A dense growth of oysters (*Pseudoperna congesta) covers the outside of the lower (left) valve of a* Volviceramus grandis *inoceramid from Ellis County, Kansas.*

In his *Notes on the Tertiary and Cretaceous Periods of Kansas,* Mudge (1876; 1877) mentions that the fossils of the Fort Hays Limestone include "*Inocerami,* fragments of *Haploscapha* [*Volviceramus grandis* (Fig. 3.1)], *Ostrea,* with occasional remains of fish and Saurians." Of mollusks in the "Niobrara proper," he noted that the "most common are *Ostrea congesta* and *Inoceramus problematicus.* Less common but still seen in many strata, are the fragments of the large *Haploscapha* [*Platyceramus platinus*], with occasionally a perfect specimen from 30 to 33 inches in length. It is thin, with a transverse fiber like the *Inocerami,* and always lies crushed flat in numerous fragments, but lying in their normal position. A few *Gryphea;* also fragments, frequently weighing ten pounds or more of a large *Hippurites* near *H. Toncasianus* [the rudist, *Durania maxima*]. Near Sheridan, we discovered a bed of *Baculites ovatus.*"

According to Williston (1897), the *Baculites* found by Mudge would have been from the Pierre Shale Formation, a geological division that had not been identified in western Kansas during Mudge's time. Elias (1933) and Gill et al. (1972) provide the most recent information regarding the occurrence of invertebrates in the Pierre Shale of western Kansas. More recently, limestone concretions containing large numbers of *Baculites maclearni* have been collected from an exposure of the Sharon Springs Member of the Pierre Shale near McAllaster Butte in northwestern Logan County (pers. obs.). They are also found preserved fairly commonly in the layer of septarian concretions near the top of the Sharon Springs Member.

Williston (1897) noted that *Ostrea congesta* was much more abundant in the "Rudistes Beds" (lower chalk) than in the overlying "Hesperornis Beds" (upper chalk). It is worth noting here that "*Ostrea congesta,*" (Fig. 3.2) the species of oysters mentioned in

*Figure 3.2. This close-up shows a crowded community of oysters (*Pseudoperna congesta*) attached to a* Volviceramus grandis *fragment. It is easy to see where the species name, "congesta," was derived.*

several early accounts, was described and named by T. A. Conrad (1803–1877). At the time, Conrad was recognized as an expert on modern and fossil shells and had published numerous articles on modern freshwater species from the northeastern United States. The Cretaceous oysters and other specimens were collected during a survey of the Missouri River by J. N. Nicollet, a French mapmaker working for the U.S. government. The name was published as a footnote (Nicollet, 1843, p. 169): "*Conrad's description of the *ostrea congesta*: Elongated; upper valve flat; lower valve venticose, irregular; the umbo truncated by a mark of adhesion; resembles a little *gryphea vomer* of Morton." *Ostrea congesta* was placed in *Pseudoperna* by Stenzel in 1971. While the description and publication of the name are not up to current standards, they are accepted. The correct citation should be listed as *Pseudoperna congesta* (Conrad, in Nicollet, 1843).

Williston (1897) was the first to note that while "several species of *Inoceramus* are found in all horizons, the Haploscaphas (*Volviceramus grandis*) are abundant only in the lower horizons." J. D. Stewart (1990; pers. comm., 1992) indicated that *V. grandis* becomes extinct in Kansas just below Hattin's (1982) marker unit 6. This has been verified by the author several times at various localities over the years and it is useful as a stratigraphic marker. If you cannot find pieces of the thick, heavy shells of *V. grandis,* you are certainly above marker unit 6. As you move above Hattin's (1982) marker unit 3, however, the shells are not as large and are less likely to be preserved intact (pers. obs.). Williston (1897, p. 241) goes on to say that on the Smoky Hill River in Trego County, near the mouth of Hackberry Creek, "there are places where these shells can be gathered by the wagon load, often distorted, but not rarely in extraordinary perfection. A very thin shelled Inoceramid [*Platyceramus platinus*] measuring in the largest specimens forty-four by forty-six or eight inches is not rare over a large part of the exposures. Invariably where exposed, as they sometimes are in their entirety on low flat mounds of shale, they are broken into innumerable pieces. For that reason, I have never known of one being collected complete or even partially complete. Not withstanding their great size, the shell substance is not more than an eighth of an inch in thickness. Fragments of Rudistes [*Durania maxima*] are not rare in some places in the lowermost horizons and I have seen specimens near the Saline River northwest of Fort Hays into which one could thrust his arm to the elbow. They are totally wanting in the Hesperornis beds."

Williston's biostratigraphic observations regarding the occurrence of invertebrates are particularly interesting because they are quite accurate even after a hundred years and have served as the basis for more recent work by Stewart (1990), Everhart (2001), and others. Thin-shelled *Platyceramus platinus* bivalves are found throughout the Smoky Hill Chalk, reaching the 1.2 m (48 in.) size noted by Williston and other early collectors at about the midpoint of the formation (middle Santonian). The smaller diameter (60 cm

[24 in.] maximum) but much more robust *Volviceramus grandis* became extinct in the lower chalk (late Coniacian). An unusual inoceramid with a thin, rippled shell that is often referred to as the "snowshoe clam" for its odd shape, *Cladoceramus undulatoplicatus* is only found in a limited zone (early Santonian) in the chalk. It occurs for a brief (geologically speaking, that is; probably no more than a couple hundred thousand years) time after the extinction of *V. grandis*. Its occurrence, however, is used as a stratigraphic marker for the beginning of Santonian-age deposits around the world. Inoceramids are much more abundant in the lower half of the chalk, suggesting that bottom conditions were more favorable for them prior to the late Santonian and early Campanian.

Rudists are represented in the chalk by a single species, *Durania maxima*. They apparently were at the northern edge of their distribution during the Late Cretaceous and were not very successful in colonizing the mud bottom of the Western Interior Sea in Kansas. From what we are able to find, it appears they lived with most of their shell beneath the mud bottom, with only the expanded top of the lower shell and a small upper shell visible on the surface (Hattin, 1988). In the chalk they are mostly found as cone-shaped masses that represent a single bivalve, but occasionally they are found in groups that look much like a giant honeycomb. Although they are an unusually modified clam, in the warmer seas to the south and around the world they were colonial reef builders, similar in that regard to modern corals. Stewart (1990) noted that they occur through the lower one-half of the Smoky Hill (late Coniacian through middle Santonian), then disappear for most of the upper chalk, then reappear in small numbers at the top of the formation. My personal observation has been that they become very rare well before the middle of the chalk (around marker unit 6). I have also collected them just below the contact with the Sharon Springs Member of the Pierre Shale. Hattin (1988) noted what he believed to be evidence of predation on *Durania,* possibly by something like the shell-crushing shark *Ptychodus* (Chapter 4). I have never observed "bite marks" or anything similar in the field, but I have noted that the thick-walled, heavy shells of *Durania* are seldom found intact. Most of the specimens that are found are fragments of larger shells which were broken up before being preserved in the chalk (Fig. 3.3). It is difficult for me to imagine something living in the Western Interior Sea that was powerful enough (and hungry enough) to crush a bivalve that has a shell 5 cm (2 in.) thick. It may well be that the larger ptychodontid sharks could do it just to get at the body of the clam inside. That is one of the mysteries that I hope to resolve in the future.

Concerning cephalopods, Williston (1897) noted that ammonites are found only rarely in the chalk, and usually only their impressions remain. He recalled seeing impressions of one about a foot in diameter. This is consistent with my observations over the years. I've only seen two fragmentary impressions of ammonite shells (*Texanites*?) that had been found by an amateur collector

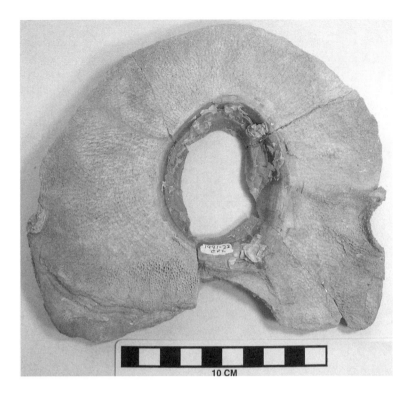

Figure 3.3. The roughly circular upper portion of the shell of a small Durania maxima *rudist from the late Coniacian chalk of Gove County. The small upper (aragonite) valve is never preserved in the chalk while the lower, cone-shaped portion of the main shell where the body of the* Durania *lived is missing in this specimen.*

(Stewart, pers. comm., 1993). Surprisingly, however, one of the fragile aptychi (jaw apparatus; see Lehman, 1979; Morton, 1981) of the ammonite was preserved on the surface of one of the molds. This occurs because the ammonite shell is made of aragonite, a calcium carbonate mineral that apparently dissolved (diagenesis) before it could be fossilized, while the thin, delicate-appearing aptychi are composed of much more stable calcite.

The presence or absence of certain species of cephalopods, particularly ammonites, scaphites, and baculites have been used by many workers to define the age of the rocks they are found in. Their use as stratigraphic markers is limited in the Smoky Hill Chalk because the aragonite shells were dissolved away and their molds or impressions were seldom preserved. Still, the limited evidence of their presence that is preserved can be useful. Hattin (1982) described the biostratigraphy of the Smoky Hill Chalk in regard to the occurrence of invertebrates and compared it with equivalent strata. Stewart (1990) used the occurrence of ammonite aptychi to define two of his biostratigraphic zones in the chalk (Fig. 3.4). The zone of "*Spinaptychus* n. sp." (new species; for a yet-undescribed type of ammonite aptychi) is found in the low chalk and is late Coniacian–early Santonian in age (ibid., p. 22). We have found another variety of aptychi which occurs just below this zone (designated *Spinaptychus* n. species "B," Stewart, pers. comm., 1992) that indicates the presence of another species of ammonite.

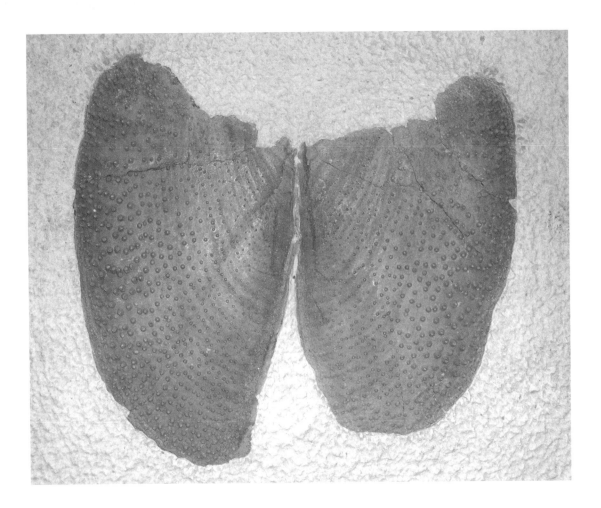

Figure 3.4. The paired aptychi (FHSM IP-528; Spinaptychus n. sp.) of an as yet unidentified ammonite found in the Smoky Hill Chalk. The delicate aptychi are composed of calcite and are preserved even though the much larger and thicker shell of the ammonite (aragonite) was dissolved.

Stewart's (1990) zone of *Clioscaphites vermiformis* and *C. choteauensis* is early Santonian and occurs below the middle of the chalk. In this case, the scaphites shells (another kind of ammonite) are represented only by the molds left in the chalk after the shells were dissolved. Hattin (1982, p. 28, also pl. 8, figs. 4, 5, 6) noted that "*Clioscaphites choteauensis* is not only common through several meters of strata that represents its zone but is also preserved excellently." The zone of *Spinaptychus sternbergi* (Stewart, 1990, p. 23) is Santonian in age and located immediately above the middle of the chalk, or Hattin's (1982) marker unit 10. Stewart (1990, p. 30) noted that the aptychi of a smaller ammonite (*Rugaptychus* sp.) occurs in the uppermost chalk (zone of *Hesperornis*).

Williston (1897, p. 242) was puzzled by fragments found occasionally in the chalk that are of a "glistening, fibrous nature." A specimen collected by H. T. Martin solved the mystery when it became apparent that the fragments represented the gladius of a "large cuttlefish, apparently different from any described species." The remains found by Martin were of a fragmentary specimen that was about six inches wide and a foot long, with a small "sepia bag"

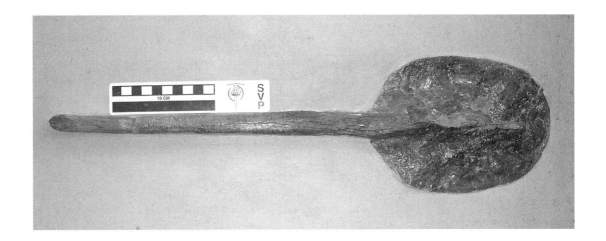

(ink sac) preserved below it. Logan (1898, p. 497) briefly described a new genus and species of squid (*Tusoteuthis longus*) from the specimen and noted that the specimen had been collected from near the top of the chalk ("Hesperornis beds"). While the root of the genus name "*teuthis*" is from the Greek, meaning "cuttlefish," it is also feminine in gender. Miller (1968) modified the masculine species name "*longus*" of Logan (1898) to the feminine "*longa*" to correct the oversight. The internal support structure (pen) of the *Tusoteuthis* (Fig. 3.5) was shaped somewhat like a tennis racquet, with a round, expanded area (gladius) at one end attached to a long central shaft (rachis). Fragmentary remains are fairly common in the chalk but are extremely fragile and seldom collected intact. The presence of large numbers of squid in the Western Interior Sea poses an interesting question regarding the marine environment since many modern squid seem to prefer deeper, colder waters than would have covered Kansas during the Late Cretaceous.

Figure 3.5. A cast of the internal support structure or "pen" of a Late Cretaceous squid, Tusoteuthis longa, in the collection of the Sternberg Museum of Natural History, Hays, Kansas.

Little additional work was done on the study of squid remains from the Smoky Hill Chalk for nearly sixty years. Miller (1957a) described a new species (*Niobrarateuthis bonneri* from a nearly complete specimen collected in Logan County by M. C. Bonner and noted that "the shape of its gladius differs materially from that of any previous described genus." The type specimen is on exhibit in the Sternberg Museum of Natural History, Hays, Kansas. Two additional new species, *Enchoteuthis melanae,* and *Kansasteuthis lindneri,* were named from single specimens discovered in Logan and Rooks Counties, respectively, and are also in the collection of the Sternberg Museum (Miller, 1968b). Stewart (1976) noted that the original description of *Tusoteuthis longa,* as well as those of *N. bonneri* and *E. melanae,* were "based on ventral views mistaken for dorsal views" and questioned the validity of the additional species. Another new species, *Niobrarateuthis walkeri,* was collected and described by Green (1977) from the lower chalk of Ellis County. Nicholls and Izaak (1987) reported on several specimens of *T. longa* from the Pembina Shale Member of the Pierre Shale

(Canada) and noted that their examination of the Kansas specimens indicated that only one species, not five, was present in the Western Interior Sea during the Late Cretaceous. In their paper regarding predation on cephalopods, Stewart and Carpenter (1990, p. 205) suggested that all teuthid species previously "described from the Niobrara Formation . . . are assignable to the taxon *Tusoteuthis longa*."

Occasionally the remains of these squid show bite marks from some unknown predator, most likely a large fish or a mosasaur. In one case (ibid.; Carpenter, 1996), the gladius of a squid was found inside the remains of a 1.5-m (5-ft.) fish called *Cimolichthys* (Chapter 5). The prey had been swallowed tail first and the open jaws of the predator suggest that the fish died with the "much too large to swallow" body of the squid lodged in its mouth. Stewart and Carpenter (1990) also reported on several coprolites containing fragments of a squid rachis (KUVP 65095; 65728; 65729). My experience indicates that fragmentary (partially eaten) squid remains, as noted by Logan (1898) and others, are common occurrences in the chalk.

The only echinoderms known from the chalk are an unusual form of crinoid called *Uintacrinus*. Related to starfish and sea urchins, these animals were apparently colonial and free-living (not attached to the sea bottom). *Uintacrinus socialis* was first found in the Uinta Mountains of Utah by Marsh's party in 1870, but was mentioned only as a "new and very interesting crinoid, allied apparently to the *Marsupites* of the English Chalk" (Marsh, 1871b, p. 195). Marsh's material from Utah was apparently so fragmentary that the species was actually described several years later largely from specimens found in the Smoky Hill Chalk of Kansas (Grinnell, 1876). The description of *Uintacrinus* provided by Charles H. Sternberg (1917, p. 156) is of some interest here because he and his sons collected many of the original specimens: "Their bodies were about the shape of half an egg, with an opening in the center, and ten arms radiating from the margin. These arms were three feet long, with feathered edges. Over the mouth, too, were smaller arms used to comb off into the mouth the tiny animal life of the sea, that was strained through, and caught in the meshes of the feathered arms." In the chalk, *Uintacrinus* appears to have been colonial because its remains indicate many individuals died at the same time and settled to the sea bottom, where they were preserved as thin layers of limestone. Miller et al. (1957) reported that the slabs are usually between a quarter inch and one inch in thickness. Williston (1897, p. 242) noted that he found a number of specimens in 1875 while collecting in the chalk with B. F. Mudge, and he indicated that some of these served as the types for the genus and species described by Grinnell.

According to Williston (ibid.), the large specimen now on display at the Museum of Natural History at the University of Kansas was found in 1891 near Elkader in Logan County. This specimen has more than a hundred individuals preserved in a slab measuring

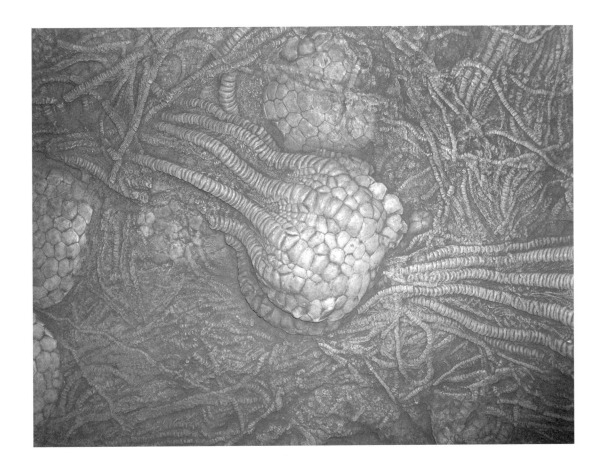

about six feet by four feet. The Sternberg Museum also has a nice specimen (Fig. 3.6) on exhibit from four miles east of Elkader. Williston (ibid.) noted that all the specimens he was familiar with came from the vicinity of Elkader (eastern Logan County). Miller et al. (1957) noted that the most eastern occurrence of *Uintacrinus* was at Castle Rock in eastern Gove County. While Miller et al. (ibid.) suggested that the specimens from Gove County were from lower in the chalk (older) that those from Logan County, more recent information (Hattin, 1982; Stewart, 1990) indicates that all the occurrences were from approximately the same stratigraphic level. *Uintacrinus* is known only from the early to middle Santonian, or just below the middle of the Smoky Hill Chalk, and is useful as a stratigraphic marker in marine rocks around the world.

Besides Williston's (1897) brief comments regarding invertebrates from the Smoky Hill Chalk, a number of other authors have also described their occurrence in some detail. Logan (1897; 1898; 1899) provided a lengthy discussion of the geology and the invertebrates of the Kansas Cretaceous. In addition to his summary of the geology, depositional environment, and invertebrate fauna, Miller (1968a; 1969) provided a review of previous work on invertebrates of the Niobrara Formation in Kansas. Most recently, Hattin (1982)

Figure 3.6. A close-up of the Late Cretaceous crinoid Uintacrinus socialis. *This specimen was found near Elkader in Logan County, and is on exhibit in the Sternberg Museum of Natural History, Hays, Kansas. The individual crinoid was preserved on the underside of a large mass of similar crinoids that died and were buried together. Such occurrences are not uncommon in some exposures near the middle of the chalk.*

described the invertebrate fauna of the Smoky Hill Chalk in relation to his stratigraphic references and provided excellent photographs of many species.

Other Invertebrates and Biostratigraphy

Other, less common invertebrates are known from the Smoky Hill Chalk but are seldom seen or collected. In addition to those mentioned above, Logan (1898, p. 481) included several other species of inoceramids, sponges (Coelenterates), serpulid worms, and arthropods (cirripeds and barnacles). In addition to listing the genera of Foraminifera that occur in the chalk, Miller (1968a) noted the occurrence of several species of oysters, *Baculites,* a pecten specimen, scaphites (*Clioscaphites*), four different kinds of ammonite aptychi, and several belemnites, including *Actinocamax walkeri* and *Belemnitella praecursor.* The occurrence of a small (23 mm) decapod crustacean (*Linuparis?*; KU 7295) is also documented (ibid., pp. 61–62). This may be the same specimen that Mudge (1876—below) had reported as a "rare crustacean" in a coprolite. I am not aware of any other crustacean specimens.

Hattin (1982, p. 71) provides a listing of invertebrates in the chalk by their method of preservation, with notes on their relative abundance. The bivalves, crinoids, and serpulids are examples of the preservation of calcareous skeletal material. Molds of scaphites, baculites, and ammonites make up the second category. Borings by sponges are the third form of fossil evidence. Squid pens represent the preservation of a combination of calcareous and organic matter. The crustacean noted above is an example of the rare fifth group, preserved chitin.

Hattin (ibid., pp. 26–29) was also the first to propose a biostratigraphy of the Smoky Hill Chalk based on the occurrence of the invertebrates. In Stewart's (1990) paper on the biostratigraphy of the Smoky Hill Chalk, he notes the occurrence of invertebrate species in each of his six biostratigraphic zones. My field experience validates Stewart's conclusions and indicates that *Volviceramus grandis* occurs only in the lower two zones, while *Platyceramus platinus* is found in the upper four zones. *Pseudoperna congesta* occurs abundantly in the lower five zones and probably is found throughout the chalk. The rudist *Durania maxima* is found in the first four zones, disappears, then reappears in the uppermost zone, just below the contact with the Pierre Shale. The squid *Tusoteuthis longa* is found throughout the chalk and, based on a specimen I collected in 1990 (Stewart, 1990, pers. comm.), also occurs just below the contact with the Fort Hays Limestone. Based on their aptychi, ammonites were probably present throughout the deposition of the chalk. The fluctuations noted in the presence or absence of various invertebrate species may be due to changes in the bottom environment (i.e., decreased oxygen), a changing climate (i.e., cooling, Stewart, 1990, p. 26), other factors which we do not understand, or lack of collecting.

Figure 3.7. A 1.6 cm (0.6 in.) fossil blister pearl attached to a fragment of a Volviceramus grandis *shell. Fossil pearls are fairly common discoveries in the Smoky Hill Chalk.*

Pearls

Brown (1940) reported on the presence of fossil pearls in the Niobrara (Smoky Hill Chalk) and Benton (Greenhorn Limestone/Carlile Shale), noting that many of the specimens had been collected by G. F. Sternberg west of Hays (Fig. 3.7). These pearls are tan to light gray in color, have lost their shiny outer layer (nacre) and occur in round, unattached forms or hemispherical shapes attached to a piece of inoceramid shell (ibid.). In some cases, the attachment occurred post-mortem when the pearl was pressed against the surface of the shell by the pressure of the overlying rocks. The largest of these pearls examined by Brown (ibid., p. 367) had a diameter of 2 cm. A clipping from the Hays newspaper in 1940 indicated that George Sternberg had donated fifty fossil pearls to the Smithsonian. These are probably the same ones reported by Brown (1940). While not rare in the chalk, they are difficult to find because they normally occur in strata where inoceramids are most common and where they are obscured by thousands of other shell fragments. Kauffman (1990, p. 66) described several giant pearls from the Smoky Hill Chalk, including a pendant pearl in the collection of the Museum of Comparative Zoology at Harvard that was 11 cm long and 6.5 cm in maximum width. The largest pearls come from the deeper, bowl-shaped shells like *Volviceramus grandis* (ibid., p. 68).

I have collected five or six pearls from the Smoky Hill Chalk over the years, but I have to admit that I wasn't specifically looking for them at any point. The largest one I have ever collected was found in the low chalk of Gove County in 2003. The specimen (FHSM IP-1451) represents an apparently huge, hemispherical pearl, 4–5 cm in diameter and 2.5 cm high, or roughly the size of half a golf ball (Fig. 3.8). With the exception of a large, very irregular specimen, all the rest of the Smoky Hill Chalk pearls in my personal collection are less than 1.5 cm in diameter. In late 2003 I collected several smaller pearls from the base of the Lincoln Limestone Member of the Greenhorn Limestone (Upper Cenomanian) in Russell County in association with numerous shark teeth. These pearls were much smaller than those collected in the Smoky Hill Chalk, averaging 4–5 mm in diameter. A polished cross-section shows concentric layers of calcite formed around a central nucleus.

Coprolites

"Coprolite" is a scientific term for the fossilized excrement, feces, or droppings of ancient animals. It was coined by Dr. William Buckland (1829). Coprolites are trace fossils, rather than body fossils (bones, teeth, etc.). Hawkins (1834, pls. 27–28) illustrated several coprolites in his book on ichthyosaurs and plesiosaurs from the Jurassic of England. In the Smoky Hill Chalk, the coprolites that we find are probably from sharks, bony fish, and marine reptiles. This fossilized waste can often tell us much about what kind

Figure 3.8. Top and bottom views of a giant inoceramid pearl (FHSM IP-1451) from the late Coniacian chalk of Gove County. This may be the second-largest pearl known from the chalk after the pendant pearl specimen cited by Kauffman (1990) in the collection of the Museum of Comparative Zoology at Harvard. (Scale = cm)

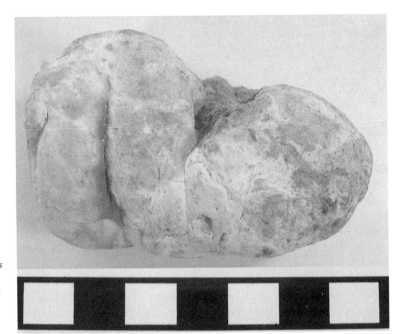

Figure 3.9. This large coprolite was found in the lower chalk of Gove County and probably is from a mosasaur, although there is no way to tell for certain. Coprolites are frequently found in the chalk and sometimes contain the bones of fish or other prey. (Scale = cm)

of prey these animals ate and even how their digestive systems worked. B. F. Mudge and S. W. Williston were among the first to discuss the presence and origin of coprolites in the Smoky Hill Chalk.

Coprolites are fairly common in the chalk and may contain partially digested bones (usually fish), teeth (Shimada, 1997), and even fragments of squid pens (Stewart and Carpenter, 1990). Sharks are thought to be the source of the so-called spiral coprolites (Stewart, 1978), but most of the other larger ones (Fig. 3.9) are likely to be from marine reptiles (mosasaurs). Coprolites tend to be more resistant to erosion than the chalk because they are composed largely of less soluble calcium phosphate instead of calcium carbonate. Fossilized gut contents (Miller, 1957b; Stewart, 1978) and regurgitated prey material (see Hattin, 1996; Everhart, 1999) are two other categories of unusual biological materials preserved in the chalk.

One of the first notes of coprolites in Kansas comes from Prof. Benjamin Mudge (1876, p. 217), the first state geologist of Kansas: "Coprolites of fish and Saurians are frequently found, containing the remains of the food of the animal. Small fish appear to be the most common food; but in one instance a rare crustacean was found preserved this way. The coprolites are not so hard as those of Europe, being a little firmer than chalk, and finer grained. . . . In some cases the undigested organic material (bones) was one-fourth the whole weight." Williston (1898, p. 214) also mentioned coprolites in his paper on mosasaurs: "Coprolites which I have always had reason to believe were from these animals are in some places very abundant, weighing from an ounce or two up to a half a pound or more. They are ovoidal in shape with sphincter or intestinal impressions upon them, and contain very comminuted parts of fish bones, fish scales, etc." Williston (1902) added that "coprolites, evidently of mosasaurs, are frequently found, with large, undigested fish bones, and fragments of fish bones."

Stewart (1990, p. 22) noted that "spiral coprolites, argued to be enterospirae by Stewart (1978), occur in this horizon [zone of *Spinaptychus* n. sp.]. McAllister (1985) has conclusively demonstrated that such structures can be found in the colon of a dissected *Scyliorhinus canicula* [a small modern shark]. Presumably, they can also be expelled intact from the rectum. Their presence in this horizon and absence in the other horizons implies the presence of some heretofore undetected shark taxon in this zone. Other types of coprolites occur in all horizons of the Smoky Hill Chalk Member."

Other things we find in the low chalk (late Coniacian) are partially digested bone fragments of fish and marine reptiles, and even limb bones of small dinosaurs. (Everhart, 1999; Everhart, 2003; Everhart, 2004a). The surface layer of the bone is usually heavily pitted, and any teeth in jaw fragments appear to have been dissolved back into their sockets. Because some of the fragments include embedded shark teeth or identifiable bite marks, I interpret them as regurgitated stomach contents. In most cases, the specimens represent pieces of the prey that were big enough to require a large shark to swallow them. Why do I think the specimens are the regurgitate from sharks and not some other large predator? With notable exceptions (Sternberg, 1922; Martin and Bjork, 1987, Everhart, 2004a), mosasaurs ate fish. However, they had digestive systems that were capable of digesting heavy bone. They were also more likely to swallow their prey whole and their teeth were not capable of shearing through large bones. Sharks, on the other hand, are known to carve up their prey, and big sharks like *Cretoxyrhina* had teeth and jaws strong enough to shear bone. We know this from pieces of mosasaur vertebrae (FHSM VP-13283) that have been severed and retain fragments of embedded shark teeth (Everhart, 1999). The digestive system of sharks, however, may not be able to completely dissolve heavy bone, and sharks are known to be able to regurgitate materials in their stomach when stressed (Stewart, pers. comm., 1994).

Sometimes we find structures that look like flattened coprolites but are composed of ground up fragments of oyster shell (Fig. 3.10). They are oval in shape and are always flattened, with average measurements of 4 cm x 2.5 cm x 1 cm (Everhart and Everhart, 1992). The smooth underside sometimes bears impressions that appear to be from the anal sphincter of some predator. Superficially, they resemble burrow structures but do not include any other debris which might be expected from winnowing of sediments by burrowing crustaceans. Interestingly, they occur contemporaneously with an unusual fish called *Martinichthys* (see Chapter 5) in the lower chalk from just below Hattin's marker unit 4 to just above marker unit 5. I suspect but have been unable to prove that

Figure 3.10. These masses of ground-up oyster shell (Pseudoperna congesta) appear to be the coprolites of some predator that apparently fed on the abundant colonies of oysters that were attached to inoceramids. They are only found in a limited stratigraphic zone in the late Coniacian chalk. (Scale = cm)

Martinichthys fed on the abundant oysters and that these shell-filled structures are the coprolites of that genus. Other, less defined masses of tiny inoceramid prisms appear to be either coprolites or regurgitates from a predator, possibly *Ptychodus,* feeding on young inoceramids.

Remains of Logs and Tree Limbs

Mudge (1877, pp. 283–284) was among the first to note the presence of "an occasional fragment of fossilized wood. . . . This wood was, in a few instances, bored before fossilization by some small animal. This might have been done by the larva of an insect (a 'borer') when the tree was living, or later by a teredo [*sic*] when the trunk floated in water. In either case it shows that the Cretaceous vegetation was subject to the same enemies as that of the present period. Some of this wood was in charred condition [carbonized] and would burn freely. Other specimens were changed into almost pure silica, the cavities studded with quartz. In one case a log, weighing about 500 pounds, had all conditions of the transformation; a portion had the appearance of soft decayed wood, which crumbled in handling, and other parts ringing like flint under a hammer. Occasionally specimens were converted into chalcedony, but the annual growth of the wood distinctly remained. In a single instance we detected the fibrous structure of the palm."

Williston (1897) noted that "fossil wood is occasionally found in the formation. A tree about thirty feet long was discovered near Elkader a year or two ago [about 1895]" and that fragments of amber had been found in association with the bark of these trees. I have occasionally come across places where small fragments of a shiny, coal-black material are eroding out of the chalk. When the source is uncovered, it turns out to be the remains of a log or tree limb that has been badly compressed by the weight of the overlying chalk. Unlike "petrified wood" that has been replaced with agate or other minerals, this wood appears to have been "carbonized" during preservation to the point that little is left but the carbon that made up the original cellulose in the wood. Carbonized wood from the chalk is fragile and breaks up readily into blocky chunks. It will burn if placed in a flame. The inside of a carbonized log when first opened smells faintly like burnt wood, but the wood was never in a fire. These are simply the remains of waterlogged trees that sank to the bottom of the sea and were buried in the chalky mud. Sometimes the logs may be 6–7 m (20+ ft.) long but are crushed to a thickness of less than 5 cm (2 in.). Inside the coal-like, carbonized portions, small pockets of delicate, fibrous wood remain. The outside of a log may be covered with patches of black calcite crystals that incorporated some of the underlying carbon as they grew. Occasionally, attached shells, mostly oysters, are visible. *Teredolite* (shipworm) borings, if any, are rare in the Smoky Hill Chalk although they are quite common in wood found the underlying Fairport Chalk Member of the Carlile Shale (pers. obs.).

Collecting fossils in the Smoky Hill Chalk is much like walking along a beach. If you look long enough, you are likely to come across the remains of just about anything that lived in the sea during the time that the chalk was being deposited. We try to collect everything that looks unusual, and if we cannot immediately identify it, we save it until we can. As our friend J. D. Stewart once told us (pers. comm., 1990), "Go ahead and pick up everything you see. You can always throw it away later. If you don't pick it up, you may never see it again." A number of our discoveries over the years, including the only known Coelacanth remains (Chapter 5) from the Smoky Hill Chalk, were collected in large part because we didn't know what they were at the time.

Four

Sharks: Sharp Teeth and Shell Crushers

The young adult mosasaur struggled wearily as it swam slowly just below the calm surface of the warm, shallow sea. It was the peak of the mosasaur breeding season. A chance encounter earlier that day with an enraged and much larger bull mosasaur had left his left front flipper crippled and useless. The six-meter marine reptile kept rolling over on its left side because the damaged flipper no longer provided the necessary steerage to keep him upright. Surfacing to breathe was difficult and his efforts to bring his head above the surface of the water were noisy. Trying to swim without the use of the flipper had tired the creature to the point of exhaustion and had also attracted the attention of other scavengers in the area.

Even though bleeding from the superficial wounds had stopped hours ago, a swarm of small *Squalicorax* sharks still cruised warily in circles around the struggling mosasaur, occasionally darting in close as if sizing up the best opportunity to attack. The largest of these sharks was less than a third the length of the mosasaur and normally would not have represented a threat. In fact, the mosasaur had killed and eaten many such sharks over the years.

Now the tables had turned and the wounded mosasaur was potential prey instead of the predator. There was no place to hide in the middle of the Western Interior Sea and no protection to be had from others of his own kind.

From deep below and far outside the circle of scavenger sharks, another much larger shark moved steadily toward the mosasaur. The shark's acute senses had detected the thrashing noises made by the wounded reptile from more than a mile away and alerted the shark to the potential feeding opportunity. Almost as long as the snakelike mosasaur and twice as heavy, the giant shark scattered the smaller scavengers in its path as it approached the mosasaur from below at high speed. The open jaws of the shark hit the mosasaur near the center of its body, on the right side, just behind the rib cage, and the impact lifted the wounded animal almost completely out of the water. As the massive jaws closed, rows of razor-sharp, two-inch teeth sliced out a dinner plate–sized chunk of skin, muscle, bones, and intestines from the mosasaur, opening a fatal wound.

Then, before the reptile could even react to the first bite, the jaws closed again, this time across the two-inch-wide bones of the spinal column. Sharp teeth sliced through muscle and vertebrae, severing the spinal cord and major blood vessels. As the powerful bite of the shark drove its teeth through the mosasaur's vertebrae, the tips of several teeth were broken off in the bone. The water quickly filled with blood as the mosasaur thrashed briefly in its death throes, its body almost cut completely in two by the shark's terrible bite.

Then, as the shark opened and closed its mouth rapidly to swallow the large chunks of flesh and bone, the other scavengers flashed into the blood-filled waters and attacked the carcass of the dead mosasaur. The big shark started to move back into the feeding frenzy, then abruptly changed directions, swimming slowly away from the swarming scavengers. Its immediate hunger satisfied, the shark turned its attention away from the battle over the remains of the mosasaur and resumed its westward journey through the warm waters. Hours later, the shark opened its mouth and regurgitated a foot-long chunk of indigestible mosasaur vertebrae.

The five vertebrae, severely eroded in places by the shark's stomach acid, but still held together by tough ligaments, sank rapidly through 400 feet of water and landed on the soft chalky mud of the sea bottom. The impact partially buried the vertebrae and protected them from further attack by scavengers. Before long, they had completely disappeared beneath the surface of the mud, where they would remain for almost 85 million years. Thousands of feet of sediment eventually covered the remains before the land rose and the ocean receded across hundreds of miles to the present Gulf of Mexico.

Over the span of millions of years, the chalky sediments were gradually compressed into layers of soft rock and then lifted nearly a mile above sea level. As this was happening, erosion wore away more than a thousand feet of shale and chalk before the vertebrae were exposed to the light of day.

Then, in 1995, a sharp-eyed, bipedal mammal wandering across the dry, eroded surface of the exposed sea bottom happened to chance upon the lumps of reddish brown bone weathering from the gray chalk. Quickly recognizing them as mosasaur vertebrae, the human took a sharp instrument and a brush out of his pack and carefully exposed the rest of the twelve-inch-long specimen. Disappointment soon clouded the human's face as he searched in vain along the chalk exposure for the rest of the mosasaur. Five vertebrae was not much of a consolation when you were hoping to find a twenty-foot-long mosasaur. Giving up, he marked and bagged each vertebra, put them in his pack, and resumed his search. It was more than a week later, when he cleaned up the specimen, that he saw the bite marks and embedded sharks' teeth and realized what he had found.

The story is fiction, of course, except for the ending where I found and examined the mosasaur vertebrae (Shimada, 1997c; Everhart, 1999). We will never know the exact circumstances regarding what happened to this particular mosasaur some 85 million years ago. Dan Varner's painting on the cover of this book shows a ginsu shark attacking a living mosasaur. In contrast, as my friend Earl Manning likes to remind me, the actual scenario may have been much less dramatic. Manning (pers. comm., 2002) suggested that it was far more likely that the mosasaur in question was a long-dead and bloated carcass floating at the surface and slowly coming apart. In such cases, sharks and other scavengers would have plenty of time to remove the more readily detachable parts like the limbs and tail, generally leaving only the skull and dorsal vertebrae to be preserved as fossils. "Bloating and floating" is certainly the case in many instances and is the only reasonable explanation for how the remains of large dinosaurs, such as *Niobrarasaurus coleii* (see Carpenter et al., 1995; Chapter 12) could have found their way into the middle of the Western Interior Sea. However, it is also reasonable to assume that these Cretaceous sharks would have behaved much like their modern cousins. Sharks are part of Mother Nature's "clean-up crew" and play a vital role in recycling organic resources back into the marine ecosystem. Sick, wounded, and otherwise vulnerable animals were likely suitable prey for ancient sharks just as they are for modern sharks.

The limited remains of this mosasaur (FHSM VP-13283) provide a number of clues that lead me to believe that the fictional story in this case is not far from the truth. The specimen consists of five vertebrae from the lower back of what was probably a young adult mosasaur named *Platecarpus tympaniticus*. The vertebrae are 5 cm (2 in.) in diameter and were still articulated when found. The three center vertebrae still have a partial rib remaining on the left side. The front and back vertebrae of the series have been severed by the shark's bite—the anterior one is cut across at an angle of about 45 degrees, the posterior one at right angles to the mosasaur's spine. Most, but not all, of the surface of the bones is corroded in a way that is consistent with the attack of stomach acids (Varricchio, 2001; Everhart, 2004b). This is especially evident on the cut ends of the two vertebrae at the front and back of the specimen that would not have had a protective covering of skin, muscle, and connective tissue and so would have been exposed to the digestive process for the longest time inside the shark's stomach.

The "smoking gun" that identifies the attacker in this case is the three teeth, or rather the broken pieces of teeth, that were left behind. Fragments of two *Cretoxyrhina mantelli* teeth are still embedded in the bones of the mosasaur (FHSM VP-13284). The tip of one tooth is in the cut and eroded surface of the last vertebrae, and a second, much larger piece is broken off in the right side of the second vertebrae. A third tooth fragment, which was apparently preserved but not recovered, left its impression pressed into the right side of vertebra number four (Shimada, 1997c, fig. 4). There are several scratches on the bones that are interpreted as the marks left by other teeth as the shark opened and closed its jaws in the process of swallowing the large piece of the mosasaur. The dorsal processes are missing on all the vertebrae. It is impossible to tell if the bite or bites that severed these five vertebrae from the spine of the mosasaur occurred while the mosasaur was living or as the result of scavenging on a rotted, floating carcass. In either case, however, the shark's ability to bite through large, solid bones is evident in this and similar remains preserved from the Smoky Hill Chalk (Fig. 4.1). Further evidence of this feeding behavior was described by Shimada and Hooks (2004) in regard to the bones of protostegid turtles found in the Mooreville Chalk of Alabama that had been bitten by *Cretoxyrhina mantelli*.

This mosasaur specimen is similar to many other partial re-

Figure 4.1. The razor-edged teeth of the ginsu shark (Cretoxyrhina mantelli) in lingual view. These teeth may reach 5 cm (2 in.) in height and were not serrated. The ginsu shark grew to lengths of over 6 m (20 ft.). (Scale = cm)

mains found in the chalk (Everhart, 1999). FHSM VP-13744 is a series of five cervical (neck) vertebrae from a medium-sized (species unknown) mosasaur. Again, the first and last vertebrae in the series have been bitten through, but there are no imbedded teeth. All the vertebrae show the effects of a fairly long exposure to stomach acids (Varricchio, 2001). In another specimen, a series of about twenty much smaller vertebrae from the end of a mosasaur's tail were found lying on the surface of the chalk (FHSM VP-13750). In this case, not only the final dismemberment of the mosasaur's tail is recorded by the presence of numerous severed vertebrae, but there is also preserved evidence of an earlier, healed bite near the end of the tail. That wound became infected and fused at least five vertebrae together into a clublike mass on the end of the tail. This "bad luck" mosasaur apparently survived one probable shark attack and lived long enough for the first wound to heal. Martin and Rothschild (1989) reported a similar fused series of mosasaur vertebrae (KUVP 1094) as a likely result of an infection following a shark bite (see also Shimada, 1997c, fig. 3). In this case, the broken tip of at least one shark tooth is still in the vertebrae.

Several years ago, my friend Tom Caggiano (see Chapter 12) found the front one-quarter or so of the skull of a small mosasaur (FHSM VP-13748), probably a young *Platecarpus tympaniticus* (Fig. 4.2). The muzzle had been sheared off right behind the third tooth in both upper jaws. The anterior ends of both the right and left maxilla and the premaxilla were still sutured tightly together, but all the teeth had been dissolved back into their sockets and the surface of the bone was corroded by stomach acids. In this case, the front of the mosasaur's skull had been bitten off by a shark and partially digested before being regurgitated. There were no other bite marks or imbedded teeth, so the identity of the attacker/scavenger is not known for certain. However, the remains are more consistent with the bone-shearing bite of a large *Cretoxyrhina* than with the flesh-slicing bite of a smaller *Squalicorax*.

Most other partially digested specimens consist of a smaller number of bones or smaller-sized remains. One or two vertebrae are fairly common, and so are small sections of mosasaur tail vertebrae. Pieces of skulls have been found with the teeth always dissolved down into the sockets. Limb bones and girdle elements (shoulder and pelvis) are also parts that are readily detachable from a mosasaur carcass. More recently, Hamm and Everhart (2001) reported on the radius and ulna of a juvenile nodosaur found in the Smoky Hill Chalk. Two long, unserrated bite marks suggest that the lower limb bones had probably been severed from a floating carcass by a large *Cretoxyrhina,* and the appearance of the surface of the bones indicated that they had been partially digested.

Another specimen is intriguing to me because it indicates an attack on a mosasaur but leaves us without knowing the immediate outcome. In 1990, I discovered a one-meter (39-in.) skull of a large *Tylosaurus* sp. (FHSM VP-13742) in Gove County. A skull of this size would have come from a tylosaur that was about seven meters

Figure 4.2. Dorsal (top) and ventral (bottom) views of the remains of the anterior end (premaxilla and both maxillae) of a juvenile mosasaur skull (FHSM VP-13748) that had been bitten off and partially digested, most likely by a large shark. Note that the teeth have been digested completely and are dissolved deeply into their sockets. (Scale = cm)

long (Everhart, 2002). The skull had weathered out the lower chalk but appeared to be relatively complete. There were, however, no post-cranial elements in the remains except for a few finger bones (phalanges). Apparently the skull had either fallen off a drifting carcass or scavengers had separated it from the rest of the skeleton before it was buried. The interesting part is that there were the tips of three broken *Cretoxyrhina* teeth embedded in the skull—one in the top of the premaxilla and one each in the outside of both dentaries. In addition, there were bite marks on the top of the premaxilla, and the tip of the premaxilla had been sheared off at an angle (Fig. 4.3). None of the bites appeared to be fatal, although the teeth embedded in the lower jaws may indicate a bite to the throat of the mosasaur. No healing of the bites was indicated. Damage from weathering precluded much further analysis of this skull, but it was evident that the large mosasaur had been bitten several times by a fairly large shark.

Figure 4.3. Left oblique view of the premaxilla of a large Tylosaurus *skull (FHSM VP-13742) that had been bitten by a ginsu shark (Cretoxyrhina mantelli). Note that the end of the premaxilla (far left) was sheared off by the bite of the shark. The broken tip of a* Cretoxyrhina *tooth is still embedded in the top of the premaxilla (T). (Scale = cm)*

Two years later, my wife, Pam, found three dorsal vertebrae from a medium-sized (5–6-m) mosasaur eroding from the edge of a gully. Our usual division of labor in the field is fairly simple: she finds them and I get to dig them. A quick check showed that at least two more vertebrae continued into the chalk. They were coming out "round end" (condyle) first, which indicated that they were pointing toward the head of the animal, not the tail. Finding articulated vertebrae is good news; it usually means that there are more to be found. I had to remove about half a meter (20 in.) of chalk that was sitting on top of the layer containing the vertebrae. It was several hours of hard digging before I was able to see what was hidden underneath. Twenty-two vertebrae were eventually uncovered, along with ribs lying on either side of the vertebral column. It was looking really good until I got to the twenty-first vertebra and found empty chalk. I dug some more and found one more vertebra, the last cervical, lying off to the side. Mosasaurs have seven cervical vertebrae. In this case, I was six vertebrae away from where the skull should have been. I dug a deeper into the chalk, looking for the skull, and found nothing. Tired and disgusted, I gave up looking for the skull and got busy recovering the vertebrae and ribs. After all that work, what I had was a headless, tailless, limbless string of vertebrae with a few ribs attached (FHSM VP-13746).

Much later, when I prepared the bones, I began to notice some obvious signs of "foul play." There was good evidence of at least three bites by a big shark along the five-foot string of vertebrae (Everhart, 2004c). The tooth marks were spaced about 3 cm (1.25 in.) apart and indicated the bites had come from a very large shark. As the result of one bite, the tip of a *Cretoxyrhina* tooth was still

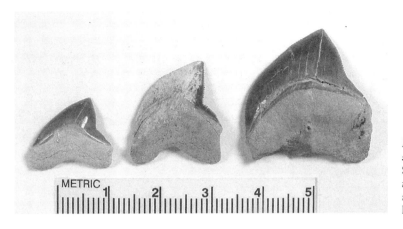

Figure 4.4. Lingual views of the serrated teeth of three species of Squalicorax (crow shark) found in the Smoky Hill Chalk (left to right): Squalicorax falcatus, S. kaupi, and S. pristodontus.

embedded in a vertebra. The ribs on one side of the carcass had been bitten off cleanly, but were reasonably intact on the other. The most anterior vertebra of the specimen, the last one I found, had been twisted to the side and showed two deep, long gouges from a ripping shark bite. And, as if to add insult to the injury, all the ribs showed the serrated bite marks of another shark, *Squalicorax falcatus*. In some areas it appeared as if the smaller shark(s) had grabbed hold of the ribs with its serrated teeth and then slid downward, peeling off the flesh and leaving long, parallel scratches on the bone. I'm sure the mosasaur was long dead at the time, but the bite marks looked particularly painful. Although it's probably not possible to establish the sequence of who was dining when, I think that one or more large *Cretoxyrhina* tore the carcass apart first and then the smaller *Squalicorax* sharks finished cleaning the bones. Or, in reverse order, it is possible that the smaller sharks may have fed first on a dead mosasaur and then were driven off by the larger predator. In any case, the remains show the interaction of three species, unusual in the chalk.

From my observations over the years, I know that *Squalicorax* probably scavenged any and all kinds of food sources (fish, turtle, mosasaur, plesiosaur, pteranodon, and even other sharks) that were available. Druckenmiller et al. (1993) reported the discovery of a *Squalicorax falcatus* specimen that was closely associated with the remains of a fish (*Ichthyodectes ctenodon*), a small turtle, and mosasaur paddle elements, which they interpreted as stomach contents of the shark. Their serrated teeth left distinctive marks on the bones of many kinds of marine animals (Fig. 4.4). In fact, at least in the low chalk, it is unusual *not* to find *Squalicorax* bite marks on larger specimens. Even the large (9 m [29 ft.]) *Tylosaurus proriger* (FHSM VP-3) specimen on exhibit at the Sternberg Museum has bite marks from serrated teeth on the right ilium, and a broken *Squalicorax* tooth was collected with the remains. Schwimmer et al. (1997) documented the evidence of *Squalicorax* scavenging on various species of vertebrates from specimens found in museum

Figure 4.5. Serrated bite marks left by the teeth of Squalicorax falcatus *on the ribs of a mosasaur* (Platecarpus tympaniticus). *In general, these sharks stripped the flesh from the bones of their prey and did not bite through bones as did the larger ginsu sharks.*

collections. As indirect evidence, their list of attributable bite marks (ibid., pp. 77–79) includes twenty-eight museum specimens, from a shark (*Ptychodus mortoni,* KUVP 59041) through a dinosaur (Nodosauridae, indet., San Diego Natural History Museum 33909). The list also includes bony fish, turtles, mosasaurs, and plesiosaurs. More directly, they also note the tip of a *Squalicorax* tooth embedded in a mosasaur vertebra and in a juvenile hadrosaur metatarsal (ibid., p. 76). Both specimens are from Alabama. During the 1997 Society of Vertebrate Paleontology meeting in Chicago, I was with David Schwimmer when we discovered serrated bite marks on a juvenile hadrosaur limb bone (FMNH PR 27383) from the Selma group of the Mooreville Formation (Late Cretaceous), Dallas County, Alabama, in the Field Museum collection. Since then, I have noted many more instances of *Squalicorax* bite marks and teeth in association with vertebrate remains in the Smoky Hill Chalk (Fig. 4.5).

Sharks in the Late Cretaceous Western Interior Sea

Although it is about as far from an ocean as you can get, Kansas is a great, if unrecognized, place to find shark teeth. The first vertebrate remains reported in the state were two shark teeth and a dorsal fin spine from three different species of shark from the "Carboniferous of Kansas" (Leidy, 1859). As settlements and the railroads moved further west, the first remains of Late Cretaceous sharks were recovered. LeConte (1868, p. 8) reported the "teeth and vertebrae of a species of shark belonging to the genus *Lamna*" from rocks near Bunker Hill station (now Bunker Hill, Kansas, in Russell County). Further west, on the bank of a small stream near Walker Creek, LeConte (ibid.) also noted that "the strata of slaty [*sic*] clay contains many teeth of fishes and sharks." Exposures in this area are mostly from the Upper Cenomanian/Lower Turonian Greenhorn Limestone and Carlile Shale.

Leidy (1868) was the first to report on a tooth of *Ptychodus* (crusher shark) from the "Cretaceous series" of western Kansas when he described a single, damaged tooth from "a few miles east of Fort Hays, Kansas." The tooth was found by Joseph LeConte during the survey for the Kansas Pacific Railroad in 1867 and donated to the Academy of Natural Sciences of Philadelphia (ANSP) upon his return to Philadelphia. Leidy considered it to be a new species and named it *Ptychodus occidentalis*. Although the tooth was later lost from the collection of the ANSP (Spamer et al., 1995), the figure shown in Leidy (1873, pls. 17 and 18) appears to me more likely to be a marginal tooth of *P. mortoni*. Leidy (1873, pl. 18, figs. 1–10 [*Ptychodus*], figs. 21–23 [*Cretoxyrhina*] and figs. 29–31 [*Squalicorax*]) was also the first to publish figures of shark teeth from the Smoky Hill Chalk, reporting on specimens collected and donated by Dr. George M. Sternberg during his tour of duty with the Army in Kansas in the late 1860s. Dr. Sternberg was the older brother of Charles H. Sternberg and was one of the first fossil collectors in western Kansas, but he is better known for his work in bacteriology and as the Surgeon General of the Army during the Spanish-American War (Sternberg, 1920).

While O. C. Marsh appears to have purposefully ignored all fish and sharks in Kansas, E. D. Cope did just the opposite, and as a consequence he named a number of new species, mostly bony fish. Possibly without realizing it, Marsh recovered two *Squalicorax kaupi* teeth (YPM 56409) in association with *Pteranodon* remains (*Pteranodon longiceps*? YPM 1169; Shimada, pers. comm., 2003) that he discovered in July 1871 (Marsh, 1871). Cope (1872a, p. 324) noted that in the "Cretaceous ocean of the West, . . . sharks do not seem to have been so common as in the old Atlantic." Later that same year, Cope (1872b, pp. 355–357) stated that "the remains of sharks and rays are far less abundant in the Cretaceous of Western Kansas than in New Jersey," but then added that "in the region near Fort Hays and Salina [probably the Greenhorn Limestone], shark's teeth are more frequently found." Following brief descriptions, Cope (ibid.) then named two new species of sharks from the Niobrara "near Fort Wallace." *Galeocerdo crassidens* Cope (*Squalicorax kaupi*) was named from two teeth, and *G. hartwelli* Cope (*Squalicorax falcatus*) from a single tooth found beneath the bones of his giant *Protostega gigas* specimen (Chapter 6). Interestingly, the species "*hartwelli*" was named for one of Cope's assistants on that dig, M. V. Hartwell. Siverson (pers. comm., 1998) suggested that *Squalicorax hartwelli* (Cope 1872b) should probably be used for the teeth found in North America instead of *S. falcatus* (Agassiz, 1843), which was originally authored to describe similar teeth found in Europe. The revision, however, is not without some disagreement by other shark experts and has not been published.

In Cope's (1875) *Vertebrata*, he again mentions the two "*Galeocerdo*" species along with naming a new species of shell-crushing shark, *Ptychodus janewayi* (named for Dr. John H. Janeway, an Army surgeon at Fort Hays who assisted Cope when he visited

Kansas). He further notes that the *P. janewayi* teeth had been found near Stockton, in Rooks County, "in a bed containing many teeth of '*Oxyrhina*' (*Cretoxyrhina*) and '*Lamna*' (*Cretolamna*)." For some reason, instead of illustrating these specimens, Cope describes and figures (ibid., p. 297, pl. 42, figs. 9–12) several teeth of *Lamna* (*Leptostyrax*) *macrorhiza* Cope from Ellis County and *L. mudgei* Cope from the "Niobrara Epoch of Kansas."

Expressing a different point of view on the abundance of shark teeth, B. F. Mudge (1876, p. 217) reported that the "teeth of Selachians are quite common. At one locality, over 400 were collected in an area of 30 inches, and apparently from the jaws of one individual—a *Ptychodus*—and all in excellent preservation." In a report to the Kansas Academy of Science at the annual meeting, Mudge (1877a, p. 5) noted discovery of "the teeth, cartilaginous jaw and vertebrae of a shark, *Galeocerdo* [*Squalicorax*] *falcatus.*" In another version of that paper, Mudge (1877b, p. 287) reported the evidence of scavenging on marine reptiles by sharks, noting that "frequently . . . the bones of Saurians were found with the marks of the serrate teeth of *Galeocerdo*, which could not have been made unless the bones were still fresh and unhardened."

A fragmentary *Xiphactinus audax* skull (KUVP 155) collected by B. F. Mudge in the early 1870s from the Carlile Shale (middle Turonian) of Russell County, shows numerous *Cretoxyrhina* bite marks on the lower jaw.

Earlier, Cope (1872a, p. 322) had also mentioned scavenging by sharks in his description of mosasaur remains he had found: "While lying on the bottom of the Cretaceous sea, the carcass had been dragged hither and thither by the sharks and other rapacious animals, and the parts of the skeleton were displaced and gathered into a small area." Williston (1898, p. 214) noted that the bones of mosasaurs "frequently bear the impression of teeth of post-mortem origin, and in many cases I have found the teeth of small sharks imbedded in them." Unfortunately, none of these specimens were documented at the time.

Williston (1900) was probably the next author to discuss in detail the occurrence of sharks in the Cretaceous of Kansas, listing eight species of *Ptychodus,* three species of Scyllidae (cat sharks) and twelve species of lamnids that had previously been identified from the state. Three of the species of *Ptychodus* (*P. anonymus, P. martini,* and *Ptychodus* sp.) were newly described by Williston himself. Only three of the *Ptychodus* species were reported from the chalk (*P. mortoni, P. martini,* and *P. polygyrus*); the rest were from the underlying "Benton Cretaceous." The "Benton Cretaceous" or "Fort Benton" is an obsolete stratigraphic term (Hattin, 1982) that includes the Graneros Shale, the Greenhorn Limestone, and the Carlile Shale (middle Cenomanian through middle Turonian). All of the "*Scyliorhinus*" specimens identified by Williston were collected from the Kiowa Shale (Early Cretaceous) in Kiowa County and are currently housed in the USNM collection. When I asked Robert Purdy (pers. comm., 2003) to re-examine them, he indicated that they were more likely to be *Leptostyrax*. More than half of the lamnid species reported by Williston (1900) were also from the Kiowa Shale (Early Cretaceous) of Clark County, Kansas. Williston (ibid.) was unable to substantiate Leidy's *Lamna sulcata* or Cope's *L. mudgei* from the specimens he examined. Four other species—*Lamna (Odontaspis?)* sp., *Lamna* sp., *L. quinquelateralis*, and *Leptostyrax bicuspidatus*—are indicated to have been found in the Kiowa Shale of Clark, Kiowa, and McPherson counties. *Corax (Squalicorax) curvatus* and *Lamna (Leptostyrax) macrorhiza* were reported from the "Benton Cretaceous" of Ellsworth and Ellis Counties, respectively. The only shark species named from the Niobrara (here meaning the Smoky Hill Chalk) were the three ptychodontids, *Isurus (Cretoxyrhina) mantelli*, *Lamna (Cretolamna) appendiculata*, and *Corax (Squalicorax) falcatus*. From its description, Williston (ibid., pp. 251–252) did not believe that the single specimen of *Scapanorhynchus raphiodon* collected by Hayden and figured by Leidy (1873, pl. 18, fig. 49) was even from Kansas. Hattin (1962) reported *Scapanorhynchus raphiodon* from the Carlile Shale (Fig. 4.6). More recently, however, *Scapanorhynchus* has been reported from the Smoky Hill Chalk (Hamm and Shimada, 2002) and Blue Hill Member of the Carlile Shale (Everhart et al., 2003).

One of the most interesting aspects of Williston's shark paper was the publication of photographs of the shark teeth in the University of Kansas collection that he was describing. So far as I am aware, this was the first time such photos had been made available in the study of shark teeth. I also find it intriguing to be able to visit the Museum of Natural History at the University of Kansas and examine many of the same teeth that were cited and photographed

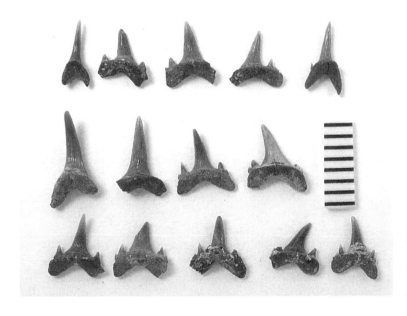

Figure 4.6. Assorted small teeth of Scapanorhynchus raphiodon *(goblin shark) in lingual view from the Blue Hill Shale Member (middle Turonian) of the Carlile Shale. (Scale = mm)*

well over a hundred years ago. Lane (1944) summarized Williston's information on sharks and reprinted several of his figures without significant modifications or additions. Schultze et al. (1982) inventoried the type and figured shark specimens in the collection of the University of Kansas Museum of Natural History, but other than modernizing the names and listing many more specimens, the list of species was essentially the same as in Williston's time. One curious omission was *Squalicorax kaupi*, a species probably collected and named more than a hundred years earlier from the chalk by Cope (1872b, pp. 355–357) as *Galeocerdo crassidens*. This is not a reflection on the work of Schultze et al. (1982) since the teeth would have to have been figured in a publication to be included. *Squalicorax kaupi* is currently well recognized in the upper chalk, and a nicely preserved set of teeth and the calcified cartilage of the jaws (FHSM VP-2213) can be found in the Sternberg collection.

Stewart (1990) was the first to define the biostratigraphic occurrence of sharks in the Smoky Hill Chalk. From his review of museum collections and experience in the field, he (ibid., p. 29) noted that *Cretoxyrhina mantelli, Cretolamna appendiculata, Scapanorhynchus raphiodon, Pseudocorax laevis,* and *Squalicorax falcatus* were found in the lower one-third of the chalk, along with the ptychodontids *P. mortoni, P. anonymus,* and *P. martini.* Stewart (ibid., p. 27) noted that he had recovered the teeth of *Rhinobatos* from acidized samples of chalk, but he did not assign them to a biostratigraphic zone. In an updated list, Stewart (pers. comm., 1993) indicated that *Rhinobatos incertae* was found in the uppermost chalk (zone of *Hesperornis*). While all the ptychodontids were apparently extinct in the chalk by the end of the Coniacian or early Santonian, the teeth of *Cretoxyrhina, Cretolamna,* and *Squalicorax* are found through the rest of the deposition of the chalk. *Squalicorax falcatus,* however, appears to have been largely replaced by *S. kaupi* by the early Campanian, which appears as early in the chalk (pers. obs.) as the middle Santonian. *S. pristodontus,* a larger species, appears right at the very top of the chalk. Two teeth of *S. pristodontus* (FHSM VP-15010 and 15011) were found recently by Pete Bussen just below the contact of the Smoky Hill Chalk with the Pierre Shale near Twin Buttes in Logan County.

Stewart (1990, p. 23) noted that the last occurrence of *Cretoxyrhina* in the middle of the continent is near the top of the chalk (early Campanian). In addition, it has been my observation that the largest shed teeth of *Cretoxyrhina* are found in the lower half of the chalk and that *Cretoxyrhina* teeth become fewer and generally smaller as you go higher in the formation. While it is intriguing that this decline coincides with the dramatic increase in the number and size of mosasaurs in the Western Interior Sea, proving that one was the cause of the other is probably not possible.

Shimada (1996) reported on the occurrence of selachians in the Fort Hays Limestone Member of the Niobrara. As might be expected, the "usual suspects" or species found in the Smoky Hill Chalk were also found in the underlying limestone, including *Pty-*

chodus mortoni. Shimada (ibid.) added a single tooth of a new and apparently rare species, *Paranomotodon* sp. (FHSM VP-12463) to the Niobrara Chalk shark species list. The following year, Shimada discussed the stratigraphic occurrence (1997a), the dentition (1997b), and the paleoecology (1997c) of *Cretoxyrhina mantelli* in three papers that resulted from extensive studies conducted as part of his graduate research. In addition, Shimada (1997d) described the very large vertebral centra (KUVP 16343) of an unknown species of shark from the Kiowa Shale (Upper Albian) of Clark County. The centra, if complete, would have been 14.5 to 18 cm (6–7 in.) in diameter. As a comparison, the largest centra from a recently discovered 5.5–6-m *Cretoxyrhina mantelli* (FHSM VP-14010) were 10 cm in diameter. The Kiowa specimen is the largest centra of a Mesozoic shark known to date and came from a shark estimated by Shimada (ibid.) to be 8.3–9.8 m (27–32 ft.) long.

Two other publications from outside Kansas regarding Cretaceous sharks should be mentioned here. Meyer (1974) did his doctoral thesis on the sharks of the Gulf Coast (Mississippi and East Texas Embayments), and Welton and Farish (1993) published a well-illustrated book on the species of sharks and rays from the Cretaceous of Texas. Both are excellent resources for identifying shark teeth from the Cretaceous of Kansas and will be even more useful as faunas in the state from other than the Niobrara are investigated.

In addition to shed teeth, the "soft parts" of some larger sharks from the Smoky Hill Chalk are well represented. The largest and most complete specimens ("shark mummies") of *Cretoxyrhina* are nearly all from the low chalk of Gove and Trego counties. These specimens usually consist of a significant portion of the shark, including calcified cartilages of the cranium and jaws, teeth, dermal denticles, vertebral centra, and even fin elements. Mudge (1877a, p. 5) was among the first to mention this unusual preservation. In 1891, Charles H. Sternberg discovered the remains of a very large *Cretoxyrhina* in the lower chalk. The specimen was acquired by the Ludwig-Maximilian University of Munich, where it was examined by Charles R. Eastman, an American student studying in Germany. Eastman (1895) wrote his doctoral dissertation on the specimen and later returned to the United States, where he studied fossil fishes.

Sternberg (1907) later commented on the discovery of this "unusual specimen of an ancient shark, *Oxyrhina mantelli*, found on Hackberry creek, Gove County, Kansas, in 1890 [*sic*]. The remarkable thing about this specimen is that the vertebral column, though of cartilaginous material, was almost complete, and that the large number of 250 teeth were in position. When Chas. R. Eastman, of Harvard, described this specimen, it proved so complete as to destroy nearly thirty synonyms used to name the animal, and derived from many teeth found at various former times." Two years later, Sternberg (1909, pp. 113–114) again described the discovery of this twenty-foot shark, probably from Gove County, south of Park, Kansas, and noted that the specimen was nearly complete, with "over 250 teeth." "[T]his is the first time and, I believe, the only time that so complete a specimen of this ancient shark has been discovered. The column and other solid parts were composed of cartilaginous matter which usually decays so easily that it is rarely petrified. I suppose my specimen was old at the time of its death, and bony matter had been deposited in the cartilage. It is not very likely that such a specimen will ever be duplicated. Dr. Eastman's study of this skeleton enabled him to make synonyms of many species which had been named from teeth alone."

Sternberg (1911, p. 71) mentions the discovery again and then describes a new find by his son George F.:

> The crowning discovery of our work here was the discovery by George Sternberg of a nearly complete skeleton of a great shark, Lamna. . . . In 1891, while employed by the Munich Museum, I discovered the first and most complete skeleton known of the shark *Oxyrhina* [*Cretoxyrhina*] *mantelli* in the same vicinity. This was made the subject of Dr. C. R. Eastman's inaugural address delivered before the Ludwig-Maximilian University of Munich for his Ph.D. degree. The specimen we collected on the south side of Hackberry creek, South of Banner post office, in Trego County, includes the plates of the mouth holding the teeth, of which some 150 are in sight, and the entire column of flattened disk-like vertebrae to within five feet of the end of the tail. The total length is about twenty feet of the preserved head and column. I think this will prove one of the greatest scientific discoveries of the year, because the sharks are cartilaginous, and consequently their skeletons are not preserved. But in the two cases mentioned enough bone was deposited to preserve the greater part of the skeleton.

Unfortunately, the specimen in the Munich museum was destroyed in World War II. The second specimen mentioned by Sternberg is currently on exhibit in the collection at the University of Kansas. The shark was mentioned again by Sternberg (1917, p. 162) when he noted that "I have preserved in the Museum of the University of Kansas a shark twenty-five feet long, and mingled with his remains were the bones of a *Portheus* [*Xiphactinus*]. . . ." Shimada (1997c, p. 927, figs. 1–2) noted that the specimen (KUVP 247) was a "nearly complete skeleton of *Cretoxyrhina mantelli*" and "is associated with many bones of *Xiphactinus audax* (KUVP 245), including the cranium, jaws, ribs and other elements throughout the matrix. They most likely represent stomach contents. . . ." It should probably be noted here that contrary to Sternberg's description, shark cartilage does not become "ossified" or filled with bone. It does in rare cases, however, accumulate enough calcium to be preserved as a fossil.

The remains of another George Sternberg specimen, a 5-m (16-ft.) *Cretoxyrhina mantelli* (FHSM VP-2187; see Shimada, 1997a;

1997b; 1997c) are on exhibit in the Sternberg Museum of Natural History. The "head" of this specimen consists of calcified cartilage (cranium and jaws) and several hundred teeth and is covered with scales (dermal denticles). The rest of the specimen is made up of a nearly continuous series of calcified vertebrae. The fins and tail are missing. The largest teeth in the front of the jaw are nearly 5 cm (2 in.) in height. In April 2002, I found a larger specimen (FHSM VP-14010) in the lower chalk (late Coniacian) of southern Gove County. The remains, including skull, pectoral fins, and vertebrae, were spread out over a distance of about 5 m (16 ft.) and did not include the missing tail. Based on comparisons of tooth and vertebrae sizes with FHSM VP-2187, Corrado et al. (2003) conservatively estimated the length of the shark to be between five and five and a half meters. However, since the posterior vertebrae were still more than 7 cm (3 in.) in diameter where it had eroded out of the hill, I believe that about two meters of the shark's tail had eroded away years earlier, and that it would have been between six and seven meters long. In either case, it was a very big shark.

Reasonably complete remains of smaller species of Late Cretaceous sharks are also known from the chalk, including a very nice Kansas specimen of *Squalicorax falcatus* (USNM 423665) on exhibit in the Smithsonian (United States National Museum) and the remains of the *S. kaupi* specimen (FHSM VP-2213) mentioned above. In 1992, I discovered another specimen in Gove County that preserved the anterior half of a 2-m (7-ft.)-long *S. falcatus*. The shark remains were upside down and had possibly been complete before eroding out of the chalk. As it was, I noticed a number of half-dollar-sized (3-cm) vertebrae lying on the surface of the chalk along the edge of a gully. They had come from the midsection of the shark. If present originally, the back half of the remains had probably been washed down the gully over the previous several years. Fortunately, the anterior half, including a complete skull and most of the teeth, was still safely buried in chalk. After picking up the loose pieces, I was able to remove the head and anterior vertebrae in one piece on a slab of chalk and carry it out of the field. The specimen (CMC VP-5722) is now in the collection of the Cincinnati Museum Center.

While these "shark mummies" provide much more information about these extinct sharks, they are relatively rare. The teeth of sharks, however, are the most common vertebrate fossils found in the chalk. Collecting shark teeth is a hobby for many of the local residents of western Kansas, and a number of extensive collections have resulted. The collection in the Fick Fossil and History Museum in Oakley, Kansas, exhibits more than ten thousand shark teeth, most of which were picked up one at a time during the lifetime of Vi Fick and her husband from chalk exposures in southwestern Gove County. A donation of fossils collected from the chalk in the 1930s by C. Y. Stout that is housed in the Sternberg Museum contains more than two thousand shark teeth. Where do all these teeth come from?

Consider that a ginsu shark (*Cretoxyrhina mantelli*) has forty-two rows of teeth in its upper jaws (twenty-one on each side) and thirty-eight in the lower jaws (Shimada, 1997b). The working teeth in the front of each row are shed and replaced continuously throughout the life of the shark, at intervals of a few weeks to several months. This means that an individual shark would shed or otherwise lose thousands of teeth during its lifetime. These teeth are replaced by what amounts to a shark tooth assembly-line process. Behind each "working" tooth in the front row is a series of replacement teeth in various stages of completion. The crowns of the teeth develop first and the roots are completed last as the tooth moves downward or upward toward the edge of the jaw. Each succeeding tooth is also slightly larger than the tooth it replaces, reflecting the continued growth in the size of the shark's jaw.

When a tooth is lost, the next tooth in the series is moved over the edge of the jaw, rotated about 180 degrees, and put to work. In *Cretoxyrhina*, there may be up to six replacement teeth in each row, so a single shark could have up to 560 (80 × 7) teeth, including developing bud-teeth, in its jaws at any one time. The actual number is probably less because the tiny symphysial teeth (eight upper, two lower) at the center of the jaws are more or less vestigial and probably aren't replaced in the same manner. Other sharks have differing numbers of teeth rows and differing rates of replacement, but all shed their teeth again and again throughout their lives.

Judging from the collections I have seen, the large, razor-edged teeth of the ginsu shark (*Cretoxyrhina mantelli*) and smaller, serrated teeth of the crow shark (*Squalicorax falcatus*) are picked up more frequently in the Smoky Hill Chalk than all the other species combined. Teeth of other species, including *Cretolamna appendiculata*, *Scapanorhynchus raphiodon*, and *Pseudocorax laevis*, are found much less often. Generally this is because of their smaller size, but it is also because these sharks occurred in smaller numbers in the deeper waters present during the deposition of the chalk near the center of the Western Interior Sea. A number of teeth of a new species of *Galeorhinus* were recovered (pers. comm. Mike Triebold, 2002, and J. D. Stewart, 2002) several years ago during the preparation of the remains of a very large (5.1-m [17-ft.]) *Xiphactinus audax* I found in 1996. As of 2004, however, the description has yet to be published. In 2003, a single tooth of a recently named genus of shark (*Johnlongia* sp.) first reported from Canada and Australia was collected from the lower chalk of Trego County by Keith Ewell (Cappetta, 1973; Siverson, 1996; Cicimurri, 2004; Shimada et al., 2004). Additional collecting will undoubtedly produce a few more species, but it is likely that most have already been discovered.

Ptychodus

The domed, shell-crushing teeth of ptychodontid sharks such as *Ptychodus mortoni* and *P. anonymus* are occasionally found one at a time like the teeth of other sharks. Sometimes, however, they occur in large numbers where they have eroded out on the surface of the chalk, or even as articulated jaw plates from a single individual (Fig. 4.7). These unusual sharks became extinct by the early Santonian in Kansas, and their teeth are only found in the lower one-third of the chalk. Due to a lack of complete specimens, we have no idea what they looked like. It is equally possible that they looked like modern rays, or that they looked like "normal" sharks. We simply do not know.

Woodard (1887, p. 128) noted that the dentition of *Ptychodus* "is that of a true Ray, and does not bear the slightest resemblance to that of the Cestraciont [hybodont] Sharks." Stewart (1980) briefly described a partial specimen of *Ptychodus mortoni* (KUVP 59041) which included calcified vertebral centra and suggested that "all living sharks and rays (including *Heterodontus*) are members of the monophyletic Neoselachii, united by synapomorphies including the presence of calcified centra. Since *Ptychodus* shares this derived state, it must be regarded as a neoselachian and not as a hybodont." One *P. anonymus* specimen (AMNH 19553) collected recently in Gove County included more than two hundred teeth, five vertebral centra, and hundreds of oral scales (denticles) from inside the mouth of the shark (Everhart and Caggiano, 2004). An examination of the vertebral centra and oral denticles from this specimen indicated that they were similar to those of *Squalicorax*.

Because their "pavement" toothed upper and lower jaws are superbly adapted for crushing hard-shelled prey like small inoceramid clams and other bivalves, it is assumed that they must have

Figure 4.7. Assorted teeth of Ptychodus mortoni *(FHSM VP-2223), a shell-crushing shark from the late Coniacian chalk of Kansas. (Scale = cm)*

fed on them. Kauffman (1972) interpreted a series of depressions along the edge of a single, 12.5-cm-long shell of *Inoceramus tenuis* from the English Chalk (Lower Cenomanian) as bite marks of *Ptychodus decurrens*. Kauffman suggested that the marks were probably made by the lateral teeth along the edge of the shark's jaw. If so, the size of the marks (about 1.5 cm across) would indicate that the shark had been quite large. It is difficult to explain why the shell was not completely crushed by the bite of the shark, but Kauffman (ibid., p. 441) suggests that the bivalve may have already been dead and had been rejected by the shark. While this scenario is certainly possible, the lack of similar remains from other localities, including the Western Interior Sea, raises questions as to the cause of these marks. Although I agree that these sharks fed on clams and other hard-shelled prey, I suspect that they avoided the larger ones in favor of the more easily cracked and readily available smaller shells. I think these sharks cruised near the bottom and scooped up mouthfuls of mud containing thin-shelled, younger clams instead of trying to attack the thicker-shelled, full-grown inoceramids.

In the Late Cretaceous of Kansas, several lines of evidence support this idea. First, from the Cenomanian through the early Coniacian when ptychodontid sharks probably had their greatest diversity, inoceramid clams were generally fairly small. By the middle of the Coniacian in the ocean over Kansas, however, one species, *Volviceramus grandis,* became larger and dominated the sea bottom. Growing to a maximum of about 50 cm (20 in.) across, their shells were quite thick (2–3 cm), and thus were probably more massive than any shell-crusher would want to routinely tackle. However, these large shells must have taken many years to grow to that size. From the time they hatched until they reached a certain size, they were vulnerable to being crunched up by the crusher shark and other predators. We find small, fragile masses of jumbled calcite prisms from tiny inoceramid shells occasionally in the low chalk. I interpret these as either the coprolites or regurgitate of a ptychodontid shark (see Hattin, 1996 for a recent discussion). It was readily apparent that something had been feeding on young inoceramids and left their indigestible remains behind.

Coincidentally with the growth in size and thickness of the shells of *V. grandis,* most of the ptychodontids became extinct in the Western Interior Sea by the end of the Coniacian. I don't believe that one was necessarily the result of the other, since the remaining species of *Ptychodus* would still have been able to feed on the younger, more fragile bivalves. However, following the extinction of *V. grandis* in the early Santonian, another species of giant inoceramid, *Platyceramus platinus,* became the dominant bivalve on the sea bottom. It is this species that is often mentioned as growing to four feet (1.2 m) in diameter (Chapter 3). However, this species was very thin-shelled and obviously had to put less energy and resources into building its large shells than did *V. grandis,* quite possibly due to the absence of predation from the ptychodontid sharks.

Although ptychodontid sharks reach their greatest diversity in

Kansas in the lower third of the Smoky Hill Chalk (late Coniacian), their first occurrence in Kansas occurred much earlier (middle Cenomanian) where *Ptychodus decurrens* is found at the transition between the Dakota Sandstone and Graneros Shale (Everhart et al., 2004). At the beginning of the upper Cenomanian, *P. whipplei*, and possibly *P. mammillaris* also occur, in addition to *P. decurrens*, in the basal Lincoln Limestone Member of the Greenhorn Limestone that overlies the Graneros Shale (pers. obs.). Williston (1900) described *P. anonymus* as a new species of ptychodontid shark from the "Benton Formation" (= obsolete term for the Graneros Shale, Greenhorn Limestone and Carlile Shale formations) but Herman (1977) later re-identified the teeth as *P. mammillaris*. Although *P. anonymus* is still regarded as a valid species (Everhart and Caggiano, 2004), the type specimens have not been relocated. More recently, however, Shimada and Everhart (2003) reported a single *P. mammillaris* tooth from the base of the Fort Hays Limestone (early Coniacian), and Everhart and Darnell (2004) reported another specimen from the Fairport Chalk (middle Turonian). While *P. anonymus* has a pre-Niobrara record in addition its occurrence in the Smoky Hill Chalk, *P. mortoni* is only found in the lower half of the chalk and is the last species of ptychodontid shark in the Western Interior Sea (Stewart, 1990; Shimada, 1996). Two species with relatively flat tooth crowns, *P. martini* and *P. polygyrus*, also occur rarely in the lower third (late Coniacian) of the chalk. *Ptychodus martini* is represented by two relatively complete jaw plates; the type specimen (KUVP 55271) described and figured by Williston (1900) and a new specimen reported by Hamm and Everhart (1999). Cope (1874) was the first to report *Ptychodus polygyrus* from Kansas, based on a single tooth "found by Professors Mudge and Merrill" in the Niobrara of Ellis County. Although the fate of that specimen is unknown, another large tooth (KUVP 55237) was reported by Williston (1900) from "the lower beds of the Niobrara Cretaceous of the Smoky Hill River" and is presently in the collection of the University of Kansas. Three other specimens from the Smoky Hill Chalk (FHSM VP-76, VP-2123 and VP-15008) are in the collection of the Sternberg Museum. Schumacher (pers. comm., 2003) showed me pictures of an associated set of 264 *P. polygyrus* teeth from Russell County in a private collection. If the specimen is from the Fairport Chalk (middle Turonian) as we suspect, it would be the best and earliest known example of the species in Kansas.

Other Times, Other Sharks

This book is generally focused on a small (geologically speaking), five-million-year window of time during the Late Cretaceous when the Smoky Hill Chalk was deposited on the bottom of the Western Interior Sea. The fauna from that interval (87–82 mya) is the most widely collected and the most thoroughly studied of any period from the Cretaceous of Kansas. The chalk is accessible in many localities, preservation is excellent and the chalk matrix is relatively easy to remove. It is also my favorite because of the occurrence of mosasaurs and the large number of their remains that have been found in the chalk since the late 1860s. But many other productive Cretaceous rocks are largely unstudied, and thus Kansas lags far behind other places in the Midwest such as Texas and South Dakota in the number of shark species that have been documented from the state.

Since I began writing this book, however, I have had a number of opportunities for further study of Kansas sharks. The discovery of the huge *Cretoxyrhina mantelli* in 2002 that is mentioned above was the beginning of a series of discoveries and studies on Kansas sharks. The opportunities even extended outside of the Cretaceous with the collection of the first known ctenacanth shark (*Ctenacanthus amblyxiphias*) remains from Kansas (Everhart and Everhart, 2003) in August 2002 from the early Permian rocks near Herington, Kansas, and the subsequent discovery in 2003 by my friend Keith Ewell of numerous Permian shark teeth and dorsal fin spines from three sites near Manhattan, Kansas. However, our work in the Cretaceous is still the focus of our research efforts.

In October 2002, my wife and I began working with a "fish tooth conglomerate" from the Blue Hill Member (Turonian) of the Carlile Shale in Jewell County, Kansas, that had been discovered in 1958 by Donald Hattin (1962). Unlike our usual field trips, however, all of our collecting was done in our fossil lab. The conglomerate dissolved very slowly and the teeth had to be collected and sorted while we peered through a microscope. While the matrix was mostly composed of thousands of tiny teleost teeth (generally *Enchodus*), over a period of several months it produced a fair number of selachian species that had never been reported from Kansas (among them, *Lonchidion* sp., *Chiloscyllium greeni*, *Ischyrhiza mira*, and *Ptychotrygon triangularis*). Many of the shark teeth were so small (1 mm or less) that they had to be picked and sorted using forceps and a binocular microscope. Imagine 500 or more *Rhinobatos* (a guitarfish) teeth in a volume about the size of an aspirin tablet! After *Rhinobatos*, *Scapanorhynchus raphiodon*, and *Squalicorax falcatus* were the most common species collected. The study generated an abstract/poster for the 2003 Society of Vertebrate Paleontology meeting (Everhart et al., 2003) and will result in a future paper.

In the late spring of 2003, I began looking at the Kiowa Shale in McPherson County with Shawn Hamm, a student at Wichita State University. Except for a thin layer of sand called the Cheyenne Sandstone in southern Kansas, the Kiowa is the oldest Cretaceous formation (Albian) exposed in Kansas. It was deposited at a time when the Western Interior Sea was advancing from the south and had not yet covered the entire state. Unlike the deeper, blue-water ocean in which the Smoky Hill Chalk was deposited, the Kiowa Shale was laid down as near-shore mud in a relatively high-energy environment. Working some ten meters below the level of the surrounding wheat fields in an active shale quarry near Marquette, we collected a much older shark fauna along with the scattered, fragmentary remains of fish, turtles, crocodiles, and plesiosaurs. Again, many of the teeth were quite small, though not as small as those from the Lovewell fauna. Most of the teeth and other remains were literally picked up from the surface of a rapidly weathering layer of pyritized sandstone. Expanding on the work of Beamon (1999), several relatively "new to Kansas" species were collected, including "*Polyacrodus* sp." (Everhart, 2004a), *Onchopristis dunklei,* and *Pseudohypolophus mcnultyi,* with *Leptostyrax macrorhiza* and *Carcharias amonensis* (Fig. 4.8) being the most common sharks.

In August 2003, a visitor (Keith Ewell) to the Sternberg Museum asked for assistance in identifying Permian-age shark teeth he had collected near Manhattan, Kansas. After making contact with Keith, I visited the site with him and saw his specimens. We then assisted him in collections made at three sites in Geary County that produced more than two hundred teeth of *Cladodus* sp., *Petalodus* sp., *Acrodus* sp., and *Chomodus* sp., dorsal spines of *Ctenacanthus, Physonemus* and *Hybodus,* and calcified cartilage (Ewell and

Figure 4.8. *Teeth of two species of sharks from the Early Cretaceous Kiowa Shale of McPherson County, Kansas:* Leptostyrax macrorhiza *(left) and* Carcharias amonensis *(both in lingual view). (Scale = cm)*

Figure 4.9. Teeth of Cretolamna appendiculata *from the basal Lincoln Limestone Member (Upper Cenomanian) of the Greenhorn Limestone, Russell County, Kansas (both in lingual view).(Scale = mm)*

Everhart, 2004). The most productive site was in the Neva Limestone (Council Grove Group, Lower Permian). While the species of sharks identified were not necessarily new to Kansas, the number of teeth, their large size, and excellent condition was certainly unexpected. The teeth of *Cladodus* sp. teeth are the most common and, in size, rival the largest of the *Cretoxyrhina* teeth we have collected from the Smoky Hill Chalk.

As a result of our meeting and subsequent conversations, Keith became interested in finding Cretaceous fossils further west. In October, as noted earlier in this chapter, Keith found the tooth of a new species of shark (*Johnlongia* sp.) in the Smoky Hill Chalk (Shimada et al., 2004) as well as numerous shark teeth in the Greenhorn Formation. Late in the year Keith discovered a rich accumulation of shark teeth, fish remains, and reptile bones (coniasaur) in the basal Lincoln Limestone Member (Upper Cenomanian) of the Greenhorn Limestone in Russell County. Together we collected samples for further examination. Again, the teeth were small (more picking with a microscope) but the numbers were large (literally hundreds of teeth), with *Carcharias amonensis* and *Squalicorax falcatus* being the most common species found, along with *Cretoxyrhina mantelli, Cretolamna appendiculata* (Fig. 4.9), at least two species of *Ptychodus*, and the ray *Rhinobatos incertus*. One of the rarest finds so far was Keith's discovery of a single 1.5-mm tooth of the primitive sawfish *Onchopristis dunklei*.

About eight miles away, in eastern Russell County, we also collected samples from a layer of poorly cemented sand in the upper Dakota Sandstone (middle Cenomanian). The sand produced a number of small *Carcharias amonensis* teeth, along with a single, large *Cretodus semiplicatus* tooth (Fig. 4.10), several *Squalicorax* sp. teeth, many *Hybodus* sp. and *Rhinobatos* teeth, and rostral denticles of the sawfish *Onchopristis dunklei*. The preservation is

Figure 4.10. Two teeth of Cretodus semiplicatus *in lingual view from the upper Dakota Sandstone (middle Cenomanian), Russell County, Kansas.*

occasionally poor due to diagenesis and the intrusion of selenite crystals, and contrasts markedly with the pristine condition of the teeth in the nearby basal Greenhorn. Few vertebrate remains (outside of the crocodile *Dakotasuchus*) have ever been reported from the Dakota Sandstone in Kansas, and even that specimen is more likely to have come from the underlying Kiowa Formation (Scott, 1970).

With the discovery of these new and productive localities, the Cretaceous (and Permian) shark faunas of Kansas appear to be quite similar to those in Texas, New Mexico, Nebraska, South Dakota, and elsewhere in the Midwest. We are looking forward to additional collecting and study of Kansas sharks in these areas.

Five
Fishes, Large and Small

The big *Xiphactinus* swam effortlessly through the clear, warm waters of the Western Interior Sea in a solitary, never-ending search for its next meal. Ten years old and nearly four meters long, it had not yet reached full size, but it was already larger than any of the other species of fish in this ocean except the giant ginsu sharks. Although still wary of the occasional prowling shark, the X-fish's only other major competitor for the larger fish that it preyed upon were the large marine lizards that shared the same waters. The X-fish had lived long enough and grown too large for any of the big marine lizards to be interested in it as a meal. Capable of taking very large prey as they were, the lizards were unlikely to attack anything larger than they could swallow, and the X-fish's massive body was well beyond what they could eat. The X-fish had on occasion taken a newly birthed mosasaur, but it generally avoided their family groups. Its favorite prey was other fish, especially some of the larger varieties that were closely related to his own kind. A large meal satisfied the X-fish's hunger and provided the energy necessary to support its constant swimming.

The X-fish's senses alerted it to a group of *Gillicus* feeding nearby and close to the surface. As it swam closer, it could see that the two-meter-long predatory fish had surrounded a school of much smaller fish, compressing them into a shimmering globe of millions of tiny forms. The *Gillicus* swam in circles around the trapped school, keeping it from spreading out or escaping. Occasionally some of the *Gillicus* would dart through the mass of little fish with their jaws wide open, gorging themselves on the sudden abundance of prey. Their attention diverted, they did not notice the X-fish's approach from the dark waters beneath them. Gauging their movements, the X-fish selected its victim and accelerated swiftly upward, meeting the other fish almost head-on. Opening its large mouth at the last moment, it seized the smaller fish's head from below. As its massive lower jaw closed, the large, conical teeth at the front punctured the thin bones covering the head of the *Gillicus* and kept it from getting away. The smaller fish struggled briefly, but the shock of impact had stunned it into submission. The X-fish held on to its prey for several moments, then quickly repositioned it so that it was pointed headfirst into the larger fish's mouth. Then it began to rapidly open and close its jaws, drawing the smaller fish deeper and deeper into its gullet. A shower of silvery scales, dislodged from the *Gillicus* by the teeth of the X-fish, glittered in the sunlight as they drifted away.

At first, swallowing the big *Gillicus* was relatively easy, but then the X-fish began to have problems. At nearly two meters in length, the *Gillicus* weighed more than a hundred pounds. It was probably the largest prey the X-fish had ever eaten. Once the head and pectoral fins of the *Gillicus* had passed into the larger fish's throat, however, there was no turning back. The bony pectoral fin rays of the *Gillicus* were folded back tightly against its body. Any attempt to reverse direction would cause them to unfold and catch in the muscular esophagus of the larger fish. The relatively large body of the *Gillicus* now filled the mouth and throat of the X-fish and made it difficult for water to reach its gills. The last two feet of the *Gillicus*, including its bony tail, still protruded beyond the jaws of the X-fish. Barely alive but tightly confined within the bony skull and forebody of the larger fish, the *Gillicus* thrashed about occasionally as it was swallowed. A final spasm caused one of its sharp fin rays to pierce the esophagus of the X-fish.

Twice as long as the *Gillicus* and much more massive, the X-fish struggled to swallow its prey. Unable to adequately replenish the oxygen in its body with the smaller fish lodged in its throat, however, it was quickly weakening. Slowly the smaller fish was moved deeper and deeper into the gullet of the larger one. Finally the bony tail moved entirely into the mouth, and the gills of the X-fish began to function normally again. Now the prey began to move more easily into the stomach of the X-fish, but something was wrong. The large fish swam in ever-slower circles as it finished swallowing the *Gillicus*. During its futile struggle to break free, a fin spine of the prey had punctured something vital in the X-fish. Within a few minutes, the X-fish stopped swimming, rolled over on its back as it died, and then sank headfirst toward the muddy bottom below.

While this story is fiction, the famous "fish-in-a-fish" specimen recovered by George Sternberg is a fact (Rogers, 1991, p. 248; Liggett, 2001, p. 64). While a group from the American Museum of Natural History was on a field trip with George Sternberg in the spring of 1952, Walter Sorenson of the AMNH found the caudal fin of a large *Xiphactinus audax* (FHSM VP-333) eroding out from the chalk in Gove County. Recovering the fossil would have taken far more time than the visitors had, so they graciously gave the specimen to Sternberg. Hours of meticulous field work under a hot summer sun by Sternberg and others from the university uncovered what was certainly one of the most complete specimens of *Xiphactinus audax* Leidy 1870 found up to that point (Bardack, 1965, p. 40). Sternberg's patient work also revealed the last meal of the larger fish, a well-preserved *Gillicus arcuatus* (FHSM VP-334). The smaller, two-meter-long fish had been swallowed headfirst and rested entirely within the ribs of the four-meter (13-ft.) *Xiphactinus* (Fig. 5.1). We know that the smaller fish had not been there long before the larger fish died because it had not yet been digested. What caused the death of this *Xiphactinus*, and a surprising number of other specimens of the species, cannot be determined from their fossils, but it most likely involved an injury that occurred when the prey was swallowed. In the case of the Sternberg specimen, a fin from the struggling *Gillicus* could have pierced the heart or a major blood vessel of the *Xiphactinus* and killed it quickly, or there may have been damage caused to the gills of the larger *Xiphactinus* that were rapidly fatal. In any case, the fossil remains have preserved a very interesting and puzzling moment in time.

While *Xiphactinus* remains as complete and well preserved as the fish-in-a-fish specimen are relatively rare, a number of *Xiphactinus* fossils have been found with a *Gillicus* as the last meal. Bardack (1965) surveyed eighteen relatively complete *Xiphactinus* specimens in museum collections and indicated that at least three (including the Sternberg specimen) had *Gillicus* remains inside. A similar-sized specimen (DMNH-1667) in the Denver Museum of

Figure 5.1. The famous "fish-in-a-fish" specimen of a large Xiphactinus audax *(FHSM VP-333) and its last meal (*Gillicus arcuatus; *FHSM VP-334) in the Sternberg Museum of Natural History, Hays, Kansas. The remains were collected from the Smoky Hill Chalk of Gove County in 1952.*

Nature and Science had lived long enough to partially digest its last meal, also a large *Gillicus*. In fact, seven of the eighteen specimens surveyed included fish as stomach contents. The giant (17 ft.) *Xiphactinus* I discovered in 1996 (NAMAL 2000-0925-009) also included the partially digested remains of another *Gillicus*. Preservation of stomach contents is rare in the fossil record (Cicimurri and Everhart, 2001) and the relatively frequent discovery of *Xiphactinus* specimens containing the remains of their last meals raises the question of why so many of them died so quickly after eating.

The remains of fish are the most common vertebrate fossils found in the Smoky Hill Chalk. Russell (1988) suggested that about 60 percent of the Late Cretaceous specimens in museum collections were fish. While this number probably includes some cartilaginous fish specimens (sharks and ptychodontids) in those collections, my experience is that the percentage of fish specimens found in the field probably approaches 80 or 90 percent of the total number. The difference can be explained by the fact that museum specimens are generally more complete, while in the field bits and pieces of fish are likely not to be collected, even though they may be identifiable to species. If your field time or storage space is limited, you tend to be a bit more choosy in what you collect.

After reviewing my field notes for the years 1988–1995 when my wife and I were vacuuming up ("hoovering") almost everything we found, I found that 78 percent of 430 vertebrate remains we collected were fish (not including isolated shark and fish teeth) and 16 percent were mosasaur. The other 6 percent were more or less equally divided between pteranodons (2 percent), turtles (2 percent) and plesiosaurs (1 percent), with a single bird bone making up a small fraction of a percent. I will admit that I picked up every mosasaur specimen that I saw, but not every fish. It is important to note, however, that these specimens ranged from complete fish to a few vertebrae or just an identifiable piece of a fin or jaw. In size, the fish specimens ranged from tiny 10-cm (4-in.) fish inside clam shells to giant *Xiphactinus* specimens that were more than 4 m (13 ft.) long. Far and away the most common remains we observed during that time, however, were fish tails. These generally consisted of a complete, mostly undamaged caudal fin with a few attached vertebrae. Mostly these were from medium-sized fish, with an occasional fairly large *Ichthyodectes* thrown in. Some fish-eating predator was apparently biting or breaking off the bony, relatively nutritionless tails while swallowing the rest of the fish. Carpenter (1996, p. 44) reported severed tails as evidence of predation on fishes in the Pierre Shale. Our unofficial count over the years has been that we find about ten tails for every one skull. Needless to say, we gave up collecting fish tails a long time ago.

Relatively few remains of fish (or anything else) are complete when found in the Smoky Hill Chalk. The multitudes of smaller fish that must have been present were most likely to be consumed whole while the larger ones were sometimes torn apart by sharks or other predators. Sometimes the fish bones we find are were partially digested (see Chapter 4 on sharks) or were found inside coprolites. Once the remains reached the bottom, they generally were not disturbed by large predators. As might be expected, in most cases, the bones of larger fish had a better chance of being preserved as fossils. The major exception to that rule occurred when schools of small fish were trapped inside giant inoceramids when they died and were preserved intact (Stewart, 1990b). In some cases, notably the "fish in a fish" *Xiphactinus* mentioned above, the remains of a last meal are identifiable inside the larger fish. Evidence of scavenging is preserved as bite marks, severed bone, and associated shed teeth (usually shark).

The ecosystem of the Late Cretaceous was probably not unlike that of the Gulf of Mexico or similar warm-water areas around the Earth today. Immense populations of "sardine"-sized fish were necessary to support larger fish and the predators that fed on them. A Late Cretaceous marine food web for the Western Interior Sea might go something like this:

1. Algae as primary producers converted sunlight, carbon dioxide, and nutrients into biological materials for growth and reproduction. The calcitic shells of the algae, called coccolithophores, and the disk-shaped, calcite coccoliths that surrounded them accumulated in vast numbers to form the limey mud on the sea floor that eventually became chalk (Hattin, 1982)

2. Microscopic single-cell or multi-cellular animals fed on the algae. Their fecal (waste) pellets containing the indigestible coccolithophores and coccoliths of their prey (ibid.) also add to the limey mud below.

3. Tiny fish, small crustaceans, and the larvae of other marine animals, including squid and ammonites, fed on the microscopic protozoa and other primary consumers. Apparently there were no large, filter-feeding vertebrates during the Late Cretaceous.

4. Small fish, young fish of many species, and invertebrates (ammonites and squid) fed on the tiny fish.

5. Medium-sized fish fed on the small fish. Pteranodons, birds, young marine reptiles, ammonites, and squid also fed at this level.

6. Large fish, sharks, plesiosaurs, and small mosasaurs fed on the medium-sized fish and on invertebrates such as ammonites and squid. Sharks, cephalopods, and possibly mosasaurs also served as scavengers to recycle biological materials from the remains of dead animals.

7. Large mosasaurs and sharks fed on everything as the top predators.

Because we find the remains of the top predators (mosasaurs) quite often, we are fairly certain that there must have been a productive ecosystem running at full tilt to support them. The coastlines, estuaries, and swamps that bordered both coasts of the Western Interior Sea may have served as semi-protected hatcheries for untold numbers of smaller fish and invertebrates that were not preserved in the fossil record but had to be there to support the base of the food web. The remains we find as fossils in the Smoky Hill Chalk are only a small fraction of the abundant life that must have existed in the Western Interior Sea during the Late Cretaceous.

The specimens that are preserved are important indicators of what was going on in the oceans of Kansas. Large predators such as mosasaurs and *Xiphactinus* are probably overrepresented as fossils because their more robust, heavier bones improved their chances for preservation. Because they are there, however, we can view them as indirect evidence of the vastly larger number of smaller fish and invertebrates that lived and died as prey and left only a scant record.

A Brief History of Fossil Fish Collecting in Kansas

Though not from Kansas, the first known fossil fish from the Niobrara Chalk was collected by the Lewis and Clark expedition in August of 1804 from the bank of the Missouri River in what is now Harrison County, Iowa. The fish jaw (ANSP 5516) is currently in the collection of the Academy of Natural Sciences of Philadelphia and is the only fossil specimen surviving from that expedition (Spamer et al., 2000). The history of this fossil is somewhat confused, however, because it was originally misidentified by Dr. Richard Harlan (1824) as the jaw of an "Enalio Saurian" (a marine lizard thought to be somewhat like ichthyosaurs and plesiosaurs). Harlan described it and gave it the name *Saurocephalus lanciformis*. It wasn't until six years later when Isaac Hays (1830) described a similar species of Cretaceous fish (*Saurodon leanus*) from New Jersey that the mistake was officially noticed and corrected. Leidy (1856, p. 302) noted that the *Saurocephalus* specimen was a fragment of a "maxillary bone with teeth of a peculiar genus of sphyraenoid fishes from the Cretaceous formation of the Upper Missouri." This group of fossil fishes would continue to give other paleontologists classification problems as more discoveries were made.

Joseph Leidy (1859) described a dorsal spine (*Xystracanthus*) and two teeth of two different species of Paleozoic sharks (*Cladodus* and *Petalodus*) found in eastern Kansas two years before Kansas became a state. The teeth had been discovered in 1858 by the U.S. Geological Survey expedition (Meek and Hayden) while traveling through Kansas during their exploration of several Midwestern states. This paper is important because it is the first description of vertebrate fossils from Kansas and is one of the first descriptions of North American vertebrates from the Permian.

Professor Benjamin Mudge (1817–1879) was one of the first paleontologists in Kansas. He began collecting fossils in the central and western parts of the state several years before Marsh and Cope arrived. As was customary for the times, he sent many of his specimens to the scientists "back East" for identification. For the most part, he was communicating with E. D. Cope during the late 1860s and early 1870s. In an 1870 letter (Williston, 1898, pp. 29–30),

Cope complimented Mudge on the scientific value of the material he had sent. In fact, many of the new species of fish and mosasaurs described by Cope were from the specimens sent to him by Mudge (Everhart, 2002). During his only trip to Kansas in late 1871, Cope (1872a) visited Mudge in Manhattan, Kansas, and examined his collection. By then, however, Mudge had begun sending some specimens, including the remains of the first known toothed bird (*Ichthyornis*) to Cope's rival, O. C. Marsh (see Chapter 11). In 1874 he began collecting fossils solely for O. C. Marsh and Yale College. Unfortunately, Mudge died in 1879.

In late 1867, Dr. Theophilus Turner, the assistant surgeon assigned to Fort Wallace in western Kansas, also began communicating with Cope about his discovery of a huge plesiosaur. Cope (1868) reported on finding the remains (scales, vertebrae, and teeth) of six species of fish, including *Enchodus,* that he believed to be stomach contents from the new plesiosaur (*Elasmosaurus platyurus*) found by Turner (See Chapter 7). When I examined the remains that had been collected by Turner (ANSP 10081) in 2002, there were still fish scales, teeth, and bone fragments visible in some of the concretionary matrix that was associated with the plesiosaur bones. However, I consider them more likely to be the normal detritus that might be expected to accumulate on the sea bottom rather than evidence of the plesiosaur's last meal. In any case, they did provide evidence of several species of fish that were living at the same time (middle Campanian) as the elasmosaur.

In addition to performing his military duties during the late 1860s, Dr. George M. Sternberg (older brother of Charles H. Sternberg) collected and sent fossils from western Kansas to the Army Medical Museum in Washington, D. C. The fossils were then forwarded to the Smithsonian Institution, where they were more readily available for study by the paleontologists of the day. Leidy (1870) described a new species of fish (*Xiphactinus audax*) from a large (40 cm [16 in.]) fragment of a pectoral fin ray (USNM 52) that Dr. Sternberg had found. After examining the same specimen, however, Cope (1871) disagreed with his mentor and renamed it *Saurocephalus audax* Leidy. Apparently, Cope's name change was not recognized by anyone else. Later that year, Cope (1872b) described a "new" genus and species from the remains of several relatively complete specimens of fish he had found near Fort Wallace on his visit to western Kansas. He called the fish *Portheus molossus* without recognizing that it was the same species that Leidy had named almost two years earlier. It is not known if Leidy recognized that it was the same species or not. Almost thirty years later, O. P. Hay (1898) published a short note in *Science* suggesting that *Xiphactinus* Leidy was the correct genus, but by then it was too late. *Portheus* Cope was too well established to be banished that easily. Even the prominent paleontologist H. F. Osborn (1904) was unaware or, even worse, ignored Leidy's name in favor of Cope's when he wrote about a newly acquired "*Portheus molassus*" specimen (AMNH 322199) at the American Museum of Natural History. Since that time, even though it is widely known to be a junior synonym of *Xiphactinus audax* Leidy, the name *Portheus molossus* Cope is much better recognized, and many museum collections still have their specimens mislabeled.

There is one further unusual twist to this story. *Xiphactinus (Polygonodon) vetus* is a sister species of *X. audax* that was named by Leidy (1856) on the basis of teeth found in the Cretaceous of New Jersey. In naming them, Leidy believed that the teeth were those of a reptile (Schwimmer et al., 1997) and apparently didn't consider that they might be from a large fish. So if you really wanted to confuse the issue, the scientific name of the big fish found in the Kansas chalk should be "*Polygonodon*" *audax* Leidy 1870. I don't think that is likely to happen, however.

Following the discovery of *Elasmosaurus platyurus* in 1867 and the first Kansas mosasaur, *Tylosaurus proriger*, in 1868, Kansas quickly became a popular place to search for fossils. In 1870, O. C. Marsh mounted the first of four organized fossil-collecting trips to the western states, including Kansas, with his students. Although they spent less than two weeks in Kansas, the Yale College scientific expedition of 1870 was an immediate success, recovering dozens of specimens, including several mosasaurs, part of the wing of a previously unknown giant pterodactyl from Kansas (Marsh, 1871), and even a fragment of an unknown bird (*Hesperornis*). Strangely, Marsh was not at all interested in fish (Shor, 1971, p. 78) and instructed his field workers to ignore them. This allowed his rival, E. D. Cope, to describe and name many of the species from the Smoky Hill Chalk. Marsh's instructions may have been the source of a statement by B. F. Mudge (1876, p. 216), who wrote, "the least interesting are the fish, which have, however, given many new species and some new genera. The small ones are near entire, but the larger are represented by well-preserved portions of the skeletons." Years earlier, Mudge had collected a number of fish specimens for Cope (see Cope, 1872a) and Cope, in turn, had credited Mudge with the discovery of a number of new species. After being hired by Marsh, however, Mudge's collecting was focused more on toothed birds, pteranodons, and marine reptiles. Chris Bennett (pers. comm., 2004) indicated that when he examined the Yale fish collection in the 1980s, "there were still unopened packages from Mudge."

Cope (1872b) published a list of the families of fishes found in the "Cretaceous Formation of Kansas" that included most of the common varieties that we are familiar with today, even though only limited collections had been made from the chalk at that time. Many of Cope's "species" were described from fragmentary remains or single teeth, and the family names would be changed again and again as more complete specimens were found. Cope (ibid, p. 357) also noted that twenty-four species had been described from the Kansas chalk and that the same genera had been found in the chalk of Europe (see Table 5.1 for an updated listing of fishes from the Smoky Hill Chalk). His list included:

1. Saurodontidae (Ichthyodectidae): *Portheus* (*Xiphactinus*), *Ichthyodectes*, *Gillicus*, and *Saurocephalus*
2. Pachyrhizodontidae: *Pachyrhizodus* and *Empo* (*Cimolichthys*)
3. Stratodontidae: *Stratodus*, *Enchodus*, and *Apsopelix*

The history of collecting, identifying, describing, and naming the many species of fish that have been found in the chalk is confusing at best. In most cases, the early paleontologists involved were working with scraps and seldom had the luxury of seeing even a large portion of the complete fish. Rather than try to relate the history of the discovery of each genus, I will go into the details of those I think are the more interesting ones.

Xiphactinus, Ichthyodectes, and *Gillicus*

The family Ichthyodectidae is represented by three species in the Western Interior Sea. Most of the history of the discovery of *Xiphactinus* (*Portheus*) has already been discussed in the preceding paragraphs. The remains of *Xiphactinus audax* Leidy, *Ichthyodectes ctenodon* Cope, and *Gillicus arcuatus* Cope are found commonly as fossils in the Smoky Hill Chalk. Bardack (1965) recognized four species of *Xiphactinus* and two species each of *Ichthyodectes* and *Gillicus* worldwide. *Xiphactinus* is the largest of the three fish in the Smoky Hill Chalk, reaching lengths of more than 5.5 m (17 ft.). Besides its extremely large size, it is distinguished from the others by teeth of unequal size in its jaws. The anterior teeth, particularly those in the premaxillae, are the largest in the jaws. *Ichthyodectes* grew to about two-thirds the size of *Xiphactinus*, reaching lengths of around 4 m (12 ft.). Its teeth are medium in size and are generally about the same size across the jaws. *Gillicus* grew to about one-third the length of *Xiphactinus*, or about 2 m (6 ft.) long. The teeth in *Gillicus* are so small that at first the jaws appear to be nearly toothless. This has led to some speculation that *Gillicus* was a filter-feeder, but I would contend that a 2-m-long fish with a big mouth does not need big teeth to feed on the large number of small fishes that were readily available. There are a number of modern marine and freshwater predatory fish that do quite well without large teeth.

The post-cranial skeletons of *Xiphactinus*, *Ichthyodectes*, and *Gillicus* are nearly indistinguishable except by size. Vertebrae (and tails) are probably the most common remains found. The remains of *Xiphactinus* are the most common of the three fish found in the chalk. This probably due to a preservational bias that favors larger animals. *Gillicus* remains are relatively uncommon, in part because they were probably completely consumed by larger predators such as *Xiphactinus* and mosasaurs. There is little fossil evidence regarding the prey of or what preyed upon *Ichthyodectes* other than the specimen reported by Druckenmiller et al. (1993) that included the remains of an *Ichthyodectes ctenodon* as stomach contents of a shark (*Squalicorax falcatus*). Shimada (1997) reported on the remains of a

TABLE 5.1.
Classification of bony fishes of the Western Interior Sea: (Adapted from Romer, 1966)

Class Osteichthyes
   Subclass Actinopterygii Klein 1885
   Incertae sedis
          Aethocephalichthys hyainarhinos (Fielitz, et al., 1999)
   Infraclass Holostei
      Order Semionotiformes
      Suborder Lepisostoidei
         Family Lepisosteidae
            Lepisosteus (Wiley and Stewart, 1977)
      Order Pycnodontiformes
         Family Pycnodontidae
            Micropycnodon kansasensis Hibbard and Graffham, 1945
      Order Amiiformes
      Suborder Amioidei
         Family Pachycormidae
            Genus *Protosphyraena* Leidy 1857
            Protosphyraena perniciosa (Cope, 1874)
            Protosphyraena nitida (Cope, 1872)
            Protosphyraena tenuis Loomis 1900
            Protosphyraena gladius (Cope, 1873)
            Protosphyreana bentonianum Stewart 1898
      Order Aspidorhynchiformes
         Family Aspidorhynchidae Woodward, 1896
            Belonostomus sp. (Lambe) 1902
   Infraclass Teleostei
      Superorder Elopomorpha
      Order Elopiformes
      Suborder Elopoidei
         Family Elopidae
            Laminospondylus transversus Springer 1957
         Family Apsopelicidae
             Apsopelix anglicus Cope 1871
         Family Pachyrhizodontidae
             Genus *Pachyrhizodus* Dixon 1850
            Pachyrhizodus minimus Stewart 1899
            Pachyrhizodus caninus Cope 1872
            Pachyrhizodus leptopsis Cope 1875
      Suborder Abuloidei
         Family Abulidae
      Order Anguilliformes
      Suborder Anguilloidei
         Family Urenchelidae
             Urenchelys abditus (Wiley and Stewart, 1981)

Superorder Osteoglossimorpha
Order Osteoglossiformes
Suborder Ichthyodectoidei
    Family Ichthyodectidae
        Genus *Icthyodectes* Cope 1870
        Ichthyodectes ctenodon Cope 1870
        Xiphactinus audax Leidy 1870
        Gillicus arcuatus Cope 1875
    Family Saurodontidae
        Saurocephalus lanciformis Harlan 1824
        Saurodon leanus Hays 1830
        Prosaurodon pygmaeus Loomis 1900
Order Tselfatiformes Nelson 1994
    Family Plethodidae Loomis 1900
        Genus *Bananogmius* Whitley 1940 (*Anogmius* Cope 1877)
        B. aratus Cope 1877
        B. favirostris Cope 1877
        B. ornatus Woodward 1923
        B. ellisensis Fielitz and Shimada 1999
        Genus *Pentanogmius* Taverne 2000
        P. evolutus (Cope 1878) = *B. polymicrodus* Stewart 1898
        Genus *Martinichthys* McClung 1926
        Martinichthys ziphioides (Cope 1877)
        Martinichthys brevis McClung 1926
        Syntegmodus altus Loomis 1900
        Thryptodus zitteli Loomis 1900
        Luxilites striolatus Jordan 1924
        Niobrara encarsia Jordan 1924
        Zanclites xenurus Jordan 1924
        Pseudanogmius maiseyi Taverne 2002
Superorder Protacanthopterygii
Order Salmoniformes
Suborder Myctophoidei
    Family Enchodontidae
        Apateodus sp.
        Genus *Cimolichthys* Leidy 1857
        Cimolichthys nepaholica (Cope 1872)
        Genus *Enchodus* Agassiz 1835
        Enchodus gladiolus Cope 1872
        Enchodus petrosus Cope 1874
        Enchodus dirus Leidy 1858
        Enchodus shumardi Leidy
    Family Dercetidae

> *Stratodus apicalus* Cope 1872
Superorder Acanthopterygii
   Order Beryciformes
   Suborder Polymixiodei
      Family Polymixiidae
>   *Omosoma garretti* Bardack 1976
   Suborder Berycoidei
      Family Holocentridae
>   *Kansius sternbergi* Hussakof 1929
>   *Caproberyx* sp. Dixon
>   Unnamed holocentrid (See Stewart, 1990)
Subclass Sarcopterygii
   Order Crossopterygii
   Suborder Coelacanthini
      Family Coelacanthidae
>   Unnamed coelacanth (Stewart, et al., 1991)

Xiphactinus as the stomach contents of a large *Cretoxyrhina mantelli* (KUVP 247) that was collected by Charles Sternberg and is on exhibit in the University of Kansas Museum of Natural History. The broken tip of a *Cretoxyrhina* tooth lodged between two vertebrae in association with the skull of another *Xiphactinus* specimen (ESU 1047) was described by Shimada and Everhart (2004). Many *Cretoxyrhina* teeth were found in association with another fragmentary *Xiphactinus* skull (KUVP 12011) which also exhibited unserrated tooth marks on the lower jaw. A relatively early *Xiphactinus* specimen (KUVP 155) collected by B. F. Mudge from the Carlile Shale of Russell County and described by Stewart (1900, p. 293, pl. 45) shows numerous, unserrated bite marks on the lower jaw.

Bardack (1965) notes that the genus *Xiphactinus* is found worldwide and may occur as early as Albian time in Europe. In North America, the genus ranges from the Upper Cenomanian (Greenhorn Limestone) in Kansas into the middle Campanian Pierre Shale (ibid, p. 10; Carpenter, 1996, p. 33). *Ichthyodectes* and *Gillicus* remains are also found from about the same time span (Bardack, 1965), although Everhart et al. (2003) reported finding the teeth fragments of an ichthyodectid similar to *Xiphactinus* in the Early Cretaceous (Albian) Kiowa Shale.

One issue that is notable for these fish, and for several other species, is that there are very few, if any, remains of juvenile or young individuals represented in collections from the Smoky Hill Chalk deposited in the middle of the Western Interior Sea. It appears that all of the remains of *Xiphactinus*, *Ichthyodectes*, and *Gillicus* found in the chalk were those of at least subadult fish. This suggests that the early stages of their life were spent elsewhere. Obviously, it could also be indicative of a preservational bias against smaller individuals, but some evidence of their presence should be preserved in the fossil record. The only juvenile specimens of *Xiphactinus* I know of are fragments of what appear to be two tiny skulls, one of which I collected.

Saurocephalus, Saurodon, and *Prosaurodon*

The family Saurodontidae includes three species from the Western Interior Sea which are closely related to the Ichthyodectidae described above (Stewart, 1999). All were small to medium-sized (1–2 m [3–6 ft.]) predatory fish and are uncommon as fossils in the chalk. As noted in the introduction to this chapter, the type specimen of *Saurocephalus lanciformis* Harlan 1824 was the first fossil fish known to be collected from the Niobrara. The most unusual character of these fish is the predentary bone that projected forward from the lower jaw. In life, it is likely that this bone was covered by something similar to the beak on a modern swordfish (Fig. 5.2). The function of this underslung sword is unknown, but it gives a different perspective to the phrase "leading with your chin." Another characteristic of these fish is their flat, bladelike teeth, set in a single row in the jaws. In *Saurocephalus* and *Saurodon,* the teeth are closely set, nearly vertical and have a keyhole-like notch at the base of each tooth on the inside of the jaws. In *Prosaurodon,* the teeth are more rounded and do not have the notch found in the other two species. In the lower jaw of *Prosaurodon,* the teeth are inclined forward except for the anterior-most three or four teeth, which are inclined posteriorly (Loomis, 1900, pl. 23, fig. 10). Stewart's (1999) phylogenetic analysis placed *Prosaurodon pygmaeus* closest to *Gillicus arcuatus.*

The three species apparently occurred at differing times in the Western Interior Sea with little overlap. According to Stewart

Figure 5.2. The upper and lower jaws of Saurodon leanus (KUVP 180) in the collection of the University of Kansas Museum of Natural History. Note the large predentary bone on the lower jaw. (Scale = cm)

(1999), *Saurocephalus lanciformis* appears only in the uppermost chalk (early Campanian) while *Saurodon leanus* Hay 1830 is known from the lower through the middle chalk (late Coniacian through the Santonian). *Prosaurodon pygmaeus* appears first about the middle of the chalk and continues upwards into the Sharon Springs Member of the Pierre Shale (ibid.; Carpenter, 1996). Pre-Niobrara records of the Saurodontidae are unknown in Kansas. However, a much older specimen from the Cenomanian Eagle Ford Group of Texas (Stewart and Friedman, 2001) suggests that family existed for some time before the deposition of the Niobrara. The extended lower jaw of saurodonts may indicate that it may have been used in digging for prey in the bottom muds, but nothing is known of their feeding habits otherwise.

Pachyrhizodus

Xiphactinus has already been mentioned as the largest of the bony fish in the Western Interior Sea. However, the skull of a large specimen of *Pachyrhizodus caninus* (FHSM VP-2189) at the Sternberg Museum is certainly as large as any *Xiphactinus* skull that I have seen. It is 56 cm (22 in.) long and has teeth 4 cm long. The entire fish, however, would have been proportionally shorter than a *Xiphactinus* with the same sized head. Still, it would have been an impressive fish.

Pachyrhizodus caninus Cope 1872 and *P. leptopsis* Cope 1874 were in the 2–3-m size range and could be considered in the middle echelon of predatory fish. They were equipped with heavy jaws that were filled with sharp, curved teeth. The teeth were noted to be short and stout by Cope (1872b, p. 343), who later said (1875, p. 220) that "they bear a superficial resemblance to those of a mosasauroid genus." Stewart and Bell (1994) reported that specimens from Texas initially described as the remains of the earliest mosasaurs (Cenomanian) by Stenzel in 1944 were, in fact, jaw fragments of *Pachyrhizodus leptopsis*. In addition, the material cited by Thurmond (1969) as the jaw of an early (Turonian) mosasaur from Texas was also re-identified by Stewart and Bell (1994) as the remains of *P. leptopsis*. The source of the confusion is that the teeth have large bases and, as Cope noted, look very much like those of a mosasaur. Stewart and Bell (1994, p. 8) noted that other specimens of *Pachyrhizodus* in museum collections have been misidentified as mosasaurs, and vice versa.

Pachyrhizodus minimus Stewart 1899 is a much smaller species (less than 1 m). The holotype (KUVP 327) is maintained in the collection of the University of Kansas. *P. minimus* has the distinction of being the most commonly found complete fish fossil in the Smoky Hill Chalk, including scales and preserved gut contents. Miller (1957) reported on a 0.8-m (34-in.) specimen (FHSM VP-326) from Trego County in the Sternberg Museum collection that includes a cast of portions of the stomach and the intestine, and an expelled coprolite. In 1990, my wife found a complete (0.6-m [24-

in.]) specimen (CMC-7552) in the early Santonian Chalk of Lane County that we donated to the Cincinnati Museum Center.

Enchodus and *Cimolichthys*

Fish remains attributable to the genus *Enchodus* Agassiz 1835 have been found worldwide and the genus apparently survives for a time past the Cretaceous/Tertiary extinction event. *Enchodus* has been erroneously referred to as the "Sabre-Toothed Herring" because of the oversized teeth located on the palatines of the skull and the ends of the lower jaw. The genus is not related to herrings but instead is placed in the same order as modern salmon. *Enchodus* is represented in the Smoky Hill Chalk by at least four species, ranging from small (*E. shumardi*) to medium-sized (*E. petrosus*) fish. However, the exact number of species is unclear even today because, while *Enchodus* remains are far from being rare, they are seldom complete. This led early workers like Cope to naming new species from fragmentary specimens.

Stewart (1990a) reported the occurrence of *Enchodus* in all but the highest level of the Smoky Hill Chalk, but their absence there is more of a collecting issue than anything else since they occur as common fossils in the overlying Pierre Shale. Goody (1976) reported on *Enchodus gladiolus* and *E. petrosus* remains in the Pierre Shale and discussed the occurrence of the genus in North America in general. In another study, *Enchodus* species represented about 25 percent of the remains observed in the Pierre Shale of Wyoming (Carpenter, 1996, p. 34). The most recent phylogenetic analysis of the family Enchodontidae was done by Fielitz (1999).

The large teeth of *Enchodus* indicate that it was certainly a predator, but as yet we are uncertain what it ate. I am unaware of any *Enchodus* remains preserved with stomach contents. While such oversized teeth might be useful in impaling soft-bodied prey such as squid, this would raise the question of how the fish got the prey off the teeth and into the mouth. Most likely the long, slender teeth were used as a "fish basket" to more effectively trap smaller fish, similar to the interlocking teeth of some plesiosaurs. While the largest species, *Enchodus petrosus*, had fangs that were nearly 6 cm (2.5 in.) long on a body that was no more than about 1.5 m (5 ft.) in length, even the smallest species had oversized fangs. They would have looked much like some of the modern deep-sea fish, which also have gaping mouths filled with long, sharp teeth. *Enchodus* teeth are usually well represented in collections of micro-vertebrate fossils, and in one instance they composed nearly all (literally millions of tiny teeth) of a thin layer of "fish bone conglomerate" found in the Blue Hill Member (middle Turonian) of the Carlile Shale of Jewell County, Kansas (Hattin, 1962, p. 102, pl. 27B; Everhart et al., 2003).

Whatever the function of these unusual fangs (Fig. 5.3), it was not apparently a deterrent to being eaten by other predators. *Enchodus* remains have been found as stomach contents of *Cimo-*

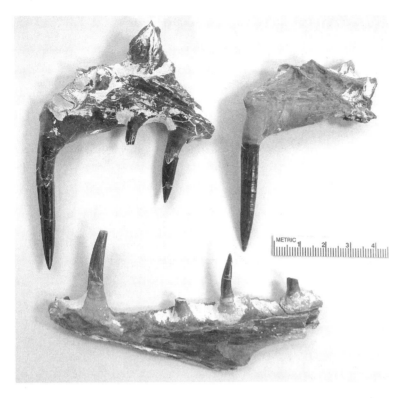

Figure 5.3. The fangs of Enchodus petrosus. *The specimen includes two palatines with large teeth and a large anterior tooth on the dentary.*

lichthys (below), mosasaurs, and plesiosaurs (Cicimurri and Everhart, 2001), and their bones are found occasionally in coprolites that are usually attributed to sharks. The skull of a large *Enchodus petrosus* at the Sternberg Museum (FHSM VP-2939) found in Trego County preserves evidence of feeding by *Squalicorax falcatus,* with serrated bite marks on the bones of the skull and a shed *Squalicorax* tooth lying against the underside of the cranium. It seems apparent that *Enchodus* existed in large numbers and were an important part of the ecology of the Western Interior Sea. In that regard, they were probably very much the equivalent of the modern herring.

One of the more common species of fish found preserved in the chalk is a medium-sized predator called *Cimolichthys nepaholica* (Cope 1872b). The genus name *Cimolichthys* (from Greek *cimoli,* a white chalky clay; and *ichthys,* fish) had been coined by Leidy in 1857 to describe fish remains (*C. levesiensis*) from the English chalk. "Cimoli" is the Latinized version of the Greek word *kimolia,* a white clay from the Aegean island of Kimolos in the archipelago of the Kyklades. Leidy apparently named the English species *C. levesiensis* for the town of Lewes in Sussex, England where the remains were found.

The genus name has been confused over the years in large part because E. D. Cope named a new genus and species, "*Empo*" *nepaholica* (1872b, p. 347), and four new species of *Cimolichthys* (ibid., pp. 351–353; *C. sulcatus, C. semianceps, C. anceps,* and

C. gladiolus) from Kansas in the same paper. Cope (1872c, p. 345) later described *E. nepaholica* as "a fish as large as pike of forty pounds." He also noted (ibid., p. 347) that *Cimolichthys*" was applied by Dr. Leidy to a fish erroneously referred by Agassiz and Dixon to *Saurodon*, Hays. He [Leidy] did not characterize it; and until the barbed palatine teeth, characteristic of it, are discovered in our species, their reference to it will not be fully established." Two years later, Cope (1874, p. 46) revised the spelling of "*nepaholica*" to "*nepaeolica*" [*nepæolica*], changed the genus name of *C. sulcata* to "*Empo*" *sulcatus*, changed *C. semianceps* to "*Empo*" *semianceps,* and added two new species of "*Empo*": *E. merrillii* and *E. contracta*. While *C. gladiolus* eventually became *Enchodus gladiolus*, *C. anceps* disappeared from the literature along with the type specimen. Goody (1976, p. 102) noted that the type specimen of *C. anceps* could not be found. Goody also determined that, based on Cope's description of the type, the specimen was more likely the ectopterygoid of *Enchodus petrosus*.

As noted above, Leidy's *Cimolichthys lavesiensis* is older (Turonian) than the specimens of "*Empo*" *nepaholica* examined by Cope from the Smoky Hill Chalk (late Coniacian–early Campanian). It doesn't appear that Cope ever accepted that *Cimolichthys* Leidy and "*Empo*" Cope were the same genus. Cope (1875, p. 230) noted that he had "formally referred some of the species of *Empo* to the genus which embraces the fish called by Leidy *Cimolichthys levesiensis;* but I find that they do not possess the same type of teeth. . . . The genus therefore takes this name [*Empo*]." Much of the confusion comes from having to work with fragmentary material. *Cimolichthys* specimens are usually poorly preserved and the teeth are quite fragile. However, Cope (1875, pls. 52 and 53) was the first to illustrate the American *Cimolichthys* and one of a few to date to figure the dermal scutes (ibid., pl. 53, fig. 9).

"*Empo*" *nepaholica* and the other new *Cimolichthys*/"*Empo*" species of Cope were later found to be fragments of the same species of fish and were recognized as such by Loomis (1900) and by Hay (1903). By virtue of being the first of the species names published in Cope's (1872b) paper, the species is rightfully called *nepaholica*. Since the genus name *Cimolichthys* Leidy 1857 takes precedence over that of "*Empo*" Cope 1872, the correct name for Cope's many "species" became *Cimolichthys nepaholica*. Like the similar problem with *Xiphactinus* Leidy and "*Portheus*" Cope, the name "*Empo*" has had a long, if ill-deserved, life of its own and is still found on labels in museum exhibits and collections. As for the other species names, Goody (1970, p. 2) indicated "there is no reason for retaining Cope's various species that are based mainly on isolated teeth and fragments of jaw bones."

Cimolichthys can be visualized as a Cretaceous barracuda (or a freshwater pike as suggested by Cope), even though it is not related to either (*Cimolichthys* is in the same order as modern salmon). The fish grew to almost 2 m (6 ft.) in length, and the skull, with triple rows of teeth set in the narrow lower jaws (Hay, 1903) is readily recognizable. Cope (1875, p. 228) noted in his description of the genus "*Empo*" that "the dentaries support several series of teeth; one of the large ones on the inner side and several smaller on the outer." His illustration (ibid., pl. 53, fig. 6) also shows three rows of teeth on the dentary.

Portions of the skull of *Cimolichthys nepaholica* have been figured by Cope (1875), Loomis (1900), Hay (1903) and Goody (1970). While their remains are frequently found, good specimens are rare because the skull was lightly constructed and tended to either come apart prior to preservation or as a result of weathering.

We can assume that *Cimolichthys* had a voracious appetite for fairly large prey because of the number of specimens that have been found with the remains of an undigested last meal inside. One of the strangest "death by gluttony" occurrences in the fossil record was reported by both Kauffman (1990) and Stewart and Carpenter (1990; see also Carpenter, 1996) regarding a specimen of *Cimolichthys* (UCM 29556) discovered in the Pierre Shale of Wyoming. This *Cimolichthys* apparently died with a large squid (*Tusoteuthis longa*—UCM 20556) lodged in its mouth. The squid had been swallowed tail first as evidenced by the rachis (squid pen) being found inside the body of the *Cimolichthys*. However, the open jaws of the fish appear to indicate that part of the head and/or tentacles of the squid were still outside the mouth. This probably meant that the fish's gills were blocked from getting oxygen from the water, causing death by suffocation.

Another unusual *Cimolichthys* specimen (FHSM VP-15065) in the Sternberg Museum was collected by Greg Winkler and Pete Bussen in the early 1990s from the lower chalk (late Coniacian) of western Gove County. In this case, the skull of a large (1.8 m) *Cimolichthys* was found eroding out of the chalk. Much of the skull was already lost, but the rest of the fish was complete back to the tip of the tail. When the specimen was initially prepared, it was found to contain not only a large *Enchodus* (FHSM VP-15066) but also the remains of another unidentified fish (FHSM VP-15067) as a last meal. It is uncertain if both fish inside the *Cimolichthys* were consumed by the larger fish or if this is a classic case of a "fish in a fish in a fish." In either case, the partially digested condition of the *Enchodus* leads me to believe that the *Cimolichthys* died within a few hours after eating it.

In a similar situation, I found a 1.3-m (4-ft.) *Cimolichthys* (FHSM VP-14024) with a 0.7-m (2-ft.)-long, partially digested *Enchodus petrosus* (FHSM VP-14025) inside in the lower chalk of southeastern Gove County in 1994. Both palatine bones of the *Enchodus*, minus the large fangs, were located near the anus of the larger fish. Besides being an indication that the prey had been swallowed headfirst, it also explains our field observation of finding an unusual number of partially digested *Enchodus* palatine bones, and no other associated remains. Being the heaviest bones in the skull of the *Enchodus*, mostly indigestible and located at the anterior end of the skull, they were probably expelled separately while the rest of the prey was still being digested. The large teeth, however, appeared to have been broken off the palatines and were not found in the rest of the remains.

It is probable that *Cimolichthys* was preyed upon by larger fish and mosasaurs. Although they were not noted by Bardack (1965) as being stomach contents in any of the *Xiphactinus audax* specimens he surveyed, I did find partially digested *Cimolichthys* vertebrae in the abdominal region of a *Tylosaurus proriger* (FFHM 1997-10) I collected in 1996–97. While their remains have not been reported as stomach contents of other predators, many of the severed tails which we commonly find in the chalk are from *Cimolichthys*. One interesting feature often observed in *Cimolichthys* remains in the field is the "cone-on-cone" calcite crystals (steinkerns) that often fill the conical hollows between the deeply cupped vertebrae.

Protosphyraena

One fairly common genus in the Smoky Hill Chalk that was noticeably missing from Cope's 1872 list was *Protosphyraena*, a primitive Late Cretaceous "swordfish" that is well represented in the fossil record from as far away as Europe and Japan. A fragment of the pectoral fin was first figured and described from the English chalk by Mantell (1822), and Leidy (1857) authored the name *Protosphyraena ferox* for the English specimens. Leidy (1865, pl. XX, figs. 7–9) also illustrated a *Protosphyraena* tooth from the Navesink Formation in New Jersey, but he erred in stating it was from a dinosaur. After examining specimens at the Smithsonian that were collected by Dr. G. M. Sternberg, Leidy (1870, p. 12) clearly described fragments of the distinctive fin of *Protosphyraena perniciosa*, but attributed them to the crusher shark *Ptychodus*. At the time, he noted (ibid.) that he was following the association made by Agassiz in 1837.

Many of the first Kansas specimens of *Protosphyraena* were collected by B. F. Mudge in the early 1870s from Rooks County. Some of them were sent to E. D. Cope, possibly as early as 1872. As the collector, however, it is apparent that Mudge had a better idea of what the fish looked like than did Cope. Mudge (1874, p. 122) wrote of the skull, "The most remarkable species of fish which we have found, the present season, are of a genus new to me, and I think to science. They are armed with a long, strong weapon at the extremity of the upper jaw, something like that of a swordfish, but round and pointed and composed of strong fibres. The jaws are provided with three kinds of teeth. On the outer edge is a row of large, flat, cutting teeth, somewhat resembling those of a shark [Fig. 5.4]. Inside, and placed irregularly, are small, blunt teeth; while in the back portion of the palate is the third set—small, sharp and needle-like in shape, forming a pavement. The jaws are also fibrous, like the snout. There are three species of this genus. Prof. Marsh has them for critical scientific examination."

Figure 5.4. A variety of flat, bladelike teeth from Protosphyraena. Although it is uncertain whether they were shed or broken off during feeding, they occur commonly in the lower chalk.

At least fourteen specimens (YPM 42137, 42138, 42152, 42200, 42285, etc.) were collected by Mudge in 1874 from Ellis and Rooks Counties and sent to Marsh. More specimens would be collected and sent to Yale in 1875 and 1876. However, the three species mentioned by Mudge had already been described by the end of the 1874 field season by Cope, and Marsh was not interested in fish.

The first species of *Protosphyraena* from Kansas was described by Cope (1873a) from teeth and skull fragments and named *Erisichthe nitida*. The type specimen (AMNH 2121) was "discovered by Prof. B. F. Mudge near the Solomon River" in Phillips County (Cope, 1875, pp. 217–218) and consisted of portions of the dentary and a pectoral fin. This specimen and other specimens of *Protosphyraena* in the Cope collection were figured by Hay (1903). "*Portheus*" *gladius* Cope was named from a fragment of a large (1 m) pectoral fin that Cope (1873b) believed to be similar to that of *Portheus* (*Xiphactinus*). The fin ray (AMNH 1859) was later figured by Cope (1875, pl. 52, fig. 3) and the genus name was changed to *Pelecopterus*. According to Cope (1873b), this specimen was also found by B. F. Mudge "near the Solomon River."

The following year, Cope (1874, p. 41) named "*Ichthyo-dectes*" *perniciosus* from fragments of saw-tooth–edged pectoral fins that were also found by B. F. Mudge. In Latin, the species name

Figure 5.5. A fragment of the saw-toothed edge of the pectoral fin of Protosphyraena perniciosa. *These fins reach 0.9 m (36 in.) or more in length. Each "tooth" marks the beginning of one of the elements (rays) that make up the fin. (Scale = cm)*

"*perniciosus*" means "destructive, harmful, or dangerous," and is certainly descriptive of the jagged leading edge of the pectoral fins (Fig. 5.5). It is apparent from Cope's descriptions at the time, however, that he did not associate the unusual pectoral fins with the swordfish-like skulls mentioned by Mudge (1874). Cope (1877b, p. 821) did note, however, that "I had already been in receipt of fragments of these beaks, associated with loose teeth of the genus *Erisichthe*, but it was Prof. B. F. Mudge who first pointed out that both belong to one and the same genus."

By then, however, Cope (1875, p. 244) had changed the name of this species to *Pelecopterus perniciosus*. In his description of another American species, Cope (1877b, p. 823) noted that another "species has been found in England, and figured by Dixon in the 'Geology of Sussex.' The portions represented in this work are the mandibles, which resemble those of the *E. nitida*, and which were supposed at that time to belong to a species of *Saurocephalus*. A muzzle, perhaps of the same species, was regarded as a Sword-fish, which was called *Xiphias dixonii* by Agassiz. It should be now termed *Erisichthe dixoni*."

In the case of Cope's three Kansas species in three different genera, and *Xiphias dixonii*, however, they were soon found all to be junior synonyms of the genus *Protosphyraena* Leidy 1857. Newton (1878, p. 788) was one of the first to comment, "Dr. Leidy in 1856 [published 1857] proposed, as already mentioned, the generic name of *Protosphyraena*, for those British specimens; and this name therefore must be adopted, and not that of *Erisichthe*, which was given by Prof. Cope in 1872 to the American specimens." Thankfully, Cope's barely pronounceable name eventually faded into oblivion although he was still referring to it as late as 1886.

Loomis (1900) was the next to work on describing the various species of *Protosphyraena*. His profusely illustrated paper in

Palaeontographica is still one of the best references available on fossil fishes from the Kansas chalk. As noted above, Hay (1903) also produced a valuable reference on these Cretaceous fishes, especially with illustrations of the specimens in the E. D. Cope collection at the American Museum of Natural History. Hay (ibid., p. 9) also noted that it was A. S. Woodward (1895) who placed Cope's *Pelecopterus perniciosus* into the genus *Protosphyraena* and first published the name *Protosphyraena perniciosa* in his Catalog of Fossil Fishes in the British Museum.

In the chalk, there are four species of *Protosphyraena* that are currently recognized: *P. nitida* (Cope 1873), *P. gladius* (Cope 1873), *P. perniciosa* (Cope, 1874), and *P. tenuis* (Loomis, 1900). *P. perniciosa* is found in the late Coniacian lower chalk, where its 90-cm (3-ft.)-long, saw-toothed fins and heavy pectoral girdle are fairly common discoveries. The skull, with its long, sword-like snout and flat, blade-shaped teeth is also found frequently, although is it difficult to identify for certain to species because is it so similar to that of *P. nitida*. The pectoral fins of *P. nitida* are readily recognizable because they lack the saw-toothed edges and are much smaller than those of *P. perniciosa*. Upon close examination of the fin of *P. nitida*, numerous fine ridges can be seen running at right angles to the edge of the fin. *P. tenuis* occurs in the early Santonian and appears to continue through the deposition of the chalk. It is a smaller species than *P. perniciosa* but has fins with a similar saw-toothed appearance. The skull of *P. gladius* is unknown, but the large, heavy fins are up to a meter in length and quite thick. Cope (1875, p. 244) noted that the fin of *P. gladius* was "a formidable weapon, and could readily be used to split wood in its fossilized condition." While this is somewhat of an exaggeration because of their relatively fragile condition as fossils, the sharp, knifelike edge of the heavy fin is quite evident.

Another, earlier species of *Protosphyraena* was described by Albin Stewart (1898, p. 27) from a specimen (KUVP 415) that had been collected by S. W. Williston in 1897 in southern Mitchell County. The remains were discovered in the Lincoln Member (Upper Cenomanian) of the Greenhorn Limestone (part of the obsolete "Fort Benton Cretaceous"), and Stewart named the new species *Protosphyraena bentonianum*. Note that the species name was first published as "*bentonia*" in 1898 by Stewart, but then was noted to be a typographical error by Stewart (1900). The specimen consists of a poorly preserved rostrum and fragmentary bones from the skull and does not appear to differ greatly from *Protosphyraena perniciosa*. In 2003, Keith Ewell and I collected the fragments of numerous *Protosphyraena* teeth (unreported data) from the upper part of the Dakota Sandstone in Russell County. The Dakota Sandstone is middle Cenomanian in age.

Stewart (1979; 1988; 1990a) was the most recent North American author to report on the occurrence of *Protosphyraena*. In doing so, however, he unintentionally introduced a change in the species name for *Protosphyraena perniciosa,* dropping the second "i" and spelling it "*pernicosa*" (Stewart, pers. comm., 2004). As he has been the only worker to study this species in the last century or so, his version of the name has come to be the accepted usage. However, the rules of the International Commission on Zoological Nomenclature (ICZN) normally require that such changes be made for good reason and be approved. The spelling of the species name authored by Woodward (1895, p. 414) is used here.

Stewart (1979) was the first to associate the occurrence of the various species of *Protosphyraena* with the stratigraphy of the chalk. He reported that *P. perniciosa* was found in the lower chalk (late Coniacian) while the other three species generally occurred during the Santonian and Campanian. Noting that the limited occurrence of *P. perniciosa* was useful as a stratigraphic marker, Stewart (1990) designated the late Coniacian lower chalk as the biostratigraphic zone of *Protosphyraena perniciosa*. However, Stewart (ibid., p. 21) also acknowledged that "consultation and field observations with Mike and Pam Everhart convince me that *Protosphyraena nitida* is a rare member of the lowest Smoky Hill Chalk fauna." The change was due to Pam's 1988 discovery of a complete *P. nitida* skull and fins (LACMNH 129752) in the lower chalk just above the contact with the Fort Hays Limestone in Ellis County (Fig. 5.6). Two other partial *Protosphyraena* sp. skulls (NJSM 15021, 15839) we collected from the same locality were donated to the New Jersey State Museum. Since then we have found a number of *P. nitida* remains in the lower chalk, and in 2003 I found another associated partial skull and fins in Gove County below Hattin's (1982) marker unit 3. It also appears that *Protosphyraena perniciosa* reappears in the upper chalk (Pete Bussen, pers. comm., 1995; pers. obs.) in the early Campanian, although it is much more common in the late Coniacian.

More than a hundred years after its discovery, it is still unusual to

Figure 5.6. A right lateral view of the skull of Protosphyraena nitida *(LACMNH 129752) found by Pam Everhart in 1988 in the lower chalk of Ellis County, Kansas. This is the only known association of a complete skull and pectoral fins (not pictured) of this species.*

find *Protosphyraena* skull and fin material together, and until recently (M. Triebold, pers. comm., 2003), a complete specimen had never been found. A nearly complete *Protosphyraena perniciosa* specimen was discovered in 2003 but is still in preparation and has yet to be reported. In contrast to some shark specimens (Chapter 4) where the cartilage became more calcified, the skeleton of *Protosphyraena* was generally poorly ossified, and their carcasses apparently tended to fall apart or were torn apart by scavengers. Only the skull and fins and a small bone (hypural) at the base of the tail were ossified enough to be preserved in most cases. Unfortunately, the skulls and pectoral fins are usually found separately. McClung (1908) described and figured the only known caudal fin of *Protosphyraena* from a specimen (KUVP 55500) found by C. H. Sternberg.

Martinichthys (Plethodidae)

Another species named by Cope (1877b), *Erisichthe ziphioides* (AMNH 2131), was eventually determined to be a completely new and unrelated genus, *Martinichthys,* which was described by McClung in 1926. The type specimen consisted of a single bone (or possibly a pair of bones that were fused) forming a blunt rostrum that Cope (ibid, p. 822) described as the "muzzle of an old individual, which has lost a good deal of its apex by attrition." Hay (1903, p. 22) believed it was simply "a species having a short and blunt snout," and renamed it *Protosphyraena ziphioides*. However, the specimen fooled both men because it was described upside down by Cope (1877b, pp. 822–823) and figured in a similar fashion by Hay (1903, figs. 13–14).

Based on differences which he observed in the shape of rostra, McClung (1926) identified six new species from a number of similar specimens as belonging to a new genus of plethodid fishes, which he called *Martinichthys* in honor of H. T. Martin at the University of Kansas. The best of these specimens, a nearly complete skull (KUVP 497) and associated vertebrae, was designated the holotype of *M. brevis.* This specimen and another specimen (KUVP 498—*M. ziphioides*) are the only remains known with skull elements and vertebrae. Since that time, several dozen isolated rostra have been collected, but there have been no additional skulls or post-cranial material found.

Martinichthys is a rare genus that has been found only in the lower Smoky Hill Chalk (late Coniacian) in Kansas and is known only from a single skull and a fairly large number of preserved, bony rostra (snouts) in the collections of the University of Kansas Museum of Natural History and the Fort Hays State University Sternberg Museum of Natural History. It now appears that, except for the heavy rostrum, the skeleton of the fish was poorly ossified and therefore unlikely to be preserved. The fact that most of the rostra found to date (Fig. 5.7), including Cope's type specimen, are heavily worn on the anterior end raises questions about the living habits of the fish.

Martinichthys is also unique in that it is only known from the chalk in western Kansas, and there only from a very limited stratigraphic interval. Most of the rostra occur within a time span of less than 150,000 years near the end of the Coniacian. Besides the ten specimens reported by McClung (1926), at least twelve other rostra had been added to the University of Kansas collection by 1993. In addition, there were a dozen or so in the Sternberg Museum collection. Almost all were from the low chalk in Gove and Trego counties. My wife and I recently donated nineteen specimens (FHSM VP-15549 to 15568) that we had collected in Gove County between 1990 and 2003. While I cannot be certain of the occurrence of the other specimens, almost all ours were found in the chalk from just below Hattin's marker unit 4 to just above marker unit 5 (Everhart and Everhart, 1993). Interestingly, for me at least, this is also the only place that an unusual coprolite composed of oyster (*Pseudoperna congesta*) fragments is known to occur (Everhart and Everhart, 1992). We are still looking for the link between these coprolites and the unusual wear found on the rostra of *Martinichthys* (and a related species called *Thryptodus zitteli*). Strange as it may seem, I am fairly certain that these fish were using their bony "noses" to batter these small shells open so that they could feed on the oysters. Nothing else I can think of can account for the obvious wear that is readily observed on most of these specimens.

After his review of McClung's ten specimens and two new specimens at the University of Kansas, Taverne (2000a) reduced the number of species to *Martinichthys ziphioides* (Cope) and *M. brevis* (McClung), and chose to validate only the two species with associated skulls (KUVP 497 and 498). Taverne (2000a, p. 10) notes that "*Martinichthys* is a valid genus of Tselfatiiformes, characterized by at least five unique characters and which is comprised of two species: *M. brevis* with a short and thick rostrum and *M. ziphioides* with a long and narrow rostrum. The other species described in this genus [McClung, 1926], *M. acutus*, *M. alternatus*, *M. gracilis*, *M. intermedius* and *M. latus* are junior synonyms of *M. ziphioides*."

While I agree with Taverne (2000a) that McClung's number of species (seven) is almost certainly too high, more work needs to be done to explain the wide variety of shapes which can be observed in the other fifty or more specimens now in museum collections. At least two "*Martinichthys*" specimens in the Sternberg Museum collection, one that I collected in 1992 (FHSM VP-15568) in Gove County and another collected by J. R. Green in Trego County in 1973 (FHSM VP-3248), may represent an undescribed taxon be-

Figure 5.7. (opposite page) The rostrum of Martinichthys brevis (FHSM VP-15567) in dorsal, left lateral, and ventral views. This specimen was collected in 2003 from the lower chalk of Gove County. Note the typical worn appearance of the anterior end of this rostrum. Only two reasonably complete skulls are known of Martinichthys. Both are in the collection of the University of Kansas Museum of Natural History. (Scale = cm)

tween *Martinichthys* and *Thryptodus* (Shimada, pers. comm., 2003).

Other Plethodids

Martinichthys is only one of a diverse family of primitive bony fish called Plethodidae which was established by Loomis in 1900. Many of the species are known from single specimens in the collection of the University of Kansas Museum of Natural History, and some of those are incomplete. As a group, they were generally medium-sized fish with many small comblike teeth in the jaws and on plates in the mouth. The skull is relatively flat and broad. It is likely that they fed on invertebrates from the sea floor, although they may have eaten small ammonites and other cephalopods nearer the surface. The most common genus is *Bananogmius* Whitley 1940, which includes four species from the chalk. Note that the name "*Anogmius*" had originally been given as a genus of *Pachyrhizodus* (Cope, 1871, p. 170) and is therefore a junior synonym of that genus. The genus *Anogmius* was thus "pre-occupied" and could not be used again, even by Cope, as he attempted to do in 1877.

While *Bananogmius* is (or was) the most common genus of plethodid in the Smoky Hill Chalk, its most common species has now been placed in another genus. According to Taverne (2000b; 2004), *Bananogmius evolutus* Cope 1878 is specifically excluded from the original genus and placed into a new genus, *Pentanogmius*. Taverne (2001a) noted that the three species remaining in the genus *Bananogmius* are *B. aratus* Cope 1877, *B. favirostris* Cope 1877, and *B. ornatus* Woodward 1923. About the same time as *B. evolutus* was placed into the new genus by Taverne, a new species (*B. ellisensis*) was described by Fielitz and Shimada (1999) from an unusual, three-dimensional specimen (FHSM VP-2118) in the Sternberg Museum from the Carlile Shale of Ellis County. This brings the total number of species within the genus *Bananogmius* back to four.

Loomis (1900) named two new plethodids: *Syntegmodus altus* and *Thryptodus zitteli*. While these are regarded as synonymous with *Bananogmius* by some workers, Taverne (2001c, p. 251) noted that the specimen of *Syntegmodus altus* (AMNH 2112) that he examined did represent a valid genus that is intermediate between *Bananogmius* and another genus called *Niobrara*. The type specimen of *Thryptodus* described by Loomis (1900, pls. 21–22) from the Kansas chalk (Fig. 5.8) was apparently destroyed in Germany during World War II, and so far as I am aware there has not been as complete a specimen found since. Shimada and Schumacher (2003) described what is probably the earliest occurrence (Upper Cenomanian) of *Thryptodus* from a specimen (FSHM VP-13996) in the Sternberg Museum. Taverne (2003) redescribed *Thryptodus zitteli* from fragmentary specimens at the University of Kansas Museum of Natural History (KUVP 456, 457, and 459)

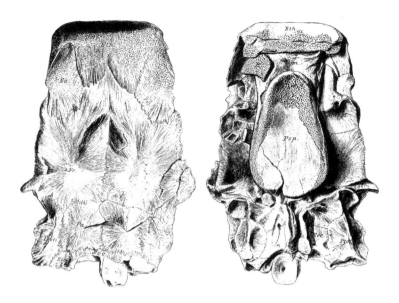

Figure 5.8. The skull of Thryptodus zitteli *as figured by Loomis (1900, pl. 21) in dorsal (left) and ventral views. This specimen was probably the most complete skull ever found of this species.*

and a specimen in the American Museum of Natural History (AMNH 19557), and characterized the species as being more closely related to *Plethodus* than to *Martinichthys*. While it is rare, the remains of *Thryptodus* are found occasionally in the lower chalk. Over the years, I have collected four fragmentary specimens in southeastern Gove County. In most cases, the remains consist only of the part of the heavy, rounded end of the snout (fused ethmoid bones), but one specimen (FHSM VP-15571) included most of the skull and several vertebrae.

Jordan (1924) added three more plethodid species to the fauna of the Smoky Hill Chalk. All are based on fairly complete specimens of small fish in the collection of the University of Kansas Museum of Natural History. His reconstruction (ibid. pl. 14) of *Niobrara encarsia* is shaped somewhat like a modern tuna. The type specimen (KUVP 179) was found in the chalk of Trego County by H. T. Martin and is 69 cm (27 in.) long. Taverne (2001b) noted that *Niobrara* was the most primitive genus within the order Tselfatiiformes (Bananogmiiformes) and that it was closely related to the genus *Bananogmius*. A similar species, *Zanclites xenurus* (KUVP 52) described by Jordan (1924) is slightly smaller (55 cm/21.5 in.) and was collected in Gove County, also by H. T. Martin. Taverne (1999) confirmed its placement in Tselfatiiformes but did not comment on the family to which it belonged. The type specimen of *Luxilites striolatus* (KUVP 295) was described from a poorly preserved skull (Jordan, 1924). Taverne (2002a) noted that *Luxilites* was a valid genus and that it belonged to a subgroup of the Plethodidae, which includes *Bananogmius, Syntegmodus,* and *Niobrara.* Taverne (2002b) also described the newest plethodid species from the Kansas chalk (*Pseudanogmius maiseyi*—AMNH 8129) from a specimen originally identified by Hay (1903) as *Anogmius* sp.

Holosteans

The holosteans are an intermediate group between primitive bony fish and the more advanced teleosts. Modern holosteans include freshwater forms such as the bowfin (*Amia*) and the gar (*Lepisosteus*) and are notable for their heavy scales. Another group of holosteans called pycnodonts lived in the oceans covering Kansas during the Cretaceous. Pycnodonts have high, narrow bodies, heavy, diamond-shaped scales, and jaws filled with small, nipping, and crushing teeth. They apparently fed on small invertebrates (epifauna) that were attached to the inoceramid bivalves on the sea floor. Most pycnodont specimens in the chalk consist of isolated jaw elements with their rows of round, smooth teeth.

A single species, *Micropycnodon kansasensis,* was described by Hibbard and Graffham (1941) from the lower chalk (late Coniacian). The type specimen (KUVP 1019) was found in Rooks County by Allan Graffham in 1939, and a more complete specimen (KUVP 7030) was collected in northeast Trego County about 1935 by George Sternberg. Dunkle and Hibbard (1946) described the second specimen and indicated that it was probably from the Fort Hays Limestone. The only *Micropycnodon* specimen that we have collected (a jaw fragment) was found by my wife in 1991 in the low chalk of Gove County. Stewart (1990a, p. 24) noted that "all the Smoky Hill Chalk Member specimens of *Micropycnodon* known to me are from the zone of *Protosphyraena perniciosa* [late Coniacian]. The single published record of *Hadrodus marshi* was probably from the zone of *Hesperornis* [early Campanian]. Within the zone of *Clioscaphites vermiformis* and *C. choteauensis* is an undetermined pycnodont that is not *Micropycnodon* and is probably not *Hadrodus.*"

While the remains of pycnodonts are rare in Kansas, they have been found almost continuously from the Albian through the late Coniacian. Jaw fragments and teeth of *Coelodus brownii* and *Coelodus stantoni* were reported by Williston (1900) from the Kiowa Shale of Kiowa and Clark counties. Isolated teeth of *Coelodus* sp. have been collected from the Kiowa Shale of McPherson County (Beamon, 1999; pers. obs.). An analysis of my 2003 collection from McPherson County indicates that pycnodont teeth make up about 21 percent of more than 500 teeth collected (unreported data). I have also recently collected pycnodont teeth (cf. *Coelodus* sp.) from the Upper Dakota Sandstone (middle Cenomanian), and was present when a pycnodont jaw plate (FHSM VP-15548; cf. *Micropycnodon*) was collected in 2003 from the base of the Lincoln Limestone Member (Upper Cenomanian) of the Greenhorn Limestone. Hibbard (1939) described a jaw fragment of *Coelodus streckeri* (KUVP 946) from the Fairport Chalk Member (Carlile Shale) of Russell County, and Zielinski (1994) reported on the discovery of the splenial tooth plate (FHSM VP-6728) of a pycnodont (cf. *Anomoeodus* sp.) from the Blue Hill Shale Member of the Carlile Shale. Everhart et al. (2003) noted numerous teeth and jaw fragments of a pycnodont-like fish, *Hadrodus priscus,* in addition to teeth of an unidentified pycnodont from a fish tooth conglomerate in the Blue Hill Shale.

Wiley and Stewart (1977) reported the discovery of remains of a gar (*Lepisosteus* sp.; KUVP 36243) from the lower chalk in western Trego County. The specimen included skull and fin elements, teeth, and scales. Gars are normally found in fresh or brackish water and this was the first report of a gar from a marine environment (ibid., p. 761). The authors were unable to explain the occurrence of the remains in a marine deposit and no additional specimens have been found. Everhart et al. (2003) reported gar scales from the upper Blue Hill Member of the Carlile Shale (middle Turonian). In the summer of 2003, I also collected gar teeth and a single ganoid scale (cf. *Lepidotes* sp.) from the Kiowa Shale of McPherson County. Williston (1900, pl. 30, fig. 4) included two teeth from the Kiowa Shale that he tentatively identified as ?*Mesodon* but noted that they probably should be referred to as *Lepidotes* sp. (ibid., p. 256). The Sternberg Museum has a section of articulated scales of a large *Lepidotes* (FHSM VP-5114) from the Kiowa Shale of Clark County, Kansas, on exhibit.

Dunkle (1969) described a new species of amioid fish (*Paraliodesmus guadagnii*—USNM 21083) from a small (17 cm) specimen found inside a *Volviceramus grandis* shell from the chalk of Gove County in 1954. Stewart (1990a, p. 24) noted that *Paraliodesmus* was also found in his biostratigraphic zone of *Clioscaphites vermiformis* and *C. choteauensis* (early Santonian). The only other amioid remains I am aware of from the Cretaceous of Kansas are seven cf. *Pachyamia* sp. teeth that I collected from the Kiowa Shale of McPherson County in 2003. A similar tooth (FHSM VP13540) was collected by Beamon (1999) and identified as "?Lepisosteid."

Other Fish Found in the Smoky Hill Chalk

Stewart (1996, p. 390) noted that "*Kansius sternbergi, Caproberyx* sp., *Trachichthyoides* sp. and one or more undescribed holocentrid genera occur within inoceramid bivalves (*Platyceramus platinus*)." These fish are all small (10 cm or less) and are generally known only from remains preserved inside the clam shells. Most are found in the Santonian when the inoceramids were large and plentiful. A list of the other species of fish found in the Smoky Hill Chalk would include *Apsopelix anglicus* Cope 1871, *Stratodus apicalis* Cope 1872, *Lepticthys agilis* Stewart 1900, *Apateodus* sp., and *Omosoma garretti* Bardack 1976. Discounting the large numbers of small fish found occasionally inside inoceramids, *Apsopelix* and *Stratodus* are the most common of this group. Just about all of them, however, would be considered rare in any collection. I have only collected one specimen of *Apsopelix* and three or four sets of fragmentary *Stratodus* remains in more than thirty years. Other than a couple of occurrences of unidentified "fish in shell," I've never seen any of the other species in the field. Many of the smaller fish documented from the Smoky Hill Chalk are single specimens or are the result of unusual preservation (e.g., inside inoceramid shells).

Small fish are particularly rare in the chalk, either because they were completely consumed as prey by larger fish or because they were literally too small to be preserved. That is, their remains (bones) were not large or solid enough to avoid being dissolved by the chemistry at work on the sea bottom. In some cases where whole schools of small fish were preserved inside clam shells (see Stewart, 1990b), they simply are yet to be described and named. It appears likely that the fish were sheltering inside the clam shells to avoid predators or possibly even feeding on parasites or other tiny fauna that was also living there. When the clam died and the shell closed, they were trapped inside where their skeletons had a better chance for preservation.

The first and only eel from the Smoky Hill Chalk, *Urenchelys abditus* (Wiley and Stewart, 1981) was found with other fish inside

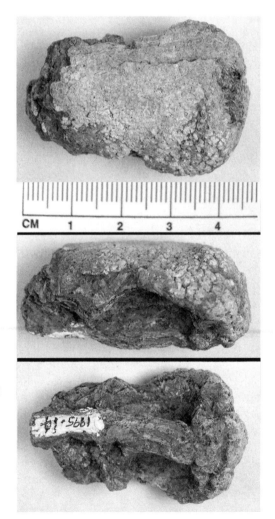

Figure 5.9. The neurocranium of an unusual and poorly known fish called Aethocephalichthys hyainarhinos *(FHSM VP-15577) in dorsal, right lateral, and ventral views. This specimen was collected from the Sharon Springs Member of the Pierre Shale in Logan County by Pam Everhart. A similar specimen at the University of Kansas Museum of Natural History was found in the upper Smoky Hill Chalk by Pete Bussen.*

a giant inoceramid. In another instance, my wife found what appeared to be the only known coelacanth remains (LACMNH 131958) from late Coniacian chalk of Gove County in 1991 (Stewart et al., 1991). Although the partial skull and lower jaws are identifiable as a coelacanth, it is probably too fragmentary to be described. We are still on the lookout for additional remains that will help establish what is probably a new genus and species of coelacanth in the Western Interior Sea.

Lastly, another strange fish, *Aethocephalichthys hyainarhinos*, was described by Fielitz et al. (1999). It has been found as far away as New Zealand, and its remains are usually limited to an odd-shaped, solidly fused neurocranium (Fig. 5.9). Although the authors (ibid.) reported one specimen from Gove County in the upper Smoky Hill Chalk, we have never found one there. My wife, however, did find a small *Aethocephalichthys* skull in the Sharon Springs Member (middle Campanian) of the Pierre Shale, near McAllaster Butte in Logan County. At least ten other specimens are in the collections of the University of Kansas and the Los Angeles County Museum of Natural History. Our friend Pete Bussen, who also collected a specimen from the Pierre Shale, refers to them as the "headlamp fish" because of the large eye sockets that occupy most of the front of the skull. According to the authors (ibid, p. 97) the etymology of the name is *Aethocephalichthys* (Greek) "strange-headed fish" and *hyainarhinos* (Greek) "pig-nosed"—not exactly a complimentary name or description.

The Smoky Hill Chalk has preserved a wide variety of fishes, large and small. It is likely that there are still new species to be found there, even after 130 years of collecting.

Six

Turtles:
Leatherback Giants

The giant turtle swam steadily through the sunlit water of the Inland Sea. Although it was early afternoon, the passing of time mattered little to her. She had been swimming almost continuously eastward now for three days on a journey that would take at least a week longer. Able to swim submerged for several minutes before surfacing to breathe, she was making good progress. Her large front limbs acted as underwater wings and enabled her streamlined body to "fly" efficiently through the water. The hundred or so eggs in her lower abdomen were nearly ready to be laid, and she was returning to the warm sands of a beach where she had hatched more than fifty years before. She seldom swam continuously like this, preferring to move more slowly and feed leisurely among the sea jellies. Instinctively, however, she knew that the best time for laying her eggs was quickly approaching. The tides would soon be at their peak and would allow her to dig a nest that was safely above the high-water mark.

She was a mature, adult *Protostega* sea turtle, with a "leatherback" shell that measured more than 2 m (6 ft.) across. She would continue to grow slowly throughout her long life, but she had already reached most of her adult size. Few predators were large enough or hungry enough to bother with something of her bulk, but it hadn't always been so. She was missing three toes on her left foot from a bite by a shark when she was much younger. After the sudden attack, the shark had disappeared as quickly as it had come, and she had recovered from the wound without further problems. Life was harsh in the Western Interior Sea and she had been lucky to survive. She did not know or care that few, if any, of the unhatched young she carried would make it through the gauntlet of hungry predators as they made their way from the nest on the beach to the relative safety of the ocean. Once they were in the water, their chances of survival improved, but not by much. There were many large fish and mosasaurs in the ocean that were quite capable of eating a baby sea turtle.

Unlike other marine reptiles, including mosasaurs and plesiosaurs, turtles had retained the ability to lay eggs, staying with a strategy of survival of the species in overwhelming numbers. So every year at about the same time, she and thousands of other females of her species made the journey to the miles of beaches surrounding the great shallow sea to lay the millions of eggs that would ensure that a few young would survive to reach adulthood.

When she surfaced to breathe the next time, there was a haze of fine dust in the air. Exhaling the stale air from her lungs and then inhaling, she noticed an acrid odor. A quick glance behind her revealed a dark cloud that stretched across the western sky. Already there were fine patches of dust floating on the water. Sensing only that something unusual was happening, she took several deep breaths and dived again, continuing to swim steadily to the east. As she swam, the water around her quickly darkened as the black cloud covered the sun and turned day into night. The dust that had at first floated on the surface soon began to clump up, and then to sink, mixing with the seawater and clouding it with fine particles.

Unknown to her, a giant volcano on the edge of the ocean more than a thousand miles to the west had exploded violently earlier in the day, sending millions of tons of volcanic ash high into the atmosphere, where it was picked up by the prevailing westerly winds. As the huge dust cloud was carried eastward over the Inland Sea, larger pieces of the ejecta began to fall out, pelting the surface of the water with a heavy rain of ash that covered and killed everything living immediately downwind of the volcano. Further east, the effects of the ash and the toxic gases that traveled with it were not immediately lethal to life, but they would have serious consequences on the local ecosystem for years to come. In the middle of the sea, the finer ash that had been carried by the wind would fall in much smaller but still significant amounts. This dust would slowly settle downward in the water and eventually completely cover the sea floor to a depth of an inch or more. Animals that depended on gills to extract oxygen from the water, such as fish and most invertebrates, would have problems with damage caused by the gritty ash, and many would die. Enough would survive to repopulate the sea, but in the path of the heaviest plume of ash it would, at the least, be a major natural disaster for sea life.

The turtle continued to swim through the now dark and muddied water between her and the distant shore. The sulfurous gases and the ash irritated her throat and eyes each time she surfaced to breathe, but she would survive. The eggs she laid eight days later would not be so lucky. Weather changes caused by the volcanic activity would create a brief chilling of global temperatures, and her eggs would fail to develop and hatch. She would not know this, however, and would return each year for the next forty-eight years to lay more eggs at the place where she first entered the ocean.

While a number of turtle species and specimens have been described from the Smoky Hill Chalk, surprisingly little is known of their occurrence in the Western Interior Sea during the Late Cretaceous. Part of the problem is the rarity of sea turtle remains as fossils, especially relatively complete specimens. We are fairly certain that turtles were never as numerous in the Western Interior Sea as they were in warmer waters along the Cretaceous Gulf Coast. Russell (1993) noted that turtles make up only about 3 percent of vertebrate specimens from the Niobrara Formation in museum collections, and are found about as often as the remains of toothed birds (Chapter 11). Another issue is the generally smaller size of most marine turtles and the lack of speed or other defensive capabilities. While turtles might seem to be more likely than other animals to be preserved because of their bony shell, it turns out that most of the remains that have been found in the Smoky Hill Chalk are skulls and limbs that appear to have come from dismembered carcasses. Having discovered the first turtle in the chalk, Cope (1872a) noted that the bones making up the shell of a very large *Protostega* were extremely thin and fragile. It is likely that the larger predators of the Cretaceous seas were well equipped to feed on sea turtles and must have done so on a regular basis. Even the relatively complete specimen of *Toxochelys latiremis* (ROM 28563) reported from the Kansas chalk by Nicholls (1988) was missing most of the readily detachable parts like the hind limbs and distal portions of the forelimbs, most likely as the result of scavenging by sharks. Turtle remains, including the complete skull of a small *Toxochelys*, were found as stomach contents of *Squalicorax falcatus* (Druckenmiller et al., 1993), and pieces of *Protostega* shell and limb material have been found with embedded teeth and bite marks attributed to *Cretoxyrhina mantelli* (Shimada and Hooks, 2004; FHSM VP-2158, pers. obs.). Cope (1872c) named a new species of shark (*G. hartwelli* Cope = *Squalicorax falcatus*) from a single tooth found beneath the bones of his giant *Protostega gigas* specimen. The tooth is a possible indication of scavenging on that specimen by sharks. Schwimmer et al. (1997, p. 78) reported bite marks from *Squalicorax* sp. on a *Toxochelys*? sp. humerus (LACMNH 50974), and two *Desmatochelys lowi* humeri (KUVP 32401 and 32405). A small *Toxochelys* specimen (FHSM VP-13449) that I collected in 1988 consisted of a complete skull, neck vertebrae, and the front edge of the carapace (upper shell). The rest had apparently been bitten off.

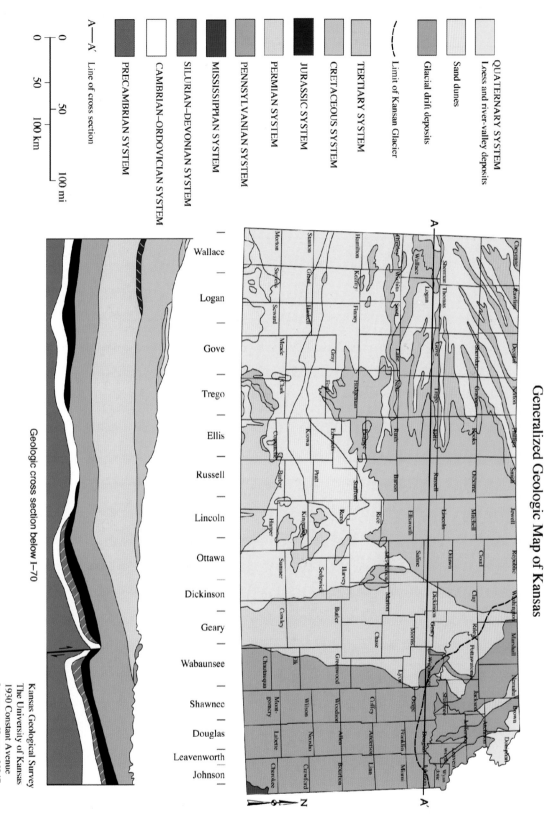

Plate 1. Generalized geologic map of Kansas. Courtesy of the Kansas Geological Survey, Lawrence, Kansas.

Plate 2. The hunter becomes the hunted, as a giant pliosaur called *Brachauchenius lucasi* attacks an early mosasaur. Although pliosaurs became extinct during the Turonian, they were still present in the Western Interior Sea when the first mosasaurs arrived. Adult mosasaurs were top predators, but their young were preyed upon by sharks and other species of mosasaurs. Life could be short for the unwary. Painting © by Dan Varner.

Plate 3. A lone polycotylid plesiosaur (*Trinacromerum osborni*) cruises in the shallow water of the middle Turonian Western Interior Sea as a much smaller *Ptychodus* shark moves out of its path. Painting © by Dan Varner.

Plate 4. Squid, such as this early relative of *Tusoteuthis longa,* were often the favored prey of plesiosaurs, including *Brachauchenius lucasi.* This giant predator was one of the last of the pliosaurs and made its final appearance in Kansas during the deposition of the Fairport Chalk Member (middle Turonian) of the Carlile Shale. Painting © by Dan Varner, courtesy of the Museum of Northern Arizona.

Plate 5. During the last few minutes of its life, a giant *Xiphactinus* struggles to swallow another fish called *Gillicus.* This *Xiphactinus* died before the smaller fish could be digested. The fossilized remains of the famous "fish-in-a-fish" specimen were collected by George F. Sternberg from the Smoky Hill Chalk, and are currently displayed in the Sternberg Museum of Natural History. Painting © by Dan Varner.

Plate 6. A giant ginsu shark (*Cretoxyrhina mantelli*) at the moment of impact in a high speed attack on a subadult mosasaur (*Tylosaurus nepaeolicus*). While we are unsure whether these sharks attacked living mosasaurs or scavenged dead ones (or both), the partially digested bones of mosasaurs are frequently found in the Smoky Hill Chalk and sometimes include the broken tips of ginsu shark teeth. Painting © by Dan Varner.

Plate 7. In this view, a *Clidastes propython* (one of the smaller mosasaurs, about twelve feet long) is about to put the bite on a marine turtle called *Calcarichelys* somewhere off the ancient gulf coast of what is now Alabama. While mosasaurs fed mostly on fish, some of them ate anything small enough to swallow. Painting © by Dan Varner.

Plate 8. It's hard to imagine the scale of this picture. The swimming birds (*Hesperornis regalis*) are about five feet long and the *Tylosaurus*—well, it's huge. Modeled after the largest specimen on exhibit, the "Bunker Tylosaur" (KUVP 5033) at the University of Kansas Museum of Natural History was at least forty-five feet long, including a skull that was nearly six feet in length. Painting © by Dan Varner.

Plate 9. This picture shows an attack by a large *Tylosaurus proriger* on a much smaller *Clidastes propython*. The largest mosasaur in the Western Interior Sea, *Tylosaurus* occasionally killed and ate other species of mosasaurs, as shown by preserved stomach contents found in the Pierre Shale of South Dakota. Painting © by Dan Varner.

Plate 10. Two *Globidens dakotensis* mosasaurs are shown feeding on clams and other shellfish found on the bottom of the shallow inland sea that covered South Dakota and much of the middle of North America during the Campanian. *Globidens* was a very specialized mosasaur with round, ball-shaped teeth and a short, heavily built skull. Painting © by Dan Varner.

Plate 11. A giant *Mosasaurus hoffmanni* just misses in an attack on a marine crocodile (*Thoracosaurus*) in the Cretaceous seas that covered present-day New Jersey. These Maastrichtian-age animals are known from both North America and the Netherlands in Europe (the North Atlantic was much smaller 68 million years ago). *Thoracosaurus* survived for some time after the K/T boundary event, while mosasaurs, plesiosaurs, dinosaurs and many other groups did not. Although first found in the Netherlands, *M. hoffmanni* is also known from the Western Interior Sea (Texas). Painting © by Dan Varner.

Plate 12. A *Mosasaurus hobetsuensis* cruises the rocky underwater shoreline of the Japanese archipelago, looking for prey. During the Late Cretaceous, the western rim of the Pacific Ocean—from Japan to New Zealand—was inhabited by many of the same genera of marine reptiles (mosasaurs and plesiosaurs) as were found in the Western Interior Sea. Painting © by Dan Varner.

The humerus of a *Protostega* (1503.57) from Logan County in the collection of the University of Wisconsin-Madison shows deep, serrated bite marks from *Squalicorax*. Even the giant *Archelon* (YPM-3000) specimen in the Yale Peabody Museum had apparently lost the lower portion of its right rear leg at some point during its life (Wieland, written communication in Williston, 1914, p. 240). While turtles have not yet been documented as stomach contents of mosasaurs in the Western Interior Sea, Dollo (1887) noted turtle remains in association with a giant *Hainosaurus* mosasaur in Europe.

Further study of the distribution of turtle species in the Smoky Hill Chalk has certainly been hampered by this lack of complete, identifiable specimens and the absence of accurate stratigraphic data. Nicholls (1988, p. 181) noted that toxochelyids, the most common turtles in the Western Interior Sea, are known mostly from skulls and that post-cranial material is rare. Over the last thirty years of working in the chalk, my success in collecting turtle remains has been limited to finding a small but nicely articulated *Toxochelys* skull (FHSM VP-13449) and a few specimens of isolated limb material or shell fragments, mostly from the lower chalk (late Coniacian). In 1994, my wife had the good fortune to discover a nearly complete right plastron (lower shell) of *Protostega gigas* (FHSM VP-13448) in the upper chalk (biozone of *Spinaptychus sternbergi*—Hattin's marker units 17–18). Unfortunately, most of the turtle material collected from Kansas during the last 130 years does not have good locality or stratigraphic data.

Cope (1872d, p. 433) noted that his *Protostega gigas* had been found "near Fort Wallace" (upper chalk) and also named a new species (1872b) from several vertebrae of the tail of a small turtle (*Cynocercus incisus*—AMNH 1582) discovered by Sgt. William Gardner in "the yellow chalk near to Butte's Creek, south of Fort Wallace." Lane (1946) regarded the second specimen as "related to *Toxochelys*" but otherwise uninformative. Cope (1873) indicated that the type specimen of *Toxochelys latiremis* (AMNH 2362) had been collected by B. F. Mudge and later noted (1875, p. 99) that the remains had been "found by Professor Mudge near the forks of the Smoky Hill River." This locality is somewhat confusing since exposures of both the Smoky Hill Chalk and Pierre Shale are found in this vicinity. At the time of the discovery, Cope was unaware of the existence of the Pierre Shale and assumed that all fossils collected in this area were from the "Cretaceous chalk." The north fork of the Smoky Hill River branches off the main stream to the southeast of McAllaster Butte and just to the south of the abandoned townsite of Sheridan. Further examination (Schultze et al., 1985, p. 25; Nicholls, 1988, p. 185) of the material has determined that the type specimen of *Toxochelys* (AMNH 2362) most likely had come from the Pierre Shale and not the chalk as Cope had assumed. Two years later, Cope (1877) noted the receipt of two nearly complete skulls of *Toxochelys* from C. H. Sternberg (AMNH 1496 and 1497), but he failed to provide any locality information. Hay (1908, p. 169) reported vaguely that both of these skulls had been "collected somewhere along the Smoky Hill River." In a similar situation, Williston noted in an 1876 letter to Marsh (Shor, 1971, p. 77) that Sternberg, who was working along the Smoky Hill River at the time, "got one or two large turtles that are good and some pretty good saurians." At the time, Williston (ibid.) was "eight miles west of Monument [Rocks?]—four miles north of the river."

In 1894, S. W. Williston at the University of Kansas described a new species of marine turtle from the "Benton Cretaceous" just across the state line near Fairbury, Nebraska. The specimen had been given to the university in 1893 by M. A. Low, and Williston named it *Desmatochelys lowi* in his honor. Although the specimen (KUVP 1200) had been damaged by "curiosity seekers" after its discovery, Williston (1894a, p. 5) noted that "the portions that were obtained are of the greatest importance," including the skull, limbs and most of the carapace. Hay (1908) indicated that portions of the back of the skull were damaged. When I examined the specimen in 2004, I was surprised to find that the skull was uncrushed, completely articulated, and in nearly perfect condition. It looked more like a modern turtle skull than one that was more than 90 million years old (Fig. 6.1). The lack of crushing most likely indicates that the specimen came from the Fairport Chalk Member of the Carlile Shale (middle Turonian). Although the number of vertebrate remains found in the Fairport Chalk is relatively small compared to the Smoky Hill Chalk, the preservation is somewhat better and there is less crushing compared to other Cretaceous strata in Kansas. The remains probably represent the oldest turtle specimen to be described from the Western Interior Sea.

More recently, Beamon (1999) reported fragmentary remains of turtles (FHSM VP-13550, 13551, and 13553) from the Early Cretaceous (Albian) Kiowa Shale of McPherson County, but he noted that they could not be identified further than to family. Williston (1894b, p. 2) reported the discovery of a "scapula-coracoid of a species as large as *Protostega*" from the Kiowa of Clark

Figure 6.1. A right lateral view of the skull of the type specimen of Desmatochelys lowi *(KUVP 1200) from the Fairport Chalk just across the state line in south-central Nebraska. This specimen is unusual for its uncrushed preservation. (Scale = cm)*

County and said that it was unlike "any turtle known to me." I was unable to relocate the specimen. In 1969, the nearly complete carapace and plastron of a medium-sized turtle (KUVP 16370) were found in the Kiowa Shale (Albian, Early Cretaceous) of Kiowa County, Kansas, by Orville Bonner. The specimen has not yet been identified or otherwise described, but it would certainly represent a much older species than *Desmatochelys lowi.*

Since the discovery of the type specimen of *Desmatochelys,* additional remains have been collected in South Dakota (Zangerl and Sloan, 1960), Arizona (Elliott et al., 1997), and possibly from Vancouver Island (Nicholls, 1992). Williston (1898, p. 354) predicted that *Desmatochelys* would also be found in Kansas. While a complete specimen has not yet been documented from the state, Schwimmer et al. (1997, p. 78) reported two shark-bitten *Desmatochelys lowi* humeri (KUVP 32401 and 32405) from the Carlile Shale of Ellis County. More recently, Bruce Schumacher (pers. comm., 2003) located the remains of a large, as yet unidentified, turtle in the Fairport Chalk (Lower Turonian) of Russell County. The age of this specimen would be slightly younger than Williston's *Desmatochelys.* At this point, the relationship of this family (Desmatochelyidae) to other Late Cretaceous turtles remains uncertain. Elliott et al. (1997) and Hooks (1998) consider it to be a separate family, while Hirayama (1997) includes it in the Protostegidae.

Much of the early work on turtles done by Cope was done hurriedly and on incomplete or otherwise damaged specimens, which resulted in significant errors that are still being corrected. Hay (1895) re-examined Cope's type specimen of *Protostega gigas* (AMNH FR 1503) and noted that Cope had mistakenly identified the left half of the plastron as being elements of the carapace. While noting that Cope's estimate of the size of the skull was wrong, Hay (ibid., p. 62) arrived at nearly the same overall length for *Protostega,* 3.92 m (nearly 13 ft.). Later, Hay (1908, p. 196) wrote that the earlier estimates of the length of *Protostega* "made by Cope and Hay were too great." In another paper, Hay (1896) also reviewed Cope's description of *Toxochelys latiremis* (Fig. 6.2) and compared it with a new skull that the Field Museum had acquired. Cope's description (1873) was based on a fragmentary specimen (part of the lower jaw, the coracoid, and some limb bones). Cope never completed a full description of the species even though he had received additional cranial material (AMNH 1476 and 1477) from C. H. Sternberg (Cope, 1877). Case (1898) followed Hay's work with a description of several new specimens and "species" of *Toxochelys* that were in the collection of the University of Kansas Museum of Natural History, and noted that it was "the single well-established genus of the sea turtles from the Cretaceous of Kansas." Not much has changed in that regard since 1898.

A total of four genera and six species of turtles were reported by Williston (1898) from the Upper Cretaceous of Kansas, including *Desmatochelys* from the "Benton Cretaceous." He further indicated that *Toxochelys* is the most common, "especially in the upper

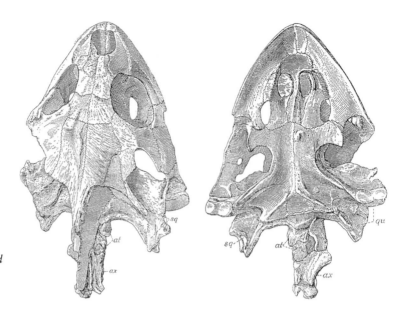

Figure 6.2. Dorsal and ventral views of the skull of Toxochelys latiremis (AMNH 1496) as figured by Hay (1908). Toxochelys is the most commonly found turtle in the Smoky Hill Chalk.

or yellow chalk." Williston (1901) described a new species of marine turtle (*Porthochelys laticeps*) from the lower chalk in Trego County near the Saline River. The skull of this new species (Fig. 6.3) is almost as wide as it is long and the jaw is massive (Hay, 1908). The new specimen was also unusual in that it included a nearly intact carapace. Unlike Cope's *Protostega* and most other marine turtles, the upper shell of the new species was completely ossified. The carapace of *Toxochelys latiremis* had not been discovered at that time, and Williston (1901, p. 198) believed that it, too, would be completely ossified. He further noted that "the relationships of *Porthochelys* are clearly with *Toxochelys latiremis*." Later discoveries, however, made it readily apparent that *Toxochelys* and *Porthochelys* represent distinctly different genera.

Possibly as the result of his 1895 discovery of the giant *Archelon ischyros*, George Wieland (1896) became the next researcher to concentrate on the study of Late Cretaceous marine turtles. The nearly complete type specimen of *Archelon* (YPM 3000) was collected from the Upper Campanian Pierre Shale of South Dakota, and thus it lived a million or so years after any of the turtles described from Kansas. Because of its completeness and because it was mostly still articulated, the specimen provided much information about protostegid turtles in general and served to deflate the exaggerated size attributed to *Protostega gigas* by Cope and others (Wieland, 1896, p. 411). According to Wieland (ibid., p. 410), the type specimen of *Archelon ischyros* exhibited at the Yale Peabody Museum was about 3.5 m (11.3 ft.) in length. Wieland was uncertain, however, that his *Archelon* represented a distinct genus, and after reconstructing the plastron and noting that it was very similar to that of *Protostega*, he (1898) renamed it *Protostega ischyra*.

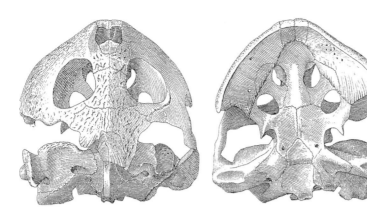

Figure 6.3. Dorsal and ventral views of the skull of Porthochelys laticeps *(KUVP 1204) as figured by Hay (1908). This specimen, including a nearly complete carapace, is on exhibit in the University of Kansas Museum of Natural History.*

After additional study, however, Wieland (1900) apparently determined that his original view was correct and resumed using the genus name *Archelon* when he described the skull and pelvis of the type specimen.

While *Protostega* was certainly a giant turtle, most of the specimens known at the time were incomplete. From the condition of the remains, it is likely that their carcasses were scavenged and torn apart by sharks or other large predators before reaching the sea bottom. Thirty years after Cope's 1871 discovery of the genus, the hind limb was still largely unknown. In looking through the collection at the University of Kansas, Williston (1902) came across a nearly complete hind limb of *Protostega* (KUVP 1201) that had been collected by Charles Sternberg near Russell Springs (Logan County) in 1902. In his brief description of the material, Williston (ibid., p. 276) commented that "the bones of the fore limb, moreover, are all much larger than those of the hind." In this case, the hind limb would have been 1.2 m (4 ft.) long and was certainly part of a very large turtle. Over a hundred years later, the specimen is still on exhibit at the University of Kansas Museum of Natural History.

Williston goes on to make an interesting comment regarding the stratigraphic occurrence of various Late Cretaceous genera that is worth repeating here. He (ibid., p. 270) noted that "the characters separating *Archelon* Wieland from *Protostega* Cope, while not very important, would seem sufficient. Nevertheless, one can derive little justification from the different geological horizons in which the forms are found. The relations between the Niobrara and Fort Pierre vertebrates are for the most part very close. I have recognized in both horizons *Tylosaurus, Platecarpus* and *Mosasaurus* (*Clidastes*), as well as *Pteranodon* and *Hesperornis,* all very typical of the Niobrara deposits, and the existence of *Claosaurus* [Marsh's dinosaur discovery in Kansas] has been recently affirmed in the Fort Pierre. On lithological grounds, there is nothing separating the two groups of deposits and I protest against the names Colorado and Montana, as perpetuating a wrong impression. On paleontological and lithological grounds, there would be much better reasons for uniting the Niobrara with the Fort Pierre than with the Fort Benton."

In the same year, Wieland (1902) described and figured the front limb of a new specimen of *Toxochelys latiremis* in the Yale Peabody collection. Wieland (ibid., p. 96) noted that this was probably "the first complete restoration of the fore flipper of an ancient marine Chelonian which has been given" and suggested that the earlier description by Cope was in error. In the same paper, Wieland also commented on the front limb of *Archelon* and (ibid., p. 107) proposed a general classification of marine turtles.

Wieland (1905) then turned his attention to the description of a new species (*Toxochelys bauri*) collected by C. H. Sternberg from near the middle of the Smoky Hill Chalk (Santonian) in western Gove County in December 1904. This was an important specimen according to Wieland (ibid., p. 326) because until then "no complete carapace of *Toxochelys* had been described." The figures of the carapace and plastron published in Wieland's paper demonstrated two important points: Williston's belief that *Toxochelys* would be found to have a solid shell like *Porthochelys* was wrong; and the carapace of *Toxochelys* was similar in construction to those of *Protostega* and *Archelon*. The holotype specimen of *T. bauri* (YPM 1786) was re-examined by Zangerl (1953) and referred to *Ctenochelys stenopora*. A check of the records at the Yale Peabody Museum in 2004 indicated that the referral had not yet been verified.

Wieland (1906) was able to report on new *Protostega gigas* material when the Carnegie Museum of Natural History purchased two specimens discovered by C. H. Sternberg. Wieland (ibid., p. 282) wrote that CMNH 1421 was collected from along Hackberry Creek in southern Gove County, Kansas, and that the two specimens were found fairly close together. If this locality had been correct, these specimens would be the earliest known examples of the species. However, Sternberg (1909, p. 116) noted in his book *Life of a Fossil Hunter* that he "should like to correct this mistake. It was found about three miles northwest of Monument Rocks, in a ravine that opens into the Smoky [Hill River], east of where Elkader once stood." The chalk in this locality is late Santonian to early Campanian in age.

CMNH 1420 was the most complete of the two specimens, but Wieland (1906, pp. 284–285) wrote that Sternberg, in his "attempt to remove and separate the bones from their matrix of chalk, mismarked some of them, and also made it impossible to determine the outlines of any of the plastral elements with exactness." He then added, "however, none of the bones are broken, and Mr. Sternberg redeemed himself by discovering and securing in such excellent condition specimen No. 1421" (a skull, lower jaw, and limbs and plastron still in the chalk matrix).

Wieland's admonition to field collectors is well worth repeating here: "As will be evident to any student of the fossil vertebrates the

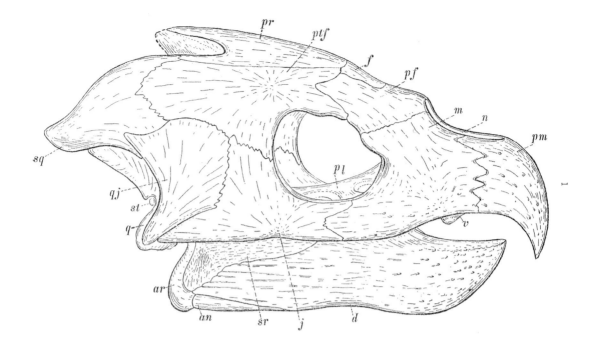

Figure 6.4. A right lateral view of the skull of Archelon ischyros as figured by Wieland (1900). Note the characteristic hooked beak of this species.

removal of the fossil from its matrix in the absence of the necessary knowledge, training and equipment, was ill advised. Such work is difficult enough in the best equipped laboratories." Sternberg might have disagreed with Wieland that he lacked the knowledge necessary to collect and prepare specimens after working almost thirty years in the chalk, but apparently he took no offense. In fact, he repeated Wieland's remarks word for word (Sternberg, 1909, pp. 116–117). There is no doubt that Sternberg was pleased with the results of his efforts and those of the Carnegie Museum. The specimen (a composite of CMNH 1420 and 1421) is one of the best examples of *Protostega gigas* currently on exhibit.

Williston (1914) provides a summary of what was then known about the "Ancient Sea Turtles," quoting extensively from his communications with Wieland. Regarding the general habits of the protostegids, Wieland suggested that there was "no doubt that *Archelon* was strictly carnivorous" and was fast enough to pursue "other than relatively slow-moving prey." Regarding the diet of *Protostega*, Williston (1914, p. 240) added that the abundance of giant clams (inoceramids) would have "afforded an almost inexhaustible source of food for these large turtles." He then noted that "perhaps the formidable beak (Fig. 6.4) was used more in social quarrels than for food-getting." While no evidence of feeding habits, pro or con, has been found since Williston's time, I think it is far more likely than *Protostega* fed near the surface, much like modern sea turtles. The toothless beak of these turtles appears far more useful for shearing flesh, like that of a modern snapping turtle, than for crushing hard-shelled invertebrates. In any case, the giant inoceramids that Williston wrote about as possible turtle food were mostly gone from the central portions of the Western Interior Sea by the early Campanian when *Protostega* arrived. We may never know what they fed upon. However, a specimen of *Toxochelys latiremis* from the Pierre Shale of Wyoming was associated with co-prolites containing fish bones (Carpenter, 1996, p. 46). Nicholls (1997, p. 219) notes that modern sea turtles feed on a variety of food sources, including sea grasses, algae, mollusks, and jellyfish, but are not known to feed on fish.

Zangerl (1953) reviewed the occurrence of marine turtles, particularly those from the Gulf Coast Cretaceous. He described and named a new species, *Lophochelys natatrix* (YPM 3606) from a specimen found by B. F. Mudge and party in the Smoky Hill Chalk. However, he did not add further information about the occurrence of *Toxochelys, Protostega,* and *Archelon* in the Western Interior Sea. Stewart (1978) reported the discovery of the remains of small toxochelyid turtles in the early Turonian Fairport Chalk (Carlile Shale Formation) of Russell and Ellis Counties at roughly the same time as *Desmatochelys*. This was long before the giant Protostegids made their appearance in the early Campanian.

Gaffney and Zangerl (1968) revised the Chelonian genus *Bothremys*. Remains of this unusual marine turtle have been found in the Cretaceous of New Jersey, Alabama, and Arkansas, and the Miocene of Maryland. The turtle is notable for its solid carapace and plastron and for pits in its upper jaws that were possibly used to position round prey, like snails, so they could be crushed. During their research, the authors also found another shattered specimen (YPM 3608) in the Yale Peabody Museum that appeared to be similar to *Bothremys*. It had been collected by B. F. Mudge and party in June 1876, apparently in what is now Logan County. Once the remains were reassembled, the specimen consisted of "a nearly complete plastron, . . . as well as about 60 percent of the carapace" and was identified as *Bothremys barberi* (ibid., p. 206). While similar to the New Jersey and Gulf forms, the authors noted several minor differences and designated it as "Sub-species C" (ibid.).

Following the collection of a nearly complete late Campanian specimen of *Toxochelys latiremis* from near Castle Rock in Gove County, Nicholls (1988) reviewed most of the known specimens and published a revision of the genus *Toxochelys*. She noted (ibid., p. 181) that the carapace was poorly known in existing specimens of this genus and that the new specimen (ROM 28563) was valuable because the carapace was complete. Whereas Zangerl (1953) noted that there were six species of *Toxochelys,* Nicholls (1988, p. 186) determined that there were really only two species: *T. latiremis* from the Western Interior Sea and *T. moorevillensis* Zangerl 1953 from the Gulf Coast.

There is, however, a downside to the story of the collection and description of this otherwise excellent specimen. The remains were discovered and collected from private property by a field crew (not including E. Nicholls) from the Royal Tyrrell Museum. While they had permission from the landowner to collect, there were some apparent misunderstandings that left the rancher feeling like the museum had not lived up to its side of the bargain. As a result, the landowner has not allowed any collecting of fossils on his property since 1988. When I talked with him in 2001, he was still adamant about refusing permission and indicated that he had never received further information about the disposition of the specimen or even a copy of the publication regarding it. I did make a point of providing him with a copy of the Nicholls paper.

Stewart (1990) noted that *Toxochelys latiremis* is found throughout the deposition of the Smoky Hill Chalk and that the holotype of another species of toxochelyid (*Lophochelys natatrix* Zangerl 1953) was found in the uppermost chalk (zone of *Spinaptychus sternbergi* or *Hesperornis*). He also indicated that the largest of the Niobrara turtles, *Protostega gigas*, had been found only in the uppermost chalk (zone of *Hesperornis*). Curiously, he did not mention any of the other species that have been described from the chalk.

Hirayama (1997, p. 234) discussed the distribution and diversity of marine turtles from the Cretaceous and suggested that the toxochelyids were bottom-dwellers and not necessarily good swimmers. While this may have been true for habitats along the margins of the Western Interior Sea, it doesn't explain the relatively large number of *Toxochelys* specimens found near midocean, hundreds of miles from the nearest shore, where the water was up to 200 meters (650 ft.) deep (Hattin, 1982). Hirayama also noted (1997, p. 236) that while "the skull of the protostegids may have been modified to feed on hard-shelled animals," the limbs were well developed for swimming and may have enabled these turtles to hunt more mobile ammonites in the water column instead of the sessile, bottom-dwelling inoceramids. Again, with no evidence available as to what these turtles actually ate, I would suspect that catching ammonites near the surface would have been a more productive method of feeding than diving for the large inoceramids at the bottom of the sea.

Hooks (1998) reviewed the Protostegidae and revised the systematics of the order based on a cladistic analysis. His cladogram indicates that *Toxochelys* is only a distant relative of the Protostegidae, sharing few characters with *Protostega* and *Archelon*. His cladogram also shows Williston's (1894a) *Desmatochelys* to be more closely related to *Toxochelys* than to the Protostegidae.

Based on the fragmentary remains of sea turtles from the earliest Cretaceous rocks in Kansas (Kiowa Shale—Albian), it is known they were generally about the size of a modern leatherback turtle. For comparison, leatherback turtles (*Dermochelys coriace*) can grow to a length of 2.4 m (8 ft.) and weigh 900 kg (2000 lb.), although most are smaller. By the Santonian, however, at least one turtle lineage began to grow much larger. *Protostega gigas* was the largest marine turtle from the early part of the Campanian, which includes the last million or so years of deposition of the chalk.

Cope (1875, p. 111), who collected and named the first specimen in 1871, estimated that *P. gigas* would have measured 3.9 m (12.83 ft.) in length and 3.4 m (11.3 ft.) across the outstretched paddles, with a skull that was about 0.50 m ("24⅝ in.") long. Note here that Cope (ibid.) repeated a math error in converting from meters to inches that he made years earlier (Cope, 1872d, p. 431). Half a meter is roughly 19⅝ inches, not 24⅝. When Hay (1895, p. 61) re-examined Cope's specimen, he didn't catch Cope's conversion error but instead noted that Cope's estimated length of the skull was too high and revised the length downward to "18 inches or 45 cm." In his comparison of Cope's type specimen with two new specimens of *Protostega gigas* (CMNH 1420 and 1420) found by C. H. Sternberg, Wieland (1906, p. 280) saw Cope's mistake and noted "a palpable numerical error in the measurements of the cranium."

Hay (1895) indicated that Cope was in error in several interpretations of the skeleton, including his misidentification of the left plastron as part of the carapace. However, he agreed almost exactly with Cope's calculated length of the turtle (3.9 m). Ten years later, having just sold a nearly complete and articulated composite specimen to the Carnegie Museum of Natural History, C. H. Sternberg (1905) strongly disagreed with Cope's reconstruction of *P. gigas* but did not refute his measurements. Sternberg had collected several reasonably complete specimens since the 1870s, including one that he sent to Cope in 1876. Wieland (1906) used the Carnegie specimens to establish a more accurate size for *Protostega,* including Cope's type specimen, noting that the actual length is far less than originally estimated. While not providing a measurement of the length of the specimens described in this paper, Wieland (1896, p. 411) had noted earlier that "*Protostega gigas* must have been less than three meters in length" when compared to his newly discovered specimen of *Archelon ischyros.*

Archelon ischyros is the largest known marine turtle from the Cretaceous or any other time (Fig. 6.5). Its size is frequently compared with that of a small car. Wieland (1896, p. 410) gave the length of the type specimen as 3.52 m (11 ft. 4 in.). K. Derstler (written comm., 1999) noted that the length of the *Protostega gigas* in the Dallas Museum of Natural History is 3.4 m, that the Yale *Archelon* specimen (YPM 3000) was 3.0 to 3.1 m, and that he had worked with a specimen of *Archelon* that was 4.6 m long. Although the remains of *Archelon* have never been found in the lower portion of the Pierre Shale in Kansas, we know that it lived in the Western Interior Sea during the latter part of the Campanian and may have survived well into the Maastrichtian. Rocks of that age in the Pierre Shale were probably exposed and eroded away in western Kansas many millions of years ago. *Archelon,* however, is well known from the upper Pierre Shale of South Dakota.

The discovery of *Protostega* is an interesting story that was well documented by E. D. Cope. While Cope (1870, p. 446) had already named a new species of Cretaceous marine turtle (*Pneuma-*

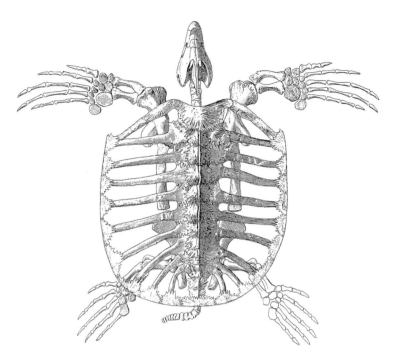

Figure 6.5. A dorsal view of the type specimen of Archelon ischyros *(YPM 3000) as figured by Wieland (1909).*

toarthrus peloreus) from an isolated vertebra found in the green sand deposits of New Jersey, he was apparently surprised by the peculiar characteristics of the much more complete specimen he discovered and collected during his only trip to the chalk of western Kansas in 1871. In a letter to Professor Leslie, dated October 9, 1871, and read at the Academy of Natural Science of Philadelphia on October 20 of that year, Cope (1872a) described and named the first turtle from the Smoky Hill Chalk:

"On another occasion, we detected unusually attenuated bones projecting from the side of a low bluff of yellow chalk, and some pains were taken to uncover them. They were found to belong to a singular reptile, of affinities probably to the Testudinata, this point remaining uncertain. Instead of being expanded into a carapace, the ribs are slender and flat. The tubercular portion is expanded into a transverse shield to beyond the capitular articulation, which thus projects as it were in the midst of a flat plate. These plates have radiating lines of growth to the circumference, which is dentate. Above each rib was a large flat ossification of much tenuity, and digitate on the margins, which appears to represent the dermo-ossification of the Tortoises. Two of these bones were recovered, each two feet across. The femur resembles in some measure that ascribed by Leidy to *Platecarpus tympaniticus* [a mosasaur specimen from Alabama], while the phalanges are of great size. Those of one series measured eight inches and a half in length, and are very stout, indicating a length of limb of seven feet at least. The whole expanse would thus be twenty feet if estimated on a Chelonian basis. The proper reference of this species cannot now be made, but both it and the genus are clearly new to science, and its affinities not very near to those known. Not the least of its peculiarities is the great tenuity [thinness] of all the bones. It may be called *Protostega gigas*."

The remains of this huge turtle were dug up and packed as well as possible under the primitive conditions that existed in western Kansas at the time. This was several years before plaster and burlap jackets were used to protect fossil bones (Sternberg, 1884), and the techniques used to remove fossils from the matrix ("pick-axe and shovel," below) were at best crude. In addition, the locality on Butte Creek was fifteen miles or so south of the nearest railroad line, and the bones had to travel more than that distance in a horse- or mule-drawn wagon over the then roadless prairie. Once packed in crates and loaded into a freight car, the remains then traveled some 1,500 miles to Philadelphia, Pennsylvania, where they were transferred again to a horse-drawn wagon and transported across town to Cope's work place. Considering the rough handling involved, it was a wonder that any of these delicate bones ever arrived in good enough condition to be useful. Cope (Almy, 1987, p. 189) was well aware of the problem and had written to Dr. Turner two years earlier asking him to pack the next shipment of *Elasmosaurus* bones more carefully (see Chapter 7).

At the March 1, 1872, meeting of the American Philosophical Society, Cope (1872b) read his complete description (eleven pages!) of the type specimen (AMNH FR 1503) collected from the chalk along Butte Creek, "near Fort Wallace," in Logan County. He also indicated that the remains were completely enclosed in the chalk when discovered and had to be excavated with "pick-axe and shovel." Apparently only the edge of the shell was visible when it was found. The bones were quite fragile, and Cope remarked that the specimen was "much fractured" by the time it had been shipped 1,500 miles to Philadelphia.

Cope also described the discovery of *Protostega gigas* in his discussion regarding the geology and paleontology of western Kansas published in the "U.S. Geological Survey of Montana and Portions of the Adjacent Territories." One has to wonder how Kansas was included with Montana, but it happened in this instance. Cope (ibid.) wrote,

> Tortoises were the boatmen of the Cretaceous waters of the eastern coast, but none had been known from the deposits of Kansas until very recently. But two species are on record; one large and strange enough to excite the attention of naturalists is the *Protostega gigas,* Cope. It is well known that the house or boat of the tortoise or turtle is formed by the expansion of the usual bones of the skeleton till they meet and unite, and thus become continuous. Thus the lower shell is formed of united ribs of the breast and of the breast-bone, with bone deposited in the skin. In the same way the

roof is formed by the union of the ribs with bone deposited in the skin. In the very young tortoise the ribs are separate as in other animals; as they grow older they begin to expand at the upper side of the upper end, and with increased age the expansion extends throughout the length. The ribs first come in contact, where the process commences, and, in the land-tortoise, they are united to the end. In the sea-turtle, the union ceases a little above the ends. The fragments of the *Protostega* were seen by one of my party projecting from a ledge of a low bluff. Their thinness and the distance to which they were traced excited my curiosity, and I straightway attacked the bank with the pick. After several square feet of rock had been removed, we cleared up one floor, and found ourselves well repaid. Many long slender pieces of two inches in width lay upon the ledge. They were evidently ribs, with the usual heads, but behind each head was a plate like the flattened bowl of a huge spoon, placed crosswise. Beneath these stretched two broad plates, two feet in width, and no thicker than binder's board. The edges were fingered, and the surface hard and smooth. And this was quite new among full grown animals, and we at once determined that more ground must be explored for further light. After picking away the bank, and carving the soft rock, new masses of strange bones were disclosed. Some bones of a large paddle were recognized, and a leg-bone. The shoulder-blade of a huge tortoise came next, and further examination showed that we had stumbled on the burial-place of the largest species of sea-turtle yet known. The single bones of the paddle were eight inches long, giving the spread of the expanded flippers as considerably over fifteen feet. But the ribs were those of an ordinary turtle just born, and the great plates represented the bony deposit in the skin, which, commencing independently in modern turtles, united with each other below at an early day, But it was incredible that the largest of known turtles should be but just hatched, and for this and other reasons it has been concluded that this "ancient mariner" is one of those forms not uncommon in old days, whose incompleteness in some respects points to the truth of the belief that animals have assumed their modern perfections by a process of growth from more simple beginnings. (1872e, pp. 323–324)

A brief description of the skeleton of *Protostega gigas* follows this section (1872e, pp. 334–335). Cope also mentions earlier discoveries of Cretaceous turtle material, including a humerus from Mississippi discovered by Dr. William Spillman (Manning, 1994, p. 67). The specimen was originally figured in Leidy (1865, p. 128, pl. VIII, figs. 1, 2) and described as "the humerus of *Mosasaurus*." While this might be considered a major mistake by today's standards, it is important to note that the limbs of both mosasaurs and giant turtles were unknown at the time. The specimen was re-identified as turtle by Cope (1872) and named *Protostega tuberosa*. Wieland (1900) examined the same bone and recognized that it was not a protostegid. He changed the genus name and called the turtle *Neptunochelys tuberosa*.

Leidy (1865, figs. 3, 4, 5) also included three views of another "gigantic humerus" from the green sand of New Jersey on the same plate. Although Leidy didn't identify it directly, he did refer to the figure (ibid., p. 128) as "supposed to be of a humerus of the *Mosasaurus.*" However, the same specimen had been previously identified by Agassiz (1849) as a turtle that he called *Atlantochelys Mortoni* (ANSP 9234) without further description. Unlike Leidy, Cope (1875, p. 113) appeared to be aware of the name given by Agassiz but chose to ignore it because "it was unaccompanied with specific or generic description." As justification, Cope indicated that he considered it to be "not only a privilege but a duty to ignore names put forward in this manner" and renamed the protostegid turtle humerus *Protostega neptunia*. To add to the confusion, Wieland (1900) indicated that Leidy (1873) had "validly rehabili-tated" the name and also indicated that Cope was in error when he considered it to be a protostegid. Wieland (1900) noted that the form of the humerus was actually closer to that of *Desmatochelys lowi* Williston and retained the name *Atlantochelys mortoni*.

The most commonly found turtle remains throughout the Smoky Hill Chalk are those of *Toxochelys,* a much smaller species than *Protostega*. Cope (1872a) mentioned the discovery of a turtle "the size of some of the large *Cheloniidae* of recent seas" which was associated with a plesiosaur skeleton (*Plesiosaurus gulo*) from the Pierre Shale near Sheridan, Kansas. In a brief note, Cope (1873) indicated that B. F. Mudge had provided him with additional mate-rial from the specimen, and which he named *Toxochelys latiremis* without further description. Although Cope spent considerable time and effort on the reconstruction (albeit in error) and descrip-tion of *Protostega gigas,* he apparently did not have sufficient ma-terial or interest to do the same with *Toxochelys*. The type speci-men is briefly noted and figured in his *The Vertebrata of the Cretaceous Formations of the West* (Cope, 1875, pp. 98–99, pl. VIII). His final note on the species (Cope, 1877) was a brief men-tion of two new skulls discovered by C. H. Sternberg.

Like mosasaurs, marine turtles certainly reached their peak in terms of body size during the Late Cretaceous. The smaller varieties were apparently far more common in the warmer waters of the Gulf Coast than they were in the north. However, the fate of the toxochelyids, desmatochelyids, and protostegids was the same as that of the dinosaurs and other species at the end of the Cretaceous.

In retrospect, marine turtles were and are the most successful of any of the reptiles that have returned to the sea, perhaps partly because they remained conservative in their evolution. Even as they evolved flippers in place of legs, they still retained a measure of flexibility and skeletal integrity in the joints that was lost in the ichthyosaurs, plesiosaurs, and mosasaurs. Unlike these other highly adapted marine reptiles, the turtles were still able to support their bodies to some extent, and to move about on land. Of the four major groups, only the turtles retained their connection with the land, returning each year to lay their eggs in the warm beach sand. They also avoided becoming apex predators. In their case, it is possible that *not* being at the top of the food chain was beneficial to their survival.

While *Archelon* and other closely related turtles did not survive the end of the Cretaceous, it is readily apparent that other marine turtles came through and prospered. If the extinction of marine reptiles at the end of the Cretaceous was largely due to a collapse of the marine ecosystem, it would be easier for a turtle feeding low on the food chain to survive than for a mosasaur or plesiosaur that depended on an abundance of fish in its diet.

Seven

Where the Elasmosaurs Roamed

The warm, silt-laden water felt strange to the young, long-necked plesiosaur as he slowly worked his way upstream against the strong current of the river. His streamlined body seemed less buoyant in the fresh water, which made raising his head above water to breathe a bit more difficult. Unlike the ocean that was his home, he could not see very far in the murky water, so he moved cautiously along the sandy bottom, trying to stay in the deeper reaches of the main channel. Dimly he remembered being here several times before with a much larger member of his species. At the time, it was all he could do to stay close to her body and not be swept away by the current. Now sensory cells in his nasal passages "tasted" strange new smells as the water passed in through his nose and exited his slightly opened mouth. Although he was hungry after his long journey, his instincts drove him up this river for something else.

Within a mile from where the river entered the ocean, the sand bottom of the river had turned to coarser gravel. The plesiosaur lowered his head and began to probe the gravel with his lower jaw. Clouds of silt and other debris rose wherever his chin touched the bottom. The sensations received through his lower jaw told him that this wasn't what he was looking for, and he began swimming again. Holding his head and neck stiffly in front of his body, he relied on alternating beats of his wing-shaped front and rear flippers to provide the thrust he needed to "fly" forward through the water.

The river narrowed as he swam upstream, and the current became faster. Swimming against a current was not something he was used to doing. Such currents were hardly ever found in the calm, shallow ocean where he had spent almost all his life. He soon began to encounter large rocks that formed eddies in the current. It was around the base of a rock larger than he was that he finally found what he was looking for. Small round stones had collected on the downstream side of the boulder. Twisting his head slightly to the side, he was able to see that there were many more here than he would need. Slowly and delicately, he picked up one smooth rock after the other with his long, slender teeth. With each, he raised his head slightly and swallowed, feeling the hard stone move into his throat and start the slow trip down his ten-foot-long esophagus to his stomach. He continued selecting and swallowing the round stones until he had satisfied his need. Raising his head above the surface briefly to breathe, his sharp underwater vision saw only a dark green blur that was the dense forest that crowded over the riverbank. Again he submerged, turned, and began his return journey back to the sea.

His movement flushed a small fish from its hiding place near the large rock. Before the fish could flee, the plesiosaur reacted and trapped it between jaws filled with long, interlocking teeth. Raising his head just above the surface, he swallowed his prey headfirst, then exhaled again and refilled his lungs. Vaguely satisfied, he submerged his head and began to swim downstream with the current with slow, deep strokes of his paddles.

Although isolated fragments of plesiosaurs, mostly vertebrae, had been described earlier in the nineteenth century by Joseph Leidy (1823–1891) and others, remains discovered by Dr. Theophilus Turner (1841–1869) near Fort Wallace in western Kansas in 1867 represented the first nearly complete plesiosaur specimen from North America and the first known elasmosaur. Dr. Turner was an Army surgeon assigned to Fort Wallace. According to his letters (Almy, 1987), Turner found the remains of the huge animal while exploring exposures of the Pierre Shale along the right-of-way of the approaching Union Pacific railroad near McAllaster Butte in present-day Logan County, Kansas. Dr. Turner gave several of the vertebrae to John LeConte, who was part of a survey crew for the railroad. LeConte in turn delivered the bones to Edward D. Cope in Philadelphia for examination (LeConte, 1868). Cope immediately recognized the bones as belonging to a large plesiosaur and wrote back to Turner (Almy, 1987) asking him to secure the rest of the specimen and ship it to him. After receiving the shipment in mid-March of 1868, Cope hurriedly unpacked and examined the remains. Less than two weeks after first seeing the

Figure 7.1. The anterior-most cervical vertebrae of the type specimen of Elasmosaurus platyurus *(ANSP 10081) in left lateral view. The specimen was collected by Dr. Theophilus H. Turner in 1867 and prepared by E. D. Cope. Note that a fragment of the occipital condyle of the plesiosaur's skull is still firmly lodged in the atlas/axis vertebrae at left. (Scale = cm)*

bones, he reported on the strange and hereto unknown extinct animal at the March 24 meeting of the Academy of Natural Sciences of Philadelphia (Cope, 1868, p. 93). Cope named the plesiosaur *Elasmosaurus platyurus* (flat-tailed, thin-plate reptile) to describe the long flat tail and the large pelvic and pectoral girdles he thought he saw in the remains. In August of the following year, Cope (1869) published a description and figures of *E. platyurus* as a "pre-print" of the *Transactions of the American Philosophical Society,* a respected scientific journal, and sent them to many of his friends and associates in the U.S. and Europe.

Unfortunately, it appears that Cope did much of his work in secret and without a review by his peers. Following the early work on American plesiosaurs by his mentor, Joseph Leidy, Cope misidentified the cervical vertebrae as caudals and believed the long neck was the tail (Fig. 7.1). As a result, he placed the head of his plesiosaur on the wrong end, creating an animal with a short neck and an extremely long tail. After reading Cope's publication and making his own examination of the specimen, Leidy, Cope's mentor, recognized the mistake. Nearly a year later, on March 8, 1870, Leidy presented his own view of the specimen, apparently without discussing it first with Cope! Leidy (1870, p. 392) noted that "Prof. Cope has fallen into the error of describing the skeleton in a reversed position to the true one, and in that view has represented it in a restored condition in his recent 'Synopsis of the Extinct Batrachia, Reptilia, and Aves,' published in the Transactions of the American Philosophical Society."

In Cope's defense, while the half-round base of the skull (occipital condyle) is still firmly attached to the first vertebra of the neck (atlas/axis), the neck of this huge animal is so long and thin that even today (pers. obs., 2002) it looks like it should be the tail. The centra of the most anterior cervical vertebrae are only 2.5 cm (1 in.) in diameter, and the small fragment from the base of the skull is not obvious. Such a long, thin neck would not have been expected on an animal that was over 12 m (40 ft.) in length. It is also possible that the atlas/axis vertebra was not fully prepared initially because it was assumed to be the end of the tail and of less taxonomic value or interest than other parts of the skeleton.

As might be expected, Cope was embarrassed by what now appears to have been an obvious oversight on his part. His immediate reaction was to try to minimize the damage by printing a notice and offering to replace all copies of the preprint (Storrs, 1984). The text of the notice was reprinted in O. C. Marsh's comments more than twenty years later in the *New York Herald* (Ballou, 1890): "An error having been detected in the letter press of the 'Synopsis of the Extinct Batrachia and Reptilia of North America,' by Edward D. Cope, it will be necessary to cancel and replace one of the forms. The author therefore requests that the recipient of this notice would please return his copy of said work to the author's address, at his expense. The volume will be returned, postpaid, with Part II, of the same work, which will be sent to those who have received the corrected Part I."

Cope's mention of an "error" in the preprint is certainly understated, but understandable from his point of view. Working quickly with the specimen and his printer, he wrote and published a revised description (Cope, 1870) in April of that year which included a corrected drawing showing the head at the end of an extremely long neck. As might be expected in the rush, however, the second version was not completely updated. Besides many errors remaining the text, it should be noted that the paper was reprinted with the original cover page from the first version, and thus is still dated August 1869. This created the perception, justified or not, that Cope was trying to cover up his original mistakes (Storrs, 1984). Although the arguments of fact between Cope and Leidy died down quickly, this incident would continue to haunt Cope for years.

There may be more to the story of *Elasmosaurus platyurus.* Cope (Almy, 1987, p. 189) noted that when he received the specimen from Turner, it was missing most of the skull (Fig. 7.2), some dorsal and cervical vertebrae, and all of the gastralia (belly ribs that

Figure 7.2. The muzzle of Elasmosaurus platyurus *Cope (ANSP 10081) in right oblique view. The anterior ends of the upper and lower jaws were found preserved together and represent the largest fragment of the skull that was recovered. The upper and lower jaws have since been separated. The muzzle was figured by Cope (1869, pl. II, fig. 8). (Scale = cm)*

are typically found in plesiosaurs). Although Turner did find and send some additional remains, most of the dorsal vertebrae and the gastralia (and the limbs) were never recovered. It appears likely that portions of the floating carcass may have been removed by scavengers or simply dropped off before it reached its final resting place on the sea bottom.

In 1991, two dorsal vertebrae and a number of very large gastroliths (stomach stones) were discovered by a local rancher (Pete Bussen) at a site about a mile northwest of where Turner found the remains of *E. platyurus* (Everhart, 2000). Larry Martin and John Chorn from the University of Kansas Museum of Natural History conducted a limited dig on the site and collected the vertebrae, rib fragments, and gastroliths (KUVP 129744). Several years later (1998), I participated in another dig on the same site with Glenn Storrs of the Cincinnati Museum Center (CMC), when the rest of the remains (CMC VP6865) were excavated. They consisted of a few more gastroliths, several complete ribs, and many gastralia. Nothing was articulated and the remains appeared to have literally been dropped in a pile on the sea floor.

That could have been the end of the story of just another incomplete plesiosaur specimen, but, while going through the collection at the Sternberg Museum in 2002, I came across seven plesiosaur dorsal vertebrae (FHSM VP-398) that were almost identical in size and preservation to the ones found by Pete Bussen in 1991. They had been donated to the museum in 1954 by a Logan County landowner. More importantly, George Sternberg's handwritten note accompanying the specimen indicated that the site was at the same locality where the University of Kansas and the Cincinnati Museum crews had recovered the other plesiosaur remains. Taken as one specimen, the three sets of remains contain most of the same bones Cope noted to be missing from *E. platyurus*. It also includes gastroliths that are now known to be present in almost all elasmosaurs (Cicimurri and Everhart, 2001). Whether or not it can be proved that they are one and the same as the holotype specimen of *E. platyurus* remains to be seen. However, I am convinced that the bloated carcass of this long-dead and decomposing elasmosaur floated along until it finally ruptured, and dropped out a large number of loose bones and gastroliths from the center of the body. Then the remaining portion of the carcass, including the head, neck, and tail, drifted farther to the southeast and sank to the bottom about a mile away at the spot where it was found in 1867 by Dr. Turner. It should be noted here that the remains of *Elasmosaurus platyurus* were found in what later became known as the Pierre Shale. At the time of their discovery, however, Cope would have assumed they were from the "Niobrara Cretaceous" just as he did in regard to the type specimen of *Toxochelys latiremis* (Chapter 6).

Besides *Elasmosaurus platyurus*, there are relatively few remains of elasmosaurs known from the Smoky Hill Chalk (Table

TABLE 7.1.

Elasmosaurids collected from the Niobrara Chalk of Kansas: by date of collection, locality and collector(s). Currently, most of these specimens are either considered to be Elasmosauridae indet. (Storrs, 1999) or *Styxosaurus snowii* (Carpenter, 1999).

Abbreviations: FFHM—Fick Fossil and History Museum; KUVP; University of Kansas Museum of Natural History; YPM—Yale Peabody Museum; USNM—United States National Museum; indet.—indeterminate/unknown because of the fragmentary nature of the specimen.

Specimen #	Original name	Collected	Locality	Collected by
YPM 1640	*Elasmosaurus nobilis* Williston (1906)	1874	Jewell County Fort Hays Limestone	B. F. Mudge
YPM 1644	*Elasmosaurus snowii* Williston (1906)	1874	Logan County	Mudge, Williston
KUVP 434	*Elasmosaurus ischiadicus* Williston (1903; 1906)	1874	Logan County	Mudge, Williston
YPM 1130	*Elasmosaurus ischiadicus* Williston (1906)	1876	Wallace County	H. A. Brous
YPM 1645	*Elasmosaurus marshii* Williston (1906)	1889	Logan County	H. T. Martin
KUVP 1301	*Elasmosaurus snowii* Williston (1890)	1890	Logan County	E. P. West
KUVP 1302	Elasmosauridae indet. (Undescribed)	1895	Logan County	C. H. Sternberg
KUVP 1312	*Elasmosaurus sternbergi* Williston (1906)	1895	Logan County	C. H. Sternberg
FFHM 2–26	Elasmosauridae indet. (Undescribed)	1926	Logan County	G. F. Sternberg
USNM 11910	*Styxosaurus snowii* (Undescribed)	1927	Logan County	C. H. Sternberg

7.1) and Pierre Shale in Kansas. Most are found in the Pierre Shale and few are as complete as Turner's *Elasmosaurus* discovery. Carpenter (1999, pp. 158–160) provides the most recent summary of the elasmosaur specimens from the Smoky Hill Chalk. Perhaps the best known of these is the skull (Fig. 7.3) and cervical vertebrae of *Styxosaurus snowii* (KUVP 1301—holotype) collected by E. P. West in 1890 from the upper chalk of Logan County (Williston, 1890). Carpenter (1999) lists several other Kansas specimens that he considers synonymous with *S. snowii*: KUVP 434; USNM 11910; YPM 1130; YPM 1644; and YPM 1645. All except YPM 1130 are from the upper chalk of Logan County. According to the records of the Yale Peabody Museum, YPM 1130 was found a little farther west in the upper chalk of Wallace County. This distinction, however, is based on a political boundary and is meaningless in terms of the actual occurrence of the specimen. Since Logan

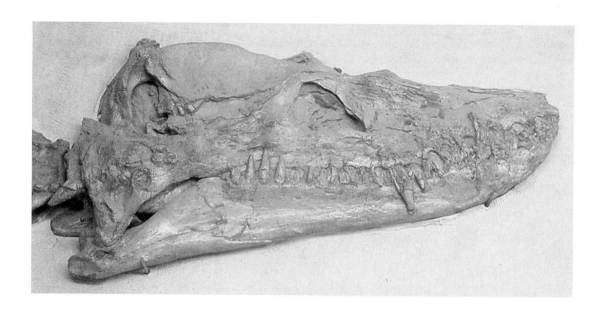

Figure 7.3. The skull of the type specimen of Styxosaurus snowii *(KUVP 1301) in right lateral view. The specimen was found by Judge E. P. West in the upper chalk of Logan County and represents the only complete elasmosaur skull known from the Smoky Hill Chalk.*

County was the eastern half of a much larger Wallace County until 1881 and called St. John County until 1885 (Bennett, 2000), it's likely that some locality information on the earlier discoveries is confused. All of these specimens, however, are from the upper Smoky Hill Chalk.

In 1893, Charles H. Sternberg collected two very large dorsal vertebrae of an elasmosaur (KUVP 1312) from the chalk at an unknown locality, originally reported to be in Gove County. In his description of the vertebrae, Williston (1906) noted that they were the largest he had ever seen and named a new species in honor of Sternberg: *Elasmosaurus sternbergi*. The two vertebral centra measured by Williston (ibid.) were 165 and 155 mm (6.5 and 6.1 in.) across, suggesting an elasmosaur "that could not have been less than 60 feet in length." By my measurements, in comparison, the largest vertebrae of *Elasmosaurus platyurus* are 130 and 120 mm (5.1 and 4.7 in.) across. Jim Martin (pers. comm., 1999) collected a single, very large elasmosaur vertebra from Vega Island, off the coast of Antarctica, that is an estimated 180 mm (7.1 in.) across.

Welles (1952) disagreed with Williston's identification of *E. sternbergi* and believed the vertebrae were more likely those of a large pliosaurid. Since that time, however, it has been determined that pliosaurids were long extinct by the time that the Smoky Hill Chalk was deposited (Chapter 8). Storrs (1999) noted, contrary to Williston, that the specimen actually consisted of one dorsal and two cervical vertebrae, but agreed with Williston (1906) that the remains were from an elasmosaur. The story might have ended there, but it didn't.

In 2004, I came across the rest of the story in a book written by the man who found the specimen. In *Hunting Dinosaurs on the Red Deer River, Alberta, Canada*, Charles H. Sternberg described the discovery of the remains of a huge elasmosaur that had been uncovered and nearly destroyed by a farmer during an excavation for a building on his farm. Sternberg (1932, p. 166; originally published in 1917) wrote: "I told Maud [his daughter] of a complete skeleton that had once been found by a farmer in the Kansas Chalk of Butte Creek, Logan County. He started to excavate a place for a stable, when he uncovered some huge vertebrae, and ribs over five feet long. He supposed they were elephant bones, and as they were broken, he thought they could not be saved, and so he dug up the bones with the chalk. They were dumped into a cow yard and beaten to powder under their feet, and could never be restored. I grieved much over the loss to science of that splendid specimen that has never been duplicated. Dr. S. W. Williston [1906], the oldest living American vertebrate paleontologist, described the few bones I was able to save from the general wreck. He did me the honor of naming it after me." The discovery of this specimen in the upper chalk (early Campanian) of Logan County makes much more sense to me than Williston's note that it came from Gove County (lower chalk). To date, more than 110 years later, nothing quite like this specimen of a truly huge elasmosaur has been found in Kansas.

Plesiosaurs have also been discovered in Kansas from below the Niobrara. In 1931, a large and fairly complete elasmosaur was discovered on an oil lease in southwestern Ellsworth County by Joe Purzer. According to George F. Sternberg's notes, the specimen was collected by M. V. Walker and himself from the Lincoln Limestone Member of the Greenhorn Limestone (middle Cenomanian) in September of that year. The remains (USNM 1195) were sold, unprepared, to the University of Nebraska State Museum around 1935 per UNSM records. Sternberg's original description notes that the specimen included many vertebrae and ribs, a complete right front paddle, portions of the pectoral and pelvic girdles including at least one ilium, both femora, most of both rear paddles, and a large assemblage of gastroliths. When I examined UNSM 1195 in 2000, the remains included 248 gastroliths, weighing 2.5 kg (5.5 lb.). The paddle currently on display in the UNSM was originally prepared and mounted by George Sternberg as indicated by archival photographs at the Sternberg Museum of Natural History. Although these remains are nearly contemporaneous with two large specimens of *Thalassomedon haningtoni* found in the Graneros Shale of Colorado (DMNH 1588) and Nebraska (UNSM 50132), the characteristics of the paddle suggest that UNSM 1195 represents an undescribed taxon (Schumacher, pers. comm., 2004). In that regard, another specimen (FHSM VP-15576) consists of an articulated series of nine cervical vertebrae from an as yet unidentified elasmosaur collected from the Graneros Shale of Ellis County that may be similar to the Colorado and Nebraska *Thalassomedon* specimens.

A bit of history of the discovery of plesiosaurs is necessary at this point to better understand the occurrence of these animals in the Late Cretaceous rocks of western Kansas. While their bones and teeth had probably been picked up as curios much earlier, the first nearly complete plesiosaur was discovered in the Jurassic rocks of Lyme Regis, England, by Mary Anning in the winter of 1820–21 (McGowan, 2001, p. 23). The name was given by the Reverend William Conybeare (De La Beche and Conybeare, 1821) and means "near-reptile," a reference to the belief at the time that plesiosaurs were closer to reptiles than were the more "fish-like" ichthyosaurs. Plesiosaurs were well known to science, and several complete specimens were already in European museums by the time the American West was being settled. Some of these Jurassic plesiosaurs, like *Plesiosaurus* and *Cryptoclidus,* had relatively long necks and small heads and would have looked very much like their Late Cretaceous elasmosaurid cousins. However, there is some evidence that many of these early plesiosaur lineages became extinct at the end of the Jurassic (Bakker, 1993), and the exact origin of the elasmosaurs is unknown. Whatever the reason, the number and diversity of plesiosaur species appears to have declined steadily from the Late Jurassic through the end of the Cretaceous.

Elasmosaurus platyurus was the first of several very-long-necked plesiosaurs (elasmosaurs) that have been discovered (1867) and described from the late Cretaceous deposits in the Midwest, and elsewhere in North America (Carpenter, 1999; Storrs, 1999). In May of 1874, B. F. Mudge collected the fragmentary remains of a large elasmosaur from the Fort Hays Limestone of Jewell County, in north-central Kansas. Later, Mudge (1876, p. 214) would describe the Fort Hays Member of the Niobrara Formation, saying, "Its fossils are *Inocerami,* fragments of *Haploscapha, Ostrea,* with occasional remains of fish and Saurians. The vertebrates are so rare that we never wasted our time in hunting them in this stratum; still, our largest Saurian, *Brimosaurus* of Leidy, was found in it in Jewell County." The specimen was later sent to the Yale Peabody Museum (YPM 1640) and described as *Elasmosaurus nobilis* by Williston (1906). Although the specimen is considered by Storrs (1999) to be "Elasmosauridae indeterminate," it does have some lasting value. Williston (1906, p. 233; see also pl. IV) wrote, "The specimen, notwithstanding what [damage] it has suffered, is of much interest since it is the only vertebrate I have any knowledge of from the [Fort] Hays limestone." Almost a hundred years later, there are still very few vertebrate remains known from the Coniacian-age Fort Hays Limestone Formation, and for that matter, very few elasmosaurs known from the Coniacian anywhere.

Later in 1874, Mudge and S. W. Williston, his assistant at the time, collected part of a very large plesiosaur (YPM 1644 ?—*Styxosaurus snowii*) from the Smoky Hill Chalk in Logan County, Kansas. Williston (1906, pp. 232–233), who was twenty-three years old at the time, wrote, "It was the first specimen of a plesiosaur that I ever saw." This was apparently the first time Mudge had seen gastroliths in association with plesiosaur remains. Mudge (1877) noted that "in the Plesiosauri, we found another interesting feature, showing an aid to digestion similar to many living reptiles and some birds. This consisted of well-worn siliceous pebbles, from one-fourth to one-half inch in diameter. They were the more curious, as we never found such pebbles in the chalk or shales of the Niobrara." In his discussion of *Elasmosaurus platyurus*, Williston (1906, pp. 226–227) indicated that "[i]t was with a specimen of an elasmosaur ("*Elasmosaurus*" *snowii*) that Mudge first noticed the occurrence of the peculiar siliceous pebbles which he described; and it was also with another, a large species yet unnamed, from the Benton Cretaceous, that the like specimens were found described by me in 1892" (see Williston, 1893).

Eventually, elasmosaurs would be found in Cretaceous marine deposits on almost every continent, including Europe (Mulder et al., 2003), South America (Welles, 1962), Australia (Long, 1998), Japan (Nakaya, 1989), New Zealand (Hector, 1874), and even Antarctica (Chatterjee and Zinsmeister, 1982; p. 66). The localities where most plesiosaur fossils have been found seem to indicate that they preferred cooler waters found at higher latitudes to those of warmer equatorial climates. In North America, they are found more often in the Cretaceous of Canada (Nicholls, 1988; Russell, 1993; Sato, 2003) than in the central United States, and they are extremely rare in the Gulf states.

Plesiosaurs are fascinating creatures, both for what we know and don't know about them. By the early part of the Late Cretaceous, they had evolved (or been reduced) to two distinct groups that are generally referred to as short-necked (polycotylids—Chapter 8) and long-necked (elasmosaurs). The giant pliosaurids (in Kansas, *Brachauchenius lucasi*, FHSM VP-321) became extinct earlier in the Late Cretaceous and were not a part of the fauna of the Smoky Hill Chalk. The short-necked polycotylids were initially fairly small, generally reaching lengths of no more than 3 m (10 ft.) during the deposition of the chalk. A much larger specimen of *Dolichorhynchops* from the Campanian Pierre Shale in South Dakota at the University of Kansas (KUVP 40001; Carpenter, 1996) suggests that they may have been as large as 6–7 m (20–23 ft.). They had long, narrow jaws filled with slender, seizing teeth (Massare, 1987) that were apparently well adapted for catching small prey. It is likely that they were capable of swimming rapidly in pursuit (Adams, 1997) of small fish and cephalopods that were their preferred prey.

Elasmosaurs, on the other hand, had shorter heads set on extremely long necks. They reached lengths of 14 m (45 ft.) or so and weighed several tons as adults. Based on work by Alexander (1989) in determining the weight of dinosaurs by the amount of water that is displaced by an accurate model, my estimate of the weight of a 12-m (40-ft.)-long elasmosaur is about 6600 kg (more than 7 tons). That is certainly much less than a modern whale would weigh (a grey whale of that length weighs about 31,000 kg or 34 tons), but it would have been a very large animal nevertheless. Among other physical limitations, their size and weight probably meant they could not have crawled ashore to rest or to lay eggs. Also, due to their unusual body plan, they were probably slow moving and not much of a threat to anything larger than a small fish.

For various reasons, plesiosaurs, and elasmosaurs in particular, seem to have captured the imagination of almost anyone interested in ancient creatures. In spite of their extinction more than 65 million years ago, their mystique lives on in Nessie (the Loch Ness monster), Champ (something seen in Lake Champlain), other sea and lake "serpents" from around the world, basking shark carcasses netted by Japanese fishing boats, and other unexplained sightings (O'Neill, 1999). All that actually remains of these plesiosaurs, however, are their bones. In this chapter, I'll try to put some flesh on these old bones.

Williston (1914, p. 77) credits the Reverend William Buckland (1784–1856) with describing a plesiosaur as "a snake drawn through the shell of a turtle" although the exact source of this quote appears to have been lost (Ellis, 2003, p. 119). The body of elasmosaurs, like other plesiosaurs, was almost totally enclosed by long ribs along the top and sides, and by large, flat pectoral and pelvic girdles connected by a continuous series of gastralia (belly ribs) across the chest and abdomen (Fig. 7.4). Four winglike paddles projected at nearly right angles, and a long neck was attached to the front of a teardrop-shaped body. Instead of lengthening the neck by increasing the length of the vertebrae, however, elasmosaurs increased the number of vertebrae. In one species, *Elasmosaurus platyurus,* the neck had more than seventy vertebrae (Welles, 1952, p. 54). Cope (1870, p. 49) originally counted 68½ cervicals but noted that some were likely to be missing. When I examined the type specimen in 2002, I noted that there was a fragmentary cervical vertebra mixed in with other bone scraps. While all the other vertebrae have been arranged and numbered at least three times (Cope, Welles, and most recently by Sachs, pers. comm., 2003), that cervical fragment had apparently never been counted with the rest. Other species of elasmosaurs, such as *Thalassomedon,* had fewer vertebrae in the neck, but their cervicals tended to be proportionately longer. In nearly all elasmosaurs that are known from relatively complete specimens, the length of the neck was usually equal to or longer than half the length of the animal.

Since the discovery of elasmosaurs, this unusually long neck has produced a number of incorrect descriptions of its "snakelike" flexibility. Descriptions and figures in earlier European publications often showed plesiosaurs on the surface of the ocean with their heads raised high in the air. These caricatures had an obvious influence on how North American plesiosaurs would be visualized. E. D. Cope (1872, p. 320; see also Webb, 1872 for an earlier, illustrated version of Cope, 1872) was certainly among the most influential when he described the appearance of swimming plesiosaurs: "Far out on the expanse of this ancient sea might be seen a huge, snake-like form which rose above the surface and stood erect, with tapering throat and arrow-shaped head, or swayed about describing a circle of twenty feet radius above the water. Then plunging

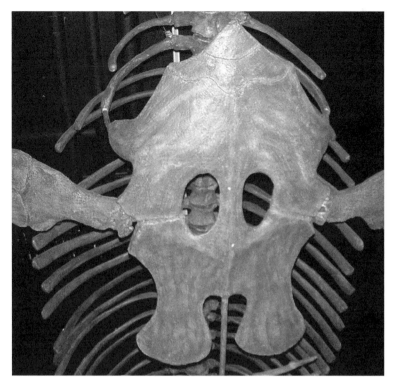

Figure 7.4. Ventral views of the pectoral (top) and pelvic (bottom) girdles of Elasmosaurus platyurus *Cope (ANSP 10081) in the exhibit at the Academy of Natural Sciences of Philadelphia as reconstructed from Cope's (1869) original drawings.*

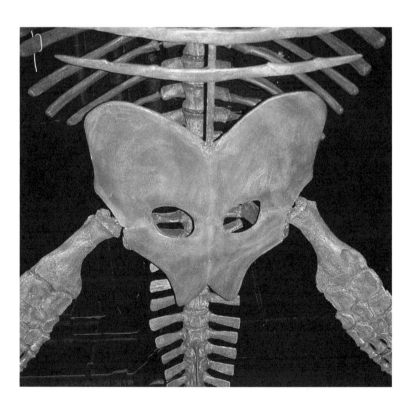

into the depths, naught would be visible but the foam caused by the disappearing mass of the life. Should several have appeared together, we can easily imagine tall, flexible forms rising to a height of the masts of a fishing-fleet, or like snakes, twisting and knotting themselves together. This extraordinary neck—for such it was—rose from a body of elephantine proportions, and a tail of the serpent-pattern balanced it behind." Later, he added (1875, p. 44), the "snakelike neck was raised high in the air, or depressed at the will of the animal, now arched swanlike, preparatory to a plunge after a fish, now stretched in repose on the water, or deflexed in exploring the depths below."

Many paleo-artists, including Charles Knight, have drawn popular but anatomically impossible representations of these fanciful (and fictional) elasmosaurs. These ideas still persist in spite of publications by Williston (1914, p. 91), Shuler (1950, pp. 24–31), Storrs (1993, p. 74), and others that have shown that the cervical vertebrae of elasmosaurs had limited movement up and down and only slightly more in a horizontal plane. For a contrasting opinion, see Welles (1943, pp. 147–152). Some of the earlier work on flexibility of the neck in elasmosaurs assumed a 1 cm space between the vertebrae (ibid.) for cartilage. Sato (2003, p. 91) suggested a 0.5 cm thickness for the intervertebral disks in a newly discovered elasmosaur from Canada (*Terminonatator ponteixensis*). An inspection of these vertebrae in articulated specimens, however, indicates that there was probably little, if any, space between them in life. The close fit, the relatively flat surface of the articulation between the vertebrae, and the lack of space between the neural spines and transverse processes on adjoining vertebrae probably meant that there was little movement between the individual vertebrae. There are several good reasons for this limited flexibility, which are discussed below.

Another form of this misconception is the view of elasmosaurs shown cruising along on the surface with their head and neck arched swanlike high above the water as they searched for prey. Two issues quickly present themselves. First, since the eyes of a plesiosaur are located on top of the skull and are generally directed upward, the plesiosaur would almost have to turn its head upside down to look down at the water in search of prey from above. In addition, their eyes were more than likely adapted for underwater vision and may have been unable to focus on prey from above the surface of the water.

A second problem is even more difficult to overcome. Unless the laws of physics were suspended on the behalf of these extinct creatures, it would have been impossible for them to lift much more than their head above water to breathe. If you would like to prove this for yourself, try lifting a heavy pole from one end while floating in water and not touching the bottom. As you try to raise the object, your legs and lower body will also rotate toward the surface to counter-balance any weight above the water. In elasmosaurs, the weight of the long neck (as much as a ton or more in a full-grown adult) dictated that the center of gravity was just behind the front flippers. Holding the head and entire neck above the surface was something that could never happen unless the rest of the plesiosaur's body were sitting on a firm bottom in shallow water (a possibly fatal situation for the elasmosaur!). Even then, the limited movement between the vertebrae, the limited musculature of the neck, and the sheer weight of the long neck would greatly limit the height to which the head could have been raised.

Still another problem that becomes evident with the long neck of elasmosaurs is what happens when the head is moved to one side or the other while the animal is moving. Located 3–6 m (10–20 ft.) or more in front of the anterior set of paddles, the head (and neck) would act as a highly effective rudder whenever it was moved away from the central axis of the animal as it swam. In other words, moving the head to the right while swimming would cause the whole body of the plesiosaur to turn in that direction. While it is possible that plesiosaurs used their head as a rudder to change directions while swimming, it would also be nearly impossible for them to swim in a straight line with their heads darting from side to side, or up and down, in search of prey as shown in many paleo-life reconstructions. In any case, it seems likely that the tiny brain of an elasmosaur would have been busy trying to maintain the position of the head even during stationary feeding activities through fine movements of the two pairs of large paddles located up to 8 to 11 m (25–35 ft.) to the rear. No disrespect for the mental capacity of these creatures is intended here since plesiosaurs, and elasmosaurs in particular, were very successful marine predators for millions of years, even if we brainy humans don't yet understand quite how they did it.

The length of the neck and the number of cervical vertebrae in elasmosaurs appears to have increased through time, conveying some adaptive advantage for this group that we also do not fully understand. *Thalassomedon haningtoni* (Welles, 1943) from the late Cretaceous (Cenomanian) Graneros Shale of Colorado and Nebraska, and *Elasmosaurus platyurus* Cope 1868 from the Campanian Pierre Shale of Kansas, are the longest elasmosaurs presently known. Both probably grew to 14 m (45 ft.) or more in length. The necks of these two species are about the same length but *Elasmosaurus* has about ten more cervical vertebrae than *Thalassomedon* (Welles, 1943). Unlike long-necked mammals and birds that evolved longer vertebrae (usually seven) through time, the neck of elasmosaurs became longer by adding more vertebrae (as many as seventy-two or seventy-three in *Elasmosaurus*). While the other large marine predators of the Late Cretaceous, including the short-necked polycotylids, snakelike mosasaurs, giant fish such as *Xiphactinus*, and huge ginsu sharks (*Cretoxyrhina mantelli*) used speed to chase down or ambush their prey, it is more likely that elasmosaurs used stealth to stalk the schools of small fish that were their primary food source. A long neck would allow the large body of the plesiosaur to remain concealed in the darkness far below a school of fish while the small head moved slowly at the end of its long neck to stealthily approach the unlucky victims from below.

Having their eyes on top of their heads, directed generally upward, as is the case in most elasmosaurs, seems to support this method of feeding. The eyes were not unusually large but were placed so that they may have been capable of stereoscopic vision, something that would be useful for accurately locating small prey (Shuler, 1950, p. 8). Because of limited movement capability in the neck, it is likely that the head was moved fairly close to the prey before striking. Precise and rapid coordination between the eyes and the distant paddles of the elasmosaur would have been essential for successful hunting of highly mobile fish and other marine life. Approaching from below may have also been used as a strategy to silhouette the prey of an elasmosaur against the sunlit surface while using the darker, deeper waters for concealment.

We know very little about the other senses of elasmosaurs. Cruikshank et al. (1991) suggested that the nostrils of plesiosaurs were well positioned to passively direct a continuous flow of water through the nasal cavity, where the senses of taste and smell may have been useful in detecting prey. Williston (1914, p. 88) notes that the inner ear (semi-circular canals) of plesiosaurs was large and well developed. This area of the brain maintains equilibrium and coordinates muscular control, and its relatively large size implies that plesiosaurs were probably quite graceful in their movements, although not necessarily fast. While there is no evidence of external ears in plesiosaurs, they may also have evolved a sensory system to locate nearby prey from the vibrations generated by their movement. It appears likely, however, that elasmosaurs relied primarily on their eyesight to find food.

Evidence from stomach contents preserved in plesiosaur remains indicates that the prey they ate was limited to small fish and invertebrates. Earlier plesiosaurs from the Jurassic and early Cretaceous appear to have preferred squid and other cephalopods, while the later and longer-necked elasmosaurs fed on small fish such as *Enchodus* (Cicimurri and Everhart, 2001). While most elasmosaurs were huge by any measure, their heads and their prey remained relatively small in relation to the size of their bodies. A 12-m (40-ft.) elasmosaur had a head length of about 0.5 m (20 in.) or about 4 percent of the total length of the animal. Unlike the kinetic (flexible) skulls of most mosasaurs (Russell, 1967), the skulls of plesiosaurs were rigidly constructed (Fig. 7.5). This means that the cross-sectional size of the prey they could swallow was limited by the distance between the fixed hinge points (quadrates) of their jaws at the back of the skull. Even in the largest of the elasmosaurs, like *Thalassomedon*, this distance was no more than about 18 cm (7 in.), and the prey they fed on was probably no longer than about eighteen inches in length (Shuler, 1950, p. 18; Cicimurri and Everhart, 2001). Their long, slender teeth also appear to argue for trapping small prey. In all cases, elasmosaur teeth appear to have been used to seize small fish or invertebrates, and not to tear flesh or otherwise dismember larger prey. Once captured, the prey would have to be swallowed whole.

Because of their unusual body plan, elasmosaurs were most likely slow-moving animals, using their paddles as wings or foils to generate the lift necessary to "fly" through the water. This type of swimming is more efficient than the rowing or paddling (e.g., like a duck) methods that are sometimes depicted. Both the front and rear limbs of plesiosaurs are similarly constructed to function as wing-like hydrofoils. Whether both were used in that manner is still open to question. While modern underwater "fliers" such as penguins use only their front limbs, many researchers now believe that plesiosaurs used a coordinated movement of both pairs of limbs. Others think the rear limbs did not have either the range of movement or the musculature necessary to effectively generate the lifting forces necessary for this method of swimming (Lingham-Soliar, 2000). Since the hind limbs were located well behind the center of gravity, they may have been more effectively used for steering. In that case, the rear limbs may have been rudders and/or stabilizers, especially to maintain the position of the body and position the head and neck while feeding.

Figure 7.5. The reconstructed skull of Styxosaurus snowii (SDSMT 451) at the South Dakota School of Mines and Technology. This specimen was found in the Pierre Shale of western South Dakota in 1945.

In elasmosaurs, both pairs of limbs are about equal in size and are highly modified into a rigid paddle. The upper end of the paddle, called the propodial (humerus/femur) is attached to the body at the shoulder and hip. The lower limb bones (radius/ulna and tibia/fibula) are reduced to flat, polygonal shapes that interlock with the smaller but similar-shaped bones of the wrist and ankle. The finger and toe bones (phalanges) were hourglass-shaped. Bones of the adjoining fingers fitted closely together and were probably held tightly within a sheath of muscles, tendons and ligaments. Although plesiosaurs had five "fingers," they exhibited a condition called hyperphalangy, which means they have many more individual finger bones than usual (humans have three bones in each finger; some plesiosaurs had as many as twenty-four). The effect of the compacted lower arm/leg and wrist/ankle structures, and the extra finger/toe bones, was to modify what had originally been a walking limb into a long, tapering paddle that had limited flexibility. In cross-section, the limb was also thicker on the leading edge than the trailing edge. It had evolved into a wing or hydrofoil for flying through the water. Movement of the paddles in a manner similar to that of a bird's wing in flight created lift and pulled the animal through the water. Unlike birds, however, the wrist was effectively "locked" and the paddles could not be folded or otherwise flexed to any great extent. At rest, a plesiosaur's limbs probably were held at right angles to the body.

Another peculiarity that has been noted among long-necked plesiosaurs since their discovery are masses of gastroliths (stomach stones) that are commonly found associated with their remains (Whittle and Everhart, 2000; Everhart, 2000; Cicimurri and Everhart, 2001). These rounded and usually highly polished rocks are normally located inside the abdomen of complete specimens. There have been as many as several hundred stones of various sizes, from egg-sized down to pea-sized or smaller, reported to be associated with ple-siosaur remains. In the case of a recently discovered specimen from Antarctica, there were thousands of small quartzite pebbles (a half-inch or less) in the stomach area (James Martin, pers. comm., 1999). Typically, most gastroliths are siliceous rocks such as quartz, quartzite, or chert. Occasionally they are granite, basalt, or other ig-neous rocks. Until recently, the largest gastroliths known were about as big as a tennis ball, but in two Kansas specimens found in the early 1990s, they were much larger. Everhart (2000) noted that the largest stones in a University of Kansas specimen (KUVP 129744) weighed about 1.4 kg (3 lb.) each and were nearly 17 cm (7 in.) in length (roughly the size of a softball!). These are the largest individual sizes and the greatest total weight of gastroliths ever documented from a plesiosaur or, for that matter, from any animal (Fig. 7.6). The total weight of the stomach stones from this plesiosaur was about 13.1 kg (29 lb.), but considering the weight of the fourteen-meter plesiosaur that carried them (estimated to weigh 6600 kg [14,500 lb.]), their rel-ative weight is insignificant. In fact, the weight of the gastroliths is probably much less than the daily food intake of the plesiosaur. More commonly, however, the total weight of gastroliths documented from individual plesiosaurs is less than ten pounds (Everhart, 2000). It ap-pears highly unlikely that these stones would have been effective in changing the buoyancy of large animals that probably had to con-sume large quantities of food daily in order to sustain themselves while living in midocean. Even buoyancy changes caused by inhala-tion and exhalation through a twenty-foot-long trachea would have been more than could have been offset by the small amount of weight represented by the gastroliths.

Figure 7.6. The gastroliths associated with a fragmentary elasmosaur specimen (KUVP 129744) from the Sharon Springs Member of the Pierre Shale in Logan County, Kansas. These stones are the largest gastroliths documented from any animal, including those of sauropod dinosaurs.

Figure 7.7. A portion of the more than 600 gastroliths recovered in 2002–03 from an elasmosaur (UNSM 1111–002) in the Mobridge Chalk Member (Maastrichtian) of the Pierre Shale in northeast Nebraska. This is the largest number of gastroliths recovered from any plesiosaur in North America.

This leads us to the question of what gastroliths were actually used for. While early fossil collectors such as Benjamin Mudge (1877) concluded from plesiosaur remains found in Kansas in the 1870s that the stomach stones were used as "an aid to digestion similar to many living reptiles and birds," Samuel Williston (1893) and others initially believed that gastroliths were swallowed for other reasons, such as a food craving or being mistaken for prey.

Currently, most researchers believe they were used in some manner for buoyancy control or for adjusting attitude/longitudinal balance (Taylor, 1981). However, a nearly complete specimen of *Styxosaurus snowii* (NJSM 15435) recently discovered in the Sharon Springs Member of the Pierre Shale of western Kansas provides a rather convincing counter-argument (Cicimurri and Everhart, 2001). These remains included about a hundred gastroliths, fish bones, scales, and teeth co-mingled in the stomach area. Unlike the skeletal material preserved as stomach contents of mosasaurs and fish which appear to have been partially digested (dissolved) by stomach acids, the remains of the fish in the stomach of this elasmosaur were broken up into nearly unrecognizable pieces. Most of the fish bones appeared to have been ground up into very small (2–3 mm [0.1 in. or less]) fragments. This specimen suggests that whatever use gastroliths may have had for buoyancy control, they were definitely useful in grinding up prey as an aid to digestion. Another elasmosaur specimen (UNSM 1111–002) discovered in the Mobridge Chalk Member of the Pierre Shale in northeastern Nebraska in 2002 had more than 600 gastroliths (Fig. 7.7) in its abdomen (unpublished data), some of which had been reduced to the size of large sand grains (less than 4 mm).

So where did these gastroliths come from? Egg-sized quartz or other igneous rocks were certainly not to be found routinely in the soft limey mud of the Cretaceous sea bottoms in Kansas. Mudge

(1877, p. 286) noted that "how far the Saurians wandered to collect them is a perplexing problem." Williston (1893, p. 122) noted that some of the quartzite stones were similar to boulders found in northwestern Iowa, or in the Black Hills of South Dakota. Each of these potential sources is 400 to 500 miles from the localities where the plesiosaur remains that contained them were found. Cobbles found in the recently discovered Nebraska elasmosaur specimen may have come from nearby granite exposures in southeastern South Dakota, or from the Quachita Mountains in western Arkansas. Wherever their origin, they certainly did not come from the soft mud bottom of the Western Interior Sea that covered Kansas. This, in itself, implies that some plesiosaurs traveled long distances to the sources of these stones. That the stones are always rounded and smooth also means that the stones were eventually worn down to sizes small enough to pass though from the crop or stomach into the gut and had to be replaced periodically throughout the life of the plesiosaur.

One interesting feature I have observed on gastroliths from a number of elasmosaur specimens from Texas to South Dakota is that the gray or black chert stones are covered with numerous arc-

Figure 7.8. Markings (conchoidal fracture scars) frequently found on elasmosaur gastroliths. These markings probably record the impact of the stones against each other in the gut of the plesiosaur. (Scale = mm)

shaped surface markings. The age of the elasmosaurs ranges from the Cenomanian through the Upper Maastrichtian. Closer examination indicates that these markings were the result of small (2–5 mm) conchoidal (bowl-shaped) fractures of the chert (Fig. 7.8). The fractures generally cover the surface of the stone, exhibit varying degrees of wear, and often cross other fractures. Similar fractures can occur naturally due to stone on stone impacts in river or beach gravels, but they do not occur on non-gastroliths in the numbers observed on the gastroliths which were examined. The more frequent occurrence of conchoidal fractures on the edges of angular-shaped stones in an experiment using a rock tumbler indicates that such damage is an important part of the mechanism for rounding and smoothing gastroliths. Gastroliths with arc-shaped markings were found within a recently described plesiosaur specimen (Cicimurri and Everhart, 2001) from Kansas in intimate association with the finely comminuted bones of small fish. These markings suggest that the conchoidal fractures occurred as the stones were ground against one another by peristaltic contractions within the plesiosaur's digestive tract. I believe that the more frequent occurrence of these markings on chert gastroliths in plesiosaurs compared to those found on similar stones from river and shore deposits is evidence of their use in processing food (Everhart, 2004b).

A final subject is the issue of plesiosaurs crawling ashore to lay their eggs. This method of reproduction has been long assumed because they are reptiles, or possibly because they were superficially similar to marine turtles. Consequently, shore-dwelling plesiosaurs have been the subject of imaginative works by paleo-artists for more than 150 years. While most modern reptiles do lay eggs, it is highly unlikely that the larger marine reptiles of the Mesozoic could have reproduced in that manner. Turtles (Chapter 6) appear to be the only major group of marine reptiles that still come ashore to lay their eggs. Unlike turtle paddles, the limbs of plesiosaurs were so modified for use in underwater flying that they would have been too rigid to be used effectively for movement on land, much less for scooping out a nest for eggs in beach sand. In addition, the length and weight of the long neck of an elasmosaur would have effectively counter-balanced the back half of the animal, lifting it off the ground and probably making the rear flippers useless for movement on land, or anything else, under the best of circumstances.

Ichthyosaurs have long been known to have given live birth to their young, as evidenced by well-preserved remains in the Jurassic shales of Germany. Live birth in the other marine reptiles (mosasaurs and plesiosaurs) has not been so easy to prove. A mosasaur (*Plioplatecarpus*) specimen found recently in the Pierre Shale of South Dakota included the remains of several young of the same species in the pelvic region (Bell et al., 1996). More recently, Caldwell and Lee (2001) reported the remains of at least four embryos preserved inside the abdomen of a mosasauroid called an aigialosaur. There is also at least one specimen of a short-necked polycotylid (*Dolichorhynchops osborni*) known with fetal material in the abdomen (Rothschild and Martin, 1993; pers. obs., see Chapter 8). It appears likely that live birth is one of the necessary adaptations that egg-laying reptiles had to make in order to successfully return to life in the ocean. If so, this raises additional questions as to how marine reptiles nurtured a fetus inside the mother's body for a relatively long period of gestation. Did they simply retain the developing embryos and yolk sacs inside the mother's body until they were ready to hatch, as do some modern snakes? Or had they developed some other method (placenta?) of nurturing baby plesiosaurs? We may never know.. Plesiosaurs were highly successful animals that clearly had to have evolved some efficient method for reproduction in the marine environment.

Giving live birth to their young probably also meant that plesiosaurs provided some form of parental care. Turtles and other reptiles which lay many eggs at one time depend on safety in numbers for the survival of a few of their young. In most cases, ichthyosaurs and mosasaurs appear to have given birth to six or fewer young. The remains of the smallest *Tylosaurus* "babies" that have been found in the Smoky Hill Chalk are about 2 m in length (Everhart, 2002). Modern animals which invest their energy in birthing a few larger babies, such as mammals, generally provide some form of protection for those young as a means of improving their chances of surviving to adulthood. Like some dinosaurs, plesiosaurs may have traveled together in small groups for that purpose. Otherwise, the survival rate for any young animals in the middle of an ocean populated with giant predatory fish like *Xiphactinus*, great white shark–sized ginsu sharks, and 10-m (30-ft.)-long, hungry mosasaurs would have been close to zero.

As it was, plesiosaurs may have been driven to the edge of extinction by the "mosasaur explosion" during the late Cretaceous. The greatest number of Late Cretaceous plesiosaur fossils in Kansas are found in the early middle Turonian (Bruce Schumacher, pers. comm., 2003). This is roughly the same time as mosasaurs make their first appearance (Martin and Stewart, 1977) in the Western Interior Sea. As mosasaurs became more numerous and larger, plesiosaurs (as judged by the number of their remains discovered to date) essentially disappeared from the middle of the Interior Sea. Their preferred habitat may have been closer to shore, or they may have been the prey of large mosasaurs such as *Tylosaurus*. One juvenile polycotylid specimen from Kansas was found as stomach contents of a large *Tylosaurus* (Sternberg, 1922; Everhart, 2004a). We do know from the bite marks on plesiosaur bones that their carcasses were scavenged by sharks and that the more easily detachable parts, such a limbs, tails, heads, and necks, were often carried away (Everhart, 2003).

For all their size and amazing adaptations to life in the ocean, it appears likely that plesiosaurs were an evolutionary dead end and were already on their way out by the end of the Cretaceous. Their greatest diversity and numbers probably occurred during the Jurassic or Early Cretaceous. By early Maastrichtian time (78 mya) there were only a few species of elasmosaurs and short-necked polycotylids left in Earth's oceans. Like the ichthyosaur "fish-lizards" that became extinct in the early part of the Late Cretaceous (Lingham-Soliar, 2003), plesiosaurs were probably losers of an evolutionary arms race that saw the rise of larger, faster teleost fish as competitors for the same prey during the Early Cretaceous and the explosive entry of the highly adaptable mosasaurs, who were both competitors and predators, in the Late Cretaceous.

Figure 7.9. Long-necked plesiosaurs, such as Elasmosaurus platyurus *and* Styxosaurus snowii, *probably used their long necks to approach prey from below. The size of their heads limited the size of the prey (mostly small fish) that they could eat. Drawing by Russell Hawley.*

Eight
Pliosaurs and Polycotylids

Just below the surface, the pod of four adult female and three smaller juvenile short-necked plesiosaurs moved steadily eastward as they migrated across the expanse of open sea toward shallower coastal waters. Their narrow, toothy heads were held stiffly in front as they moved their long paddles up and down rapidly like the wings of a bird. With little apparent effort, they were literally flying through the clear, warm water. At regular intervals, almost in unison, they would break through the surface and come completely out of the water as they quickly exhaled and inhaled. Their smooth, scaleless skin, blue-black on top and a pale cream color underneath, was briefly visible when they were above the surface.

The bright midday sun illuminated the water around them but did not penetrate far into the depths below. There was only a gentle breeze, and the surface of the dark blue waters around them was almost calm. Normally, they would have avoided the deeper waters near the center of the sea, preferring the relative safety nearer to shore. It was late in the season, however, and the females were about to give birth to their young. The leader of the group, a pregnant female, was taking the most direct route across the sea to a sheltered nursery area near the mouth of a large river. The stubby but well-streamlined bodies of the polycotylids were making little splash or other noise as they came out of the water periodically to breathe.

The noise was, however, enough to draw the attention of a large tylosaur hunting nearby. His acute hearing picked up the sounds from nearly a mile away and alerted him to the passage of the short-necked polycotylids. Normally slower than these fish-eating plesiosaurs, the tylosaur swam rapidly with wide sweeps of his long tail at nearly right angles to the pod on a course that would intercept them. He would have only one chance to attack from below before they outdistanced him.

As the pod approached, the tylosaur surfaced, inhaled, and dived again. This time his descent took him under the path of the pod. Nearly motionless now, he looked up and watched as they approached, silhouetted against the sunlit surface. Then, using his front paddles to orient his body upward, he lashed his tail suddenly and accelerated toward the surface. His target was one of the smaller juveniles in the center of the group, and he timed his approach perfectly. Coming up underneath the polycotylid, he opened his jaws at the last moment, then closed them savagely around the head and neck of his prey. Bones crunched as the plesiosaur's lightly built skull was crushed. The momentum of the tylosaur carried his upper body and that of his prey out of the water, then he fell over sideways with a great splash. The remaining plesiosaurs scattered quickly in all directions, then reformed, moving quickly away from the scene of the ambush.

Still holding the much smaller polycotylid tightly between his jaws, the tylosaur began to position his prey so that it was pointed headfirst into his throat. Attracted to the noise of the attack and the blood of the plesiosaur that was now spreading into the water, a group of small sharks swam in wary circles around the big tylosaur. They remained at a safe distance and the tylosaur ignored them for the moment. Satisfied with the position of his prey, the tylosaur opened his mouth wider and surged forward, lodging the narrow head of the short-necked plesiosaur into his throat. Then he began using his double-jointed lower jaws to slowly ratchet the body into his gullet. Each time the prey was moved backward, the sharp, hooked teeth on the roof of the tylosaur's mouth grabbed hold of the plesiosaur's skin, allowing the lower jaw teeth to be released and moved forward. Slowly the plesiosaur's body was swallowed, much like a large rat being eaten by a snake. The tylosaur rested briefly while floating at the surface as it began to digest its large meal.

Cretaceous plesiosaurs have traditionally been divided into two groups: the long-necked, small-headed variety, called elasmosaurids (i.e. *Elasmosaurus*); and the short-necked, large-headed variety, called pliosaurids. Pliosaurids such as *Pliosaurus* (Owen, 1842) and *Liopleurodon* from the Jurassic, and Early Cretaceous varieties like *Kronosaurus* (Australia) and *Brachauchenius*

(Kansas/Texas/Utah), were true "sea monsters" of their day (Ellis, 2003), with skulls that were as long as 3 m (9 ft.). Smaller "short-necks," or polycotylids, from the Late Cretaceous, like *Trinacromerum, Dolichorhynchops,* and *Polycotylus,* have been frequently lumped into the pliosaurid group because of the obvious resemblance of their body plan to that of their extinct cousins. Carpenter (1996), however, determined that these Late Cretaceous "short-necks" were actually more closely related to the longer-necked elasmosaurids than to pliosaurids on the basis of similarities in their skulls. Assuming this determination is correct, the pliosaurid lineage became extinct for unknown reasons during the early part of the Late Cretaceous, along with the last of the ichthyosaurs. In a review of stomach contents preserved with specimens, Cicimurri and Everhart (2001) noted that the food preferences of plesiosaurs in general appeared to change from cephalopods (squid, ammonites, etc.) to bony fish during that same period. The polycotylids, including *Trinacromerum* and *Dolichorhynchops,* may have evolved as a smaller, faster version of the elasmosaur lineage, possibly to fill the ecological niche left vacant by the extinction of these two earlier types of marine reptiles.

A more likely ending to the story at the beginning of this chapter would have been the polycotylids sensing the approach of the giant tylosaur and dodging at the last moment. Then, as now, predators probably failed in capturing prey more often than they succeeded. In either case, while the story is fiction, the fact is that a large *Tylosaurus proriger* was discovered by Charles Sternberg in 1918 with the remains of a partially digested polycotylid within its ribcage. Did adult tylosaurs hunt and kill plesiosaurs as the story suggests? Or did they simply scavenge the bloated remains of dead plesiosaurs that happened to become available from time to time, competing with sharks as scavengers? Unfortunately, we don't know the answer to those questions. About all we are certain of is that a juvenile polycotylid was eaten by a nine-meter-long *Tylosaurus proriger,* and that the *Tylosaurus* died before completely digesting its meal.

In the summer of 1918, Charles Sternberg and his sons were prospecting for fossils in the Smoky Hill Chalk along Butte Creek in Logan County, Kansas. According to Sternberg (1922):

> I was so fortunate as to find a fine tylosaur skeleton the second day in the field. There were twenty-one feet of the skeleton present in fine chalk. The complete skull was crushed laterally, nearly the complete front arches and limbs were present, as was also the pelvic bones and both femora. All the vertebrae to well into the caudal region beyond the lateral spines were continuous, with the ribs in the dorsal region. Between the ribs was a large part of a huge plesiosaur with many half-digested bones, including the large humeri, part of the coracoscapula, phalanges, vertebrae, and, strangest of all, the stomach stones, show-

ing that this huge tylosaur, that was about twenty-nine feet long, had swallowed this plesiosaur in large enough chunks to include the stomach. How powerful the gastric juice that could dissolve these big bones! This specimen I sent to the United States National Museum.

After reading Sternberg's account in 2001, I was amazed that the intimate association of these two marine reptile specimens had never been mentioned again in the literature. I contacted Bob Purdy and Michael Brett-Surman at the Smithsonian and learned that the *Tylosaurus* (USNM 8898) was, in fact, their exhibit specimen, and had been so since it was obtained from Charles Sternberg in 1919. A photograph of the preparation of the *Tylosaurus* exhibit was even featured in the *Scientific American* magazine (see Gilmore, 1921). The plesiosaur material (USNM 9468), however, had been stored safely away in the collection, where it had been mostly gathering dust for more than eighty years. A review of the Smithsonian's curation records clearly showed the association, but otherwise the plesiosaur remains had been essentially ignored by paleontologists since their discovery.

I made arrangements to visit the collection, and early on the morning of September 11, 2001, my wife and I drove up the Kansas Turnpike toward Kansas City for a flight to Washington, D.C. Needless to say, it was one of the longest days of our lives as we listened to reports of the terrorist attacks on the World Trade Center and the Pentagon on National Public Radio. We made it as far as Lawrence, Kansas, before realizing that we weren't going anywhere, and like the contrails of the commercial airliners in the clear blue sky overhead, we turned around and went home.

In November 2001, I visited the Smithsonian, where I was able to examine the plesiosaur specimen. Unfortunately, after reading and rereading Sternberg's note, I had set my expectations a little too high. The plesiosaur remains were contained in a single drawer and there wasn't much to see. The bones that Sternberg had found were badly corroded, apparently by the mosasaur's stomach acid. Unlike the mosasaur specimen in the exhibit, much of the chalk matrix had not been removed from the bones of the little plesiosaur (Fig. 8.1). The specimen is described more fully by Everhart (2004b) but it's worth mentioning here that it wasn't a "huge plesiosaur" as indicated by Sternberg. Rather, the remains appeared to be those of a juvenile polycotylid, probably no more than 2–2.5 m (6–7 ft.) long. Most of the plesiosaur's skeleton was missing and it is likely that some of the bones had been removed by sharks that had scavenged the mosasaur carcass (there was a single *Squalicorax falcatus* shark tooth included in the box of plesiosaur bones). As noted by Sternberg (1922), the fact that the plesiosaur bones were found inside the heavy ribs of the mosasaur probably meant that the much smaller sharks were unable to reach them. In any case, the evidence shows that mosasaurs did eat plesiosaurs, and for that matter, probably anything else they could swallow.

Figure 8.1. The partially digested upper limb bones (humeri?) from a small polycotylid plesiosaur (USNM 9468) found as stomach contents of a nine-meter *Tylosaurus proriger (USNM 8898)* in the collection of the United States National Museum (Smithsonian) in Washington, D.C. The remains were found in 1918 in Logan County, Kansas, by Charles Sternberg.

Almost fifty years earlier, shortly after the discovery of *Elasmosaurus platyurus* by Dr. Theophilus Turner (Cope, 1868), another "new" kind of plesiosaur was found in the upper Smoky Hill Chalk "about five miles west" of Fort Wallace (Cope, 1871, p. 386). The locality given by Cope (and probably provided to him) is suspect because there are no exposures of "yellow cretaceous limestone" to the west of Fort Wallace; most likely the locality was along the Smoky Hill River five or so miles to the east. A land agent and part-time fossil hunter named William E. Webb from Topeka had discovered the fragmentary remains of a plesiosaur that Cope (1869) called *Polycotylus latipinnis* (USNM 27678 *and* AMNH 1735). Note that the specimen has two numbers because portions of it are curated in both the United States National Museum in Washington, D.C., and the American Museum of Natural History in New York City (Storrs, 1999).

Peterson (1987, p. 228) wrote, "[A]lthough there were no organized scientific expeditions into the area [western Kansas] in 1868, the Kansas Pacific Railway promoted several well attended excursions to the end of the line at Sheridan near Fort Wallace. One small group organized as a hunting and adventure party by William E. Webb, on reaching the end of the line in September was told of a large fossil near the fort [for a fictionalized account of this expedition, see Webb, 1872; also Davidson, 2003]. Webb, with the help of a 'professor' in the group, collected the fossil and shipped it to

Cope who placed it in a new genus and described it as the first true Plesiosauroid found in America." While Cope (1871, p. 34) did use those exact words in his publication, the remains of many other plesiosaurs had been discovered in North America prior to *Polycotylus*. It was, however, the first specimen of an unknown (at the time) group of Late Cretaceous plesiosaurs that would be called polycotylids.

The genus and species were described essentially from a pelvic arch and twenty-one vertebrae. The genus name (*Polycotylus*) refers to the deep "cupping" of the anterior and posterior surfaces of the vertebrae, a characteristic that was quite different from other plesiosaur vertebrae seen by Cope until then. According to Carpenter (1996), however, "the holotype material is scrappy and of questionable value."

Williston (1906) apparently held the same opinion, and redescribed *Polycotylus* from a second, more complete set of remains (YPM 1125) that was collected personally by O. C. Marsh from the chalk of Logan County in November of 1870. Considering the growing rivalry between Cope and Marsh, it is interesting that Marsh did nothing with this much more complete specimen and did not even mention it in his report (1871) of the fossils collected by the 1870 Yale scientific expedition. Perhaps even more interesting in view of Marsh's belated criticism (Ballou, 1890) of Cope's mistake regarding the reconstruction of *Elasmosaurus platyurus*, Marsh did not publish a single paper on Cretaceous plesiosaurs during his long career. I think this is odd considering that the Yale Peabody Museum has at least eleven plesiosaur specimens collected from the chalk of western Kansas from the 1870s.

While the specimens of *Polycotylus latipinnis* Cope found to date are fragmentary (Carpenter, 1996, p. 268), two other species within the family are well represented in the fossil record. In the past the genus names *Trinacromerum* and *Dolichorhynchops* have often been used interchangeably, most recently by Adams (1997), who mistakenly named a new species of a large *Dolichorhynchops* from the Campanian of Canada as *Trinacromerum bonneri*. In his discussion of the occurrence of short-necked plesiosaurs, Carpenter (1996, p. 284) determined that *Trinacromerum bentonianum* (Cragin, 1888) has a 3.3-million-year range, from Upper Cenomanian into the Turonian, and *Dolichorhynchops osborni* (Williston, 1902) has a four-million-year range, beginning in the earliest Campanian. This suggests that there is an eight- to nine-million-year gap in the current fossil record of the Western Interior Sea—between the middle Turonian and the beginning of the Campanian—that is unoccupied by polycotylids. In their review of the occurrence of pliosaurids, polycotylids, and elasmosaurids in the old "Benton Formation" in Kansas (Graneros Shale, Greenhorn Limestone, and Carlile Shale), Schumacher and Everhart (unpublished data) found that *Trinacromerum* was relatively abundant through the middle of the Fairport Chalk Member (middle Turonian) of the Carlile Shale.

Figure 8.2. The mounted type specimen of Dolichorhynchops osborni *(KUVP 1300) in the collection of the University of Kansas Museum of Natural History. Note that the skull is a model constructed by H. T. Martin because the original was crushed and too fragile to place in the exhibit.*

Within the Smoky Hill Chalk, the gap noted by Carpenter (1996) until the first occurrence of *Dolichorhynchops* and *Polycotylus* extended from the beginning of the chalk (late Coniacian) through the end of the Santonian, a period of almost 3.5 million years. This interval also corresponds roughly to the greatest expansion of the Western Interior Sea (Hattin, 1982, p. 59), and the beginnings of the subsequent regression, a period in which nearshore, shallow-water environments, possibly the preferred habitats of polycotylids, were the furthest away from the depositional area of the Smoky Hill Chalk in western Kansas. Recently, Everhart (2003) reported the discovery of several specimens of fragmentary plesiosaur remains from the lower chalk (late Coniacian to early Santonian) and suggested that polycotylids were present throughout the deposition of the chalk, "albeit in small numbers." Although not previously reported, the discovery of such specimens should not be considered unusual in light of the fact that the polycotylid lineage would not have disappeared and suddenly reappeared. The most likely explanation is that polycotylids preferred a different environment somewhere away from the deep water in the middle of the Western Interior Sea.

The holotype of *Dolichorhynchops osborni* (KUVP 1300, Fig. 8.2) was discovered by a young George F. Sternberg in 1900 and collected by his father, Charles H. Sternberg (Williston, 1903). The skull of KUVP 1300 is crushed laterally but is well preserved and complete (Fig. 8.3). Storrs (1999, p. 9) indicated that the specimen came from the Smoky Hill Chalk (Campanian) of Logan County, Kansas. Carpenter (1996, p. 271) concurred, noting that KUVP 1300 was collected "east of Wallace" in Stewart's (1990, p. 23) zone of *Hesperornis*. Bonner (1964, p. 41) reported that the specimen of *Trinacromerum osborni* (FHSM VP-404; now *D. osborni*, per Carpenter, 1997, p. 193) on exhibit in the Sternberg Museum of Natural History at Fort Hays State University was collected from Logan County, a mile southwest of Russell Springs. A third, partial *D. osborni* specimen (MCZ 1064) collected by G. F. Sternberg in 1926 (Sternberg and Walker, 1957, p. 57; Everhart, 2004a) was also discovered in Logan County (Carpenter, 1996, p. 271). The partially digested bones of the plesiosaur (USNM 9468) reported by Charles Sternberg (1922; p. 119; Everhart, 2004b) as gut

Figure 8.3. A left lateral view of the crushed skull of Dolichorhynchops osborni (KUVP 1300). This is the same view as published in plate II of Williston (1903). The slender teeth and delicate construction of the skull suggest that Dolichorhynchops was feeding only on small prey.

contents within a large *Tylosaurus proriger* (USNM 8898) mosasaur and the subject of the story at the beginning of this chapter, came from an exposure along (Twin) Butte Creek in Logan County. The remains of this plesiosaur were tentatively identified as a juvenile polycotylid (O'Keefe, pers. comm., 2001). Although their exact stratigraphic occurrence is unknown, the horizon of all these Logan County localities is in the upper one-third (Upper Santonian through Lower Campanian) of the Smoky Hill Chalk.

One of the fragmentary late Coniacian plesiosaur specimens (FHSM VP-13966) that I found in 1992 consisted of the partially digested remains of the back of a skull and lower jaws of *Dolichorhynchops* (Everhart, 2003). The bones were scattered across a wide area of the chalk at one of our favorite sites in Gove County. The specimen now consists of more than twenty small fragments of bone, all of which appeared initially to be badly weathered. I have since determined that they were actually partially digested, most likely by a large shark (see Chapter 4 in regard to bones regurgitated by sharks). At the time of discovery, however, I was unable to identify the remains. When J. D. Stewart examined the material in 1992, he concluded that the bone fragments were from the skull of a plesiosaur, noting that two of the larger fragments appeared to be the hinge joints (right and left articular, surangular, and angular) of the lower jaws.

This was really good news as far as I was concerned because we had found little plesiosaur material in the chalk. Two years earlier, Stewart (1990, p. 25) had written that he knew of no "lower chalk specimens" other than a "few juvenile propodials." Few things can ever be said to be certain in paleontology. As luck would have it, I happened to be with J. D. Stewart in May of 1990 when he found a fairly complete hind limb of a *Dolichorhynchops* (LACMNH 148920) in the lower chalk. Unfortunately, it was too late to add it to his paper (Stewart, 1990) in the *Niobrara Chalk Excursion Guidebook* that was published for the fiftieth anniversary meeting of the Society of Vertebrate Paleontology (SVP) in Lawrence, Kansas, in October.

However, the plesiosaur skull fragments and our other plesiosaur specimens sat around for almost ten years before I got around to working with them. In 2001, I sent the skull fragments (FHSM VP-13966) to Ken Carpenter at the Denver Museum of Nature and Science. He had recently published on his studies of short-necked plesiosaurs (Carpenter, 1996) from the Western Interior Sea and is familiar with polycotylids. Ken was able to identify the fragments, with a reasonable amount of certainty, as elements of the skull, braincase, and lower jaws of a plesiosaur, most likely *Dolichorhynchops*. After talking to Ken and rereading his 1996 paper, I realized that this specimen and five other sets of remains from the low chalk represented the first real evidence that plesiosaurs were living in the Western Interior Sea during late Coniacian time. While this was certainly not unexpected, it was the first time it could be actually documented from this portion of the chalk. Ugly and uninformative as they were, I published a short paper on them (Everhart, 2003) and added them to the record of plesiosaur remains from the Smoky Hill Chalk.

Several notable specimens occur outside of the Smoky Hill Chalk formation in Kansas. Certainly one of the first discoveries was ten articulated vertebrae found in Russell County by B. F. Mudge in 1872 in the "Benton Cretaceous." The stratigraphic occurrence was determined by Schumacher and Everhart (unpublished data) from Mudge's locality information to be the middle of the Fairport Chalk (middle Turonian). Mudge's handwritten note is still affixed to the specimen and indicates that he believed he had found the remains (KUVP 1325) of an "*Ichthyosaurus*." Later, the specimen was described as the type of *Trinacromerum anonymum* by Williston (1903) and then reidentified as *T. bentonianum* by Carpenter (1996). A more complete specimen of *T. anonymum/bentonianum* (YPM 1129) was found a year later in Osborne County by Joseph Savage and is now in the Yale Peabody Museum.

The type specimen of *Trinacromerum bentonianum* was described by Cragin (1888) from two skulls (USNM 10945 and 10946) that were collected from the upper part of the Fairport Chalk (middle Turonian) in Osborne County, Kansas. Riggs (1944) described another specimen (KUVP 5070) that had been found in 1936 on a road cut through the Hartland Shale Member of the Greenhorn Formation (Upper Cenomanian) along U.S. Highway 81 a few miles south of Concordia in Cloud County. The construction workers had apparently taken great care to protect the specimen, and the skull (Fig. 8.4) was recovered intact along with much of the rest of the skeleton. Riggs (1944) named the plesiosaur *Trinacromerum willistoni* in honor of S. W. Williston, but it was later determined by Carpenter (1996) to be yet another example of *T. bentonianum*.

In talking to Dr. Paul Johnston (pers. comm., 2003), recently retired from the geology department at Emporia State University, I

Figure 8.4. A right lateral view of the crushed skull of the type specimen of Trinacromerum willistoni Riggs (KUVP 5070, now T. bentonianum per Carpenter, 1999) in the collection of the University of Kansas Museum of Natural History.

learned of the discovery of another, fairly complete *Trinacromerum bentonianum* specimen that turned into a paleo-crime story. According to Dr. Johnston, in the summer of 1971 he found plesiosaur bones eroding from a road cut at the top of the Greenhorn Limestone, about a mile south of Wilson Lake in eastern Russell County. He covered up the bones and brought help back with him the next week. They uncovered the skeleton of a large plesiosaur coming out of the hillside headfirst at an angle. The skull and most of the cervical vertebrae were already gone, probably taken off by heavy equipment when the road was built. He remembers seeing one of the front paddles, the pectoral girdle, an articulated vertebral column, ribs, and at least one of the rear paddles in their excavation. They were on a public road and had quite a bit of traffic past the dig, including two men on motorcycles who stopped and looked things over pretty closely. He remembered that they asked a lot of questions. Near the end of the excavation, Dr. Johnston and his crew covered the remains and returned to Emporia for a few days. When they returned, the plesiosaur bones had been hacked out of the ground. All that was left was a few bone fragments and one complete front paddle that had still been buried in the hillside. The paddle is currently on exhibit (Fig. 8.5) in the Paul Johnston Museum of Geology at Emporia State University and has recently been identified as *T. bentonianum* (B. Schumacher, pers. comm., 2003).

There are two important pliosaurid specimens (not polycotylids) from below the chalk in central Kansas that should be mentioned here to avoid any further confusion. Williston (1903) described and named a new species of plesiosaur, *Brachauchenius*

Figure 8.5. A forepaddle of Trinacromerum bentonianum (ESU 5000) in the collection of the Paul Johnston Museum of Geology at Emporia State University, Emporia, Kansas. The specimen was found in Russell County in 1971.

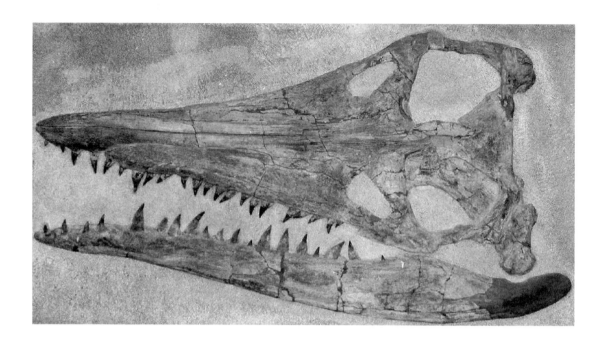

Figure 8.6. The skull of a giant
pliosaur (Brachauchenius lucasi—
FHSM VP-321) in the collection
of the Sternberg Museum of
Natural History. The specimen
was collected in 1950 from the
Fairport Chalk Member (middle
Turonian) of the Carlile Shale in
Russell County, Kansas, and
represents one of the last known
occurrences of a pliosaur in North
America.

lucasi, from a 90-cm (nearly 3-ft.)-long skull and thirty-seven vertebrae in the Smithsonian collection (USNM 4989) collected from
the "Benton Formation" in Ottawa County, Kansas. According to
Robert Purdy (pers. comm., 2003), the Smithsonian purchased the
specimen from Charles Sternberg in the spring of 1884. At the
time, the skull was prepared and exhibited upside down, along
with the articulated cervical and dorsal vertebrae. It has since been
removed from the original mounting so that the damaged upper
portions of the skull can be examined (McHenry, pers. comm.,
2004). The skull of a second specimen (FHSM VP-321) is much
larger, more complete, and better-preserved than the specimen described by Williston (Fig. 8.6). The remains were found in 1950
near the town of Fairport in Russell County by Robert and Frank
Jennrich while they were looking for shark teeth. Although the
skull was collected by George Sternberg in October 1950 and
placed on exhibit soon afterwards in the Sternberg Museum, it was
not actually reported until it was described by Carpenter (1996).

The FHSM VP-321 skull is currently on exhibit in the Sternberg
Museum of Natural History at Fort Hays State University. A cast of
the skull is also on display at the University of Kansas Museum of
Natural History, Lawrence, Kansas. The skull is about five feet (152
cm) in length along the midline and must have come from a creature
that was truly huge. Williston (1907) and Carpenter (1996) agree
that Brachauchenius is closely related to the Jurassic pliosaur Liopleurodon ferox. After his examination of the Sternberg specimen,
McHenry (pers. comm., 2004) suggested that there were also many
similarities between Brachauchenius lucasi and an Early Cretaceous
pliosaur from Australia called Kronosaurus queenslandicus.

Liggett et al. (1997) reported the discovery of a partial paddle (FHSM VP-13997) of a giant plesiosaur from the base of the Lincoln Limestone Member of the Greenhorn Limestone in western Russell County. If complete, the paddle would have measured more than 2 m (6 ft.) in length (Schumacher, pers. comm., 2003). The specimen was tentatively identified as the limb of *Brachauchenius lucasi*. Conservatively, this paddle would represent a pliosaur that was about 5 m (16 ft.) wide from paddle tip to paddle tip, and 7 m (22 ft.) long from nose to tail. The Sternberg Museum also houses two other specimens that have been attributed to *B. lucasi* (a partial propodial, FHSM VP-2149; and vertebrae, VP-2150). Additional remains of *Brachauchenius* have been discovered in Russell County (J. D. Stewart, pers. comm., 1999) but are as yet unreported.

Even with the discovery of three nearly complete specimens of *Dolichorhynchops osborni* in the past century or so, plesiosaur remains of any kind are rare occurrences in the Smoky Hill Chalk. *Dolichorhynchops* persisted past the end of the deposition of the chalk (about 82 mya) and lived well into the deposition of the Pierre Shale during Campanian time. Although rarely found in the Pierre Shale of Kansas, its remains have been collected more often in South Dakota and Wyoming. One specimen in the University of Kansas collection (KUVP 40001—from the Sharon Springs Member of the Pierre Shale, Fall River County, South Dakota) appears to show that the species was getting much larger (Carpenter, 1996). The skull length of this specimen is 98 cm (38 in.) as compared with 51 cm (20 in.) for the specimen (FHSM VP-404) in the Sternberg Museum. Scaling up suggests a pliosaur that was nearly 6 m (20 ft.) long, very close to the size of the much earlier *Brachauchenius* specimens from Kansas. This also approximates the body size (not including the neck and head) of *Elasmosaurus platyurus*.

In 1992, we participated in a dig by the New Jersey State Museum on a nearly complete *Styxosaurus snowii* in the Sharon Springs Member of the Pierre Shale of Logan County that had been found by our friend, Pete Bussen. At the time, my wife Pamela was having trouble "seeing" vertebrate remains in the selenite-filled dark gray shale. She asked Pete for help and he went up the hill with her to show her what to look for. Within thirty minutes, she had picked up a podial (finger bone of a paddle) and asked him if it was from a plesiosaur. It was, and she discovered the remains of one of the largest specimens of *Dolichorhynchops osborni* currently known from Kansas. Most of the remains were contained in a concretion, badly damaged and far from complete. Elements recovered included both rear paddles, part of the pelvis, and a number of caudal vertebrae. One of the femora was about 40 cm (16 in.) long, compared with the 33-cm (13-in.)-femur of FHSM VP-404 (Bonner, 1964), and much more massive. The material was identified by Ken Carpenter in 1994 and later donated to the Cincinnati Museum Center (CMC VP-7055).

There is one other specimen of *Dolichorhynchops osborni* that should be mentioned. The remains were found about 1987 by Orville Bonner in the Sharon Springs Member of the Pierre Shale in Logan County, just above the contact with the Smoky Hill Chalk. It was headless and otherwise unremarkable, *except* for what was preserved inside the abdomen. In their book on paleopathology, Rothschild and Martin (1993, p. 294) cited a personal communication with O. (Orville) Bonner in 1991 which alluded to the discovery of "young within the body" of "short-necked, polycotylid plesiosaurs." The specimen (LACMNH 129639) was obtained by the Los Angeles County Museum of Natural History in 1988 (McLeod, pers. comm., 2000). In September 2000, while visiting with Orville's younger brother Chuck, I was shown pictures of the plesiosaur dig and provided with some additional information. The scattered remains were removed in several jackets. A photograph of one of the partially prepared jackets showed adult-sized limb material next to what appeared to be the remains of at least one baby plesiosaur. The dorsal processes of the small vertebrae were not fused to the centra and the limb girdles appeared to be incompletely formed. The remains appear to represent a small plesiosaur but one that was much too large to have been consumed by the adult *Dolichorhynchops*. This, and the fact that the bones of the smaller individual do not appear to have been damaged by stomach acids, pretty much rules out cannibalism. The specimen has not been formally described, but it appears to represent a pregnant female *Dolichorhynchops* that was carrying at least one near-term fetus.

Lastly, it is worth noting that the earliest remains of plesiosaurs found in Kansas are from the Kiowa Shale (Early Cretaceous–middle Albian). The Kiowa Shale is exposed along the southern border of Kansas in Clark, Kiowa, and Comanche counties, where it overlies rocks of Permian age. It is also found in the central part of Kansas (mainly in Saline, McPherson, and Ellsworth counties). According to Scott (1970), much of Kansas was on the edge of a shallow sea at the time, and these interbedded shale and sandstone deposits represent a near-shore, high-energy environment. Consequently, the remains of the plesiosaurs and other vertebrates recovered from the Kiowa are generally disarticulated and incomplete. Although the Kiowa has been collected since the 1870s, little is known about the plesiosaurs or other animals of that time in Kansas. F. W. Cragin, who found the type specimen of *Trinacromerum bentonianum,* was among the first collectors of plesiosaur material from the Kiowa Shale. In about 1893, the remains of a plesiosaur (*Plesiosaurus mudgei*—KUVP 1305), consisting of nine vertebrae, a fragmentary femur, and about two hundred gastroliths, were collected and named by Cragin (1894) from the Kiowa Shale in Clark County (see also Schultze et al., 1985). The specimen is briefly mentioned in the text by Williston (1903) who apparently believed it was synonymous with another fragmentary specimen (*Plesiosaurus gouldii*). An excellent black-and-white photograph (ibid., pl. 29) shows the remains, including the many gastroliths.

Kansas has produced the remains of many plesiosaurs since the discovery of *Elasmosaurus* and *Polycotylus* in the late 1860s. They are found in every Cretaceous rock unit in the state, including the Dakota Sandstone (pers. obs.), and are especially common in the Kiowa Shale, the Greenhorn Limestone, and the Fairport Member of the Carlile Shale. I have no doubt that major new specimens will be discovered in the near future.

Figure 8.7. The short-necked polycotylids, Polycotylus latipinnus *and* Dolichorhynchops osborni, *depended upon their speed and long, toothy jaws to capture small prey. As teleosts evolved into larger and faster forms during the Late Cretaceous, the competition for food increased. The young of these plesiosaurs were occasionally prey for mosasaurs. Drawing by Russell Hawley.*

Nine

Enter the Mosasaurs

The female mosasaur swam slowly through the calm, warm water of the Inland Sea. Three of her week-old young swam cautiously on either side and slightly above her rear flippers. A fourth baby, the smallest and weakest of the litter, had not been able to keep up with the steady pace of the mother. When it dropped behind two days before, it had been quickly swallowed whole by a large fish. The surviving three babies were instinctively alert to any signs of predators and kept as close as possible to their mother's scaly side. They also stayed well behind her head and out of the reach of her toothy jaws. Even though she was their mother, her feeding instinct was very strong. A momentary mistake in the recognition of her offspring could be instantly fatal to them.

For the past several days, the mother mosasaur had guided her young through areas of the ocean that teemed with swarms of small, soft-bodied prey. The young mosasaurs were accomplished hunters almost from the moment of birth and had easily caught enough of the little squid to keep their bellies full. If they were to survive, it was essential that they eat as much and as often as they could. They were growing quickly but still would be vulnerable to attacks by other predators in the Western Interior Sea for many months to come.

The mother mosasaur paused briefly in the water to rest, floating motionless in the calm sea. Her young slithered part way up on her narrow back to take advantage of the relative safety of her large body. Few other animals in the ocean, except mosasaurs of her species, could match her 10-m (30-ft.) length.

For her young, it was a rare chance to absorb warmth from the sun that shone hotly in the clear, blue-white sky. While the mother mosasaur rested with her eyes and nose barely above water, her predatory senses were active. Hunger gnawed at her belly. She was used to eating often and well, and her recent pregnancy had all but depleted her body's reserves. Dimly aware that she couldn't hunt her usual prey without endangering her offspring, she had curbed her urge to stalk the schools of large fish that she sensed were nearby. She had eaten only twice in the last three days, first the bloated carcass of a small shark and then a large, swimming bird that had blundered across her path.

As she watched and waited, her senses detected the commotion caused by several large fish feeding on a trapped school of smaller fish. She could "feel" vibrations that carried through the water as the smaller fish leaped into the air to avoid the jaws of their attackers. For a few moments, she didn't move, but the urge was too great. With a powerful flick of her long, sinuous tail, she started her large body moving in the direction of the feeding frenzy.

Startled by her sudden movement, her young splashed awkwardly into the water and then swam swiftly to catch up and return to her side. As she swam beneath the surface, her movements were efficient and silent. Years of experience had taught her that hunting was more successful when her prey was unaware of her presence until it was too late to get away. The noise created by the smaller fish in their panic to escape would conceal her approach from the larger fish. They were her intended prey. One would fill her belly and satisfy her hunger for a day or so.

As she got closer to the commotion, she sensed rather than saw the larger fish darting into the turbulence caused by the panic of their smaller prey. Moments later, a long, silver torpedo shape emerged from the cloud of bubbles and almost ran into her open mouth. Too late, the fish recognized the danger and tried to turn away. With a swift lunge, the mosasaur's jaws snapped closed just behind the head of the fish. The water turned red with blood and a shower of scales glittered in the sunlit water as the fish struggled briefly and then went limp.

Waiting a moment to make certain that the fish was dead, the mosasaur then opened and closed her jaws in rapid movements to position her victim to be swallowed headfirst. When the head of the fish was inside the mosasaur's mouth, two rows of sharp teeth on the roof of the mouth helped hold the fish in place as the lower jaw flexed and pulled it deeper inside. Once the fish was securely started down the mosasaur's throat, she raised her head out of the water and used gravity to help her swallow her large meal. When the fish's large bony tail was the only part remaining outside her

jaws, she closed her mouth and shook her head sharply. With a snap, the tail broke off at the base and skipped across the water.

Again, the mosasaur floated almost motionless in the water as she finished swallowing the large fish. Her young milled nervously around her flanks, uncertain what to do in all the confusion. She had almost forgotten about them in her drive to satisfy her hunger.

The feeding frenzy had attracted other predators to the area. A large shark moved swiftly toward the fray, sensing the same signals that the mosasaur had reacted to. As it got closer, the shark detected the faint traces of blood in the water. Seeing three small objects that struggled at the side of the larger stationary one, it raced upward from the depths.

By the time the mother mosasaur and her young sensed the pressure wave generated by the approaching shark, it was too late to react. As the babies turned to flee, the shark's jaws closed viciously across the muzzle and neck of the largest one. The sharp, bladelike teeth of the shark sliced easily through the smaller animal's skull and cervical vertebrae, killing it instantly.

With the body of the young mosasaur in its mouth, the momentum of the shark carried it into the side of the mother mosasaur. The shark was less than half the length of the mosasaur and no match for the larger animal. As the mosasaur turned almost double on itself, slashing with open jaws at the flank of the shark, the shark flicked its tail and swam swiftly away with its victim held securely in its mouth. The mother mosasaur pursued the shark for a short distance, then slowed so that her two remaining young could catch up. Swimming quickly out of danger, the shark wolfed down the carcass of the little mosasaur. Hours later, it would regurgitate a few of the indigestible bones from the mosasaur's skull.

It is important to note here that we know virtually nothing about the social or parenting behavior of mosasaurs. Their modern relatives (monitor lizards and snakes) do not display the sort of parental involvement that I have imagined for this mosasaur mother and her young. Mosasaurs appear to have been solitary animals for the most part, at least on the basis of thousands of individual mosasaur specimens that have been found to date. In fact, the only associations of two or more mosasaurs that I am aware of

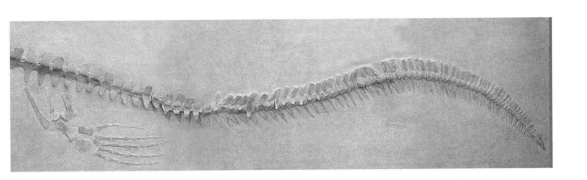

are: 1) The *Plioplatecarpus* specimen with at least four embryonic young reported by Bell et al. (1996); 2) the *Tylosaurus proriger* specimen described by Martin and Bjork (1987) that contained a much smaller *Clidastes* as stomach contents; and 3) the co-mingled and partially digested bones of two juvenile *Platecarpus* skulls, (FHSM VP-14846 and VP14847; Amy Sheldon, pers. comm., 1994) that I discovered in 1990. Other examples of interaction (fighting) between mosasaurs include a *Mosasaurus conodon* skull from the Pierre Shale with bite marks and the tooth of another *M. conodon* embedded in its left quadrate (Bell and Martin, 1995), and a *Tylosaurus* sp. skull (FHSM VP-2295) from the lower (late Coniacian) chalk with major bite marks across the top of the skull and right lower jaw that could have only been caused by a larger mosasaur. Beyond these few examples, the fossil record is largely mute regarding questions of whether mosasaurs may or may not have lived together.

While the above story is fiction, it is based on a fragment from the front of the skull of a young mosasaur (FHSM VP-13748) found in 1997 by Tom Caggiano in Gove County. It had been severed cleanly across the muzzle, behind the third tooth in both maxillae, and partially digested. The premaxillae and the anterior ends of both maxillae were still joined together. I knew it was partially digested because of the eroded condition of the surface of the bone (Everhart, 1999; Varricchio, 2001) and because the teeth were completely dissolved down into their sockets. Most likely this was the work of a large ginsu shark (*Cretoxyrhina mantelli*). Other partial remains of mosasaurs, including several with the broken tips of *Cretoxyrhina* teeth still embedded in them, are commonly found as fossils in the lower (late Coniacian) chalk. One such specimen, FHSM VP-13283 (Shimada, 1997, p. 928, fig. 4; Everhart, 1999), is an articulated series of five vertebrae from the lower back of a large (est. 6 m) mosasaur that has two shark teeth embedded in it. The anterior and posterior vertebrae (approximate diameter 5 cm [2 in.]) in the series are both severed across the centra and most of the surface of all the vertebrae has an eroded appearance. When I discovered this specimen in 1995, it was still mostly in the chalk, so any erosion of the bone surface is not due to weathering. There is no way of telling, however, whether or not these specimens were

TABLE 9.1.
Mosasaurs from the Western Interior Sea/North America

Genus *Halisaurus*	Marsh 1869
Halisaurus sternbergii	(Russell, 1970); See *Clidastes sternbergi* Wiman 1920
Genus *Clidastes*	Cope 1868
Clidastes liodontus	Merriam 1894
Clidastes propython	Cope 1869
Genus *Globidens*	Gilmore 1912
Globidens alabamaensis	Gilmore 1912
Globidens dakotensis	Russell 1975
Globidens "Kansas"	See Everhart, 1996
Genus *Mosasaurus*	Conybeare 1822 (*Mosasaurus hoffmanni* Mantell 1829)
Mosasaurus conodon	Cope 1881
Mosasaurus ivoensis	Persson 1863 (KUVP 1024)
Mosasaurus missouriensis	Harlan 1834 (*Mosasaurus Maximiliana* Goldfuss 1845)
Mosasaurus maximus	Cope 1869
Genus *Platecarpus*	Cope 1869
Platecarpus tympaniticus	Cope 1869; includes *P. ictericus* and *P. coryphaeus*
Platecarpus planifrons	Cope 1874 (Originally *Clidastes planifrons* Cope)
Genus *Ectenosaurus*	Russell 1967
Ectenosaurus clidastoides	Merriam 1894
Genus *Plioplatecarpus*	Dollo 1882
Plioplatecarpus primaevus	Russell 1967
Plioplatecarpus depressus	Cope 1869
Genus *Prognathodon*	Dollo 1889 (*Prognathodon solvayi* Dollo)
Prognathodon crassartus	(Cope 1872) See Williston 1898, p. 180
Prognathodon overtoni	Williston 1897
Genus *Tylosaurus*	Marsh 1872
Tylosaurus proriger	Cope 1869
Tylosaurus nepaeolicus	Cope 1874
Tylosaurus n. sp.	(Stewart, 1990)

the result of a shark attack on a living mosasaur or from scavenging of a carcass. I suspect that most of these specimens were from sharks feeding on already dead mosasaurs, much like the behavior of modern sharks. However, there is no reason to doubt that Cretaceous sharks would have missed an opportunity for an easy meal on an injured or otherwise vulnerable mosasaur.

Mosasaurs were marine lizards that lived only during the Late Cretaceous (Fig. 9.1). They were the top predators of the Earth's oceans at roughly the same time as *Tyrannosaurus rex* lived in western North America. Their closest modern relatives are probably monitor lizards (varanids) like the Komodo dragon and quite possibly snakes (Caldwell, 1999), although the exact relationships are still a matter of debate among mosasaur workers. The ancestors of mosasaurs were probably related to small terrestrial lizards called aigialosaurs that lived close to the ocean from the Late Jurassic through the Cretaceous. Carroll and Debraga (1992) reported three mosasaur-like aigialosaur specimens from Cenomanian–Turonian (93 mya) deposits in Yugoslavia.

About 95 million years ago, well into the second half of the Cretaceous, the ancestral, shore-dwelling mosasaurs began to evolve rapidly. Within a relatively short time, geologically speaking, they had adapted to life in the sea to the point that they could no longer leave the water. Mosasaurs were not the first terrestrial reptiles to return to the sea, but they were probably the most successful in terms of their diversity and eventual dominance (Table 9.1). Then, for reasons we don't fully understand, they became extinct after about 25 million years of existence, along with dinosaurs and many other groups at the end of the Cretaceous.

Mosasaurs suddenly appear in the fossil record of the Western Interior Sea in the second half of the Cretaceous period (early Turonian), several million years prior to deposition of the Smoky Hill Chalk. However, very few mosasaur fossils prior to the beginning of Coniacian time (89 mya) have been reported. Some of the specimens of "early mosasaurs" that had been reported turned out to be the jaws and teeth of a large predatory fish called *Pachyrhizodus* (Stewart and Bell, 1994; see Chapter 4).

Figure 9.2. Dorsal and ventral views of a "Platecarpus-like" mosasaur frontal (top of the skull) from the Fairport Member of the Carlile Shale of Ellis County, Kansas. This specimen (KUVP 97200) and several vertebrae from the same formation are some of the earliest known remains of mosasaurs from North America. (Scale = cm)

Martin and Stewart (1977) described two sets of vertebrae and a jaw fragment from the middle Turonian Fairport Chalk Member of the Carlile Shale Formation and noted their affinities with *Clidastes*. Another skull element (a *Platecarpus*-like frontal; KUVP 97200) is also in the KUVP collection (Bell, pers. comm., 2004) from the same strata in Ellis County, Kansas (Fig. 9.2). Lingham-Soliar (1994) reviewed mosasaur remains from the Upper Turonian of Angola in western Africa. Bell and VonLoh (1998) reported on new records of mosasauroids from the Greenhorn Formation (early Turonian) of South Dakota and the Boquillas Formation of western Texas. A more detailed discussion of the stratigraphic occurrence of mosasaurs is provided by Bell (1997b).

Ichthyosaurs and plesiosaurs had inhabited the oceans for millions of years after the Triassic, evolving into many forms and surviving several major extinction events (Ellis, 2003). For unknown reasons, the ichthyosaurs declined significantly after the beginning of the Cretaceous (Bakker, 1993; Russell, 1993; Lingham-Soliar, 2003) and are thought to have been extinct by the time that the earliest mosasaurs re-entered the water. Shimada (1996) reported the only known specimen of an ichthyosaur from Kansas, a single vertebra (FHSM VP-2169) from the Early Cretaceous (Albian) Kiowa Shale. Russell (1993) suggested that mosasaurs were able to fill many of the ecological niches left vacant by the demise of ichthyosaurs. According to Bakker (1993), plesiosaurs replaced the ichthyosaurs and mosasaurs may have replaced the marine crocodilians that had also disappeared by the beginning of the Coniacian in the Western Interior Sea. The giant crocodilian *Deinosuchus rugosus* lived along the coast during the Campanian (Schwimmer, 2002), but probably didn't venture far out to sea. It is quite possible that both the ichthyosaurs and the plesiosaurs had been losing the evolutionary battle of "who eats what" to competition from faster, larger, and more advanced varieties of bony fish such as *Xiphactinus* (Massare, 1987; Lingham-Soliar, 1999a) and the giant ginsu sharks (*Cretoxyrhina mantelli*). Other groups of reptiles, including marine crocodiles, teleosaurs, placodonts, and turtles (Chapter 5) had also enjoyed limited successes in the oceans of the Mesozoic, but none approached the worldwide domination of the seas that mosasaurs would attain in the Late Cretaceous.

Plesiosaurs also appear to be less numerous in the Late Cretaceous than during the Jurassic (there is no Jurassic record in Kansas) and had evolved into specialized forms like the slow-moving, long-necked *Elasmosaurus* and the short-necked, fast-swimming *Trinacromerum, Polycotylus,* and *Dolichorhynchops* (Carpenter, 1996; Adams, 1997). Initially, the short-necked plesiosaurs (polycotylids) found in Kansas were much smaller than their distant Jurassic cousin, *Liopleurodon,* an early Cretaceous relative, *Kronosaurus,* from down under, and a large pliosaurid from Kansas called *Brachauchenius.* They would grow much larger during the Campanian. During deposition of the chalk in the Late Cretaceous, however, mosasaurs outnumbered the remaining elas-

Table 9.2.
Approximate duration of ages that make up the Upper Cretaceous in the Western Interior Sea of North America, with remarks regarding mosasaurs.

Age	Time Span	Remarks
Maastrichtian	71.3–65.4 mya (5.9 my)	Greatest diversity and distribution; invasion of freshwater habitats. Beginning of "Third Wave" mosasaurs.
Campanian	83.5–71.3 mya (12.2 my)	"Second Wave" mosasaurs; *Hainosaurus, Mosasaurus, Globidens, Plioplatecarpus, Prognathodon*
Santonian	85.8–83.5 mya (2.3 my)	Mosasaurs get much larger; worldwide distribution
Coniacian	89.0–85.8 mya (3.2 my)	"First Wave" mosasaurs; *Tylosaurus, Platecarpus,* and *Clidastes*
Turonian	93.5–89.0 mya (4.5 my)	Early mosasaurs; *Clidastes* and *Tylosaurus / Platecarpus* precursors
Cenomanian	99.0–93.5 mya (5.5 my)	Ancestral mosasaurs (Aigialosaurs?) return to the ocean

mosaurs and polycotylids by more than ten to one, at least in the number of museum specimens recorded from the Niobrara (Russell, 1988). In support of Russell's data, my wife and I collected seventy mosasaur specimens and just five plesiosaur specimens between 1988 and 1995, mostly in the lower chalk of Gove County.

The early ancestors of mosasaurs probably fed in shallow coastal waters and returned to land to rest and breed much like the marine iguanas that inhabit the Galapagos Islands today. Several species of modern monitor lizards spend much of their time in the water, but are otherwise terrestrial. Over the space of a few million years, these ancestral mosasaurs became much larger and more specialized, evolving rapidly into a diverse assortment of highly successful marine predators (Table 9.2). By middle Coniacian time (87 mya), the "first wave" of mosasaurs (*Tylosaurus, Platecarpus,* and *Clidastes*) was well established in the Western Interior Sea that covered Kansas and most of the central portions of North America (Williston, 1898a; Russell, 1967; Everhart 2001). As mosasaurs continued to evolve, growing larger and diversifying rapidly, a "second wave" of genera and species showed up about the beginning of the Campanian (83.5 mya). Following a possible near-extinction event near the middle of the Campanian reported by Lindgren and Siverson (2004), mosasaurs rebounded, and the "third wave" was just getting underway during the final years of the Cretaceous, shortly before their final extinction.

The meanings of mosasaur names are of some interest because they give some insight into certain characteristics that early workers saw in the first fossil remains. *Tylosaurus* (Marsh, 1872b) means "knob (or snout) lizard" after the elongated bony muzzle that projected beyond the last teeth of the upper jaws (Fig. 9.3). There are three species of *Tylosaurus* known from the chalk: *T. proriger* (Cope 1869—"prow-bearing"); *T. nepaeolicus* (Cope

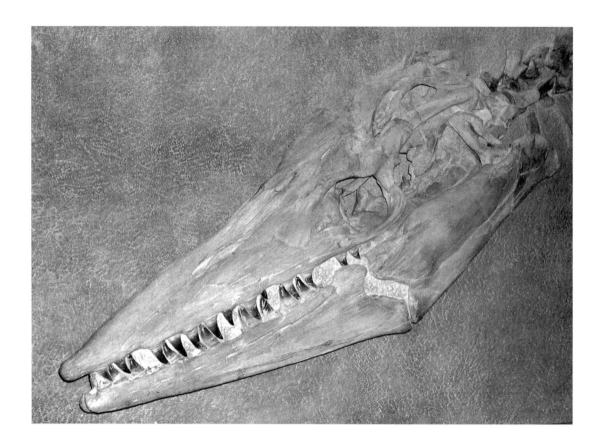

Figure 9.3. This skull of Tylosaurus proriger *(FHSM VP-3)* is about 1. 2 m (4 ft.) in length. Note the extended rostrum in front of the front teeth in the upper jaw.

1874) (from the Nepaholla River, an old name for the Solomon River in north-central Kansas; Everhart, 2002a); and a third, as-yet unnamed species (Stewart, 1990; Bell, 1997b; Everhart, 2004b).

Platecarpus Cope 1869 means "oar (or paddle) wrist" for the flatness and relative immobility of the wrist joint. Several species of *Platecarpus* were named initially, some from fragmentary remains, but most workers (Russell, 1967; Stewart, 1990; Bell, 1997b; Everhart, 2001) now agree that most of the earlier names (*P. coryphaeus* Cope and *P. ictericus* Cope) are synonymous with *Platecarpus tympaniticus* Cope 1869 (referring to the difference in the form of the tympanic bone or quadrate). *P. tympaniticus* was originally discovered in Mississippi by Dr. William Spillman (Manning, 1994) and was later recognized in the fauna of the Western Interior Sea (Russell, 1967). Although many mosasaur workers have agreed that Cope's original descriptions are inadequate for separating the various species of *Platecarpus,* there has been little effort since Cope's time to straighten out the problems. A study is currently in progress to sort out and re-describe the genus (Takuya Konishi, pers. comm., 2004). Fortunately, there are many excellent specimens available for study.

The fragmentary type specimen (AMNH 1491) of *Platecarpus planifrons* (Cope 1874) (meaning "flat forehead/frontal bone") was collected by Professor B. F. Mudge. It was originally identified

and named as *Clidastes planifrons* by Cope (1874). In the first comprehensive review of mosasaur species, Williston (1898a) determined that the remains were not those of a *Clidastes* and renamed it *Platecarpus planifrons*. However, because the type specimen was so fragmentary, Russell (1967) considered the specimen to be of "uncertain taxonomic position." *P. planifrons* was again recognized in the Smoky Hill Chalk by Bell (report of pers. comm. by Schumacher, 1993; Bell, 1997b). The more recent (1996) discovery of a complete skull of *P. planifrons* (FHSM VP-13910) by Steve Johnson of Wichita, Kansas (Everhart and Johnson, 2001), raises new questions concerning the placement of this species in *Platecarpus* (Fig. 9.4). The species is characterized in part by a shortened skull and the fewest number of teeth in the maxilla (ten) and dentary (eleven) known in any mosasaur (Table 9.3). Note that Burnham (1991) reported that an as-yet unnamed species of *Plioplatecarpus* (UNO 8611–2) from the Demopolis Formation in Alabama had eleven teeth in one dentary and twelve in the other.

Clidastes Cope 1868 roughly translates as "one who locks up" and describes the bony processes (called the zygosphene and zygantrum) on the vertebrae of this genus which restrict movement and limit the bending of the dorsal portion of the vertebral column (Russell, 1967). This was a necessary adaptation to stiffen the back of the animal for more efficient swimming. *Clidastes propython* Cope 1869 ("before-python," refers to the snakelike affinities Cope saw when he classified mosasaurs into a separate order called Pythonomorpha) was a relatively late arrival (mid-Santonian) in the Western Interior Sea and was far more common along the Gulf Coast. *Clidastes liodontus* Merriam 1894 (smooth-toothed) first appears in the Smoky Hill Chalk during late Coniacian time (FHSM VP-13909; Everhart, 2001). *Clidastes sternbergii* was described by Wiman (1920) but was later transferred to the genus

Figure 9.4. The reconstructed skull (length = 38 cm [15 in.]) of Platecarpus planifrons (FHSM VP-13910) in the collection of the Sternberg Museum of Natural History. This specimen was found in Gove County by Steve Johnson of Wichita, Kansas, in 1996, and is the only known complete skull of this rare species.

Table 9.3.
Mosasaur Tooth Counts

	Maxilla	Dentary	Pterygoid	Citation/Specimen #
Mosasauridae				
Halisaurus sternbergii	unk	22	unk	RMM 6890 (per J. Lindgren)
Russellosaurinae				
Tylosaurus novum sp.	12	13–14	10-11	Everhart, in prep)
Tylosaurus nepaeolicus	13	13	10	FHSM VP-7262
Tylosaurus proriger	13(12)	13	10	(Russell, 1967)
Ectenosaurus clidastoides	17	16?	9-11	FHSM VP-401
Platecarpus planifrons	10	11	9	FHSM VP-13910
Platecarpus tympaniticus	12	12(11 rarely)	10–12	(Russell, 1967)
Plioplatecarpus primaevus	11	12	13	(Holmes, 1996)
Plioplatecarpus UNO 8611–2	12	11-12	10–13	(Burnham, 1991)
Mosasaurinae				
Clidastes liodontus	14–15	16	15	(Russell, 1967)
Clidastes propython	18	17–18	14	(Russell, 1967; Williston, 1898)
Globidens alabamaensis	12–13?	unk	unk	(Gilmore 1912)
Globidens dakotensis	13	unk	0	(Russell, (1975)
Globidens "Kansas"	unk	13–14	unk	(Everhart and Everhart, 1996)
Prognathodon overtoni	12	14	6	SDSMT 3393
Prognathodon stadtmani	11–12	14	6	(Kass, 1999)
Plesiotylosaurus crassidens	13	16-17		(Camp, 1942)
Plotosaurini				
Mosasaurus missouriensis	14	15	8-10	(Russell, 1967)
Mosasaurus conodon	15	17	10	(Russell, 1967)
Plotosaurus bennisoni	18	17	15	(Camp, 1942)
Plotosaurus tuckeri	18	17+/-	12–13	(Camp, 1942)
European Species				
Mosasaurus hoffmanni	15	13	8	(Cuvier, Dollo)
Mosasaurus hoffmanni	14	14	8	(Lingham-Soliar 1995)
Mosasaurus maximus	13-14	14	8	(Russell, 1967)
Hainosaurus bernardi	12	13	8?	(Lingham-Soliar 1992)
Prognathodon solvayi	12	13	8	(Lingham-Soliar/Nolf, 1989)
Plioplatecarpus marshi	12/13	12	11	(Lingham-Soliar, 1994)
Plioplatecarpus houzeaui	12	12	?	(Lingham-Soliar, 1994)
African Species				
Goronyosaurus nigeriensis	11	12	8	(Lingham-Soliar 1999)
Pluridens walkeri	unk	30	unk	(Lingham-Soliar 1998)

Abbreviations: FHSM—Fort Hays Sternberg Museum, Hays, Kansas; RMM—Red Mountain Museum, Birmingham, Alabama; SDSMT—South Dakota School of Mines and Technology, Rapid City, South Dakota; UNO—University of New Orleans, New Orleans, Louisiana; unk—unknown.

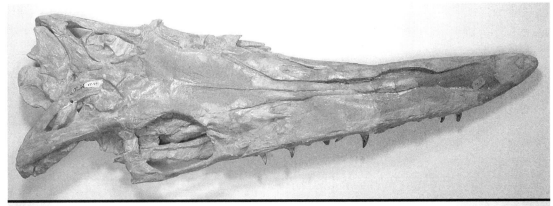

Halisaurus (Marsh 1869—"sea lizard") by Russell (1970) on the basis of similar material found in Alabama. The only reasonably complete Kansas specimen was found in Logan County in 1918 by Levi Sternberg (Sternberg, 1922), and purchased by Uppsala University in Sweden. Bardet and Suberbiola (2001) re-examined the Kansas remains (UPI R 163) and concluded that it is the oldest representative of the genus *Halisaurus* and the most primitive mosasaurid (see also Bell, 1997b; Lindgren and Siverson, 2004).

Another unusual and apparently rare species in the Smoky Hill Chalk is *Ectenosaurus clidastoides* Merriam 1894. Merriam (1894, p. 30) originally named the species "*Platecarpus clidastoides*" because of its *Clidastes*-like appearance. However, Russell (1987, pp. 156–158) disagreed with its placement within *Platecarpus* and erected a new genus (*Ectenosaurus*—"drawn-out lizard") to describe the unusually long, narrow snout. The type specimen was probably destroyed in WWII (ibid.). The most complete specimen known (FHSM VP-401) (Fig. 9.5), missing only the tail and hind limbs, and including skin impressions, was found by George Sternberg northwest of WaKeeney in Trego County in 1963 (ibid.). A reconstruction of this skull is figured by Russell (ibid., fig. 86). One of the distinguishing features of this species is its long, narrow skull and relatively large number of teeth, especially when compared to *Platecarpus* (seventeen versus twelve in the maxilla—see Table 9.3).

Figure 9.5. Dorsal and ventral views of the long and slender skull of Ectenosaurus clidastoides *(FHSM VP-401) in the collection of the Sternberg Museum of Natural History. This is the most complete specimen known of this rare species. The specimen also includes skin impressions.*

Tylosaurs were the largest of the early mosasaurs, reaching about 9 m (29.5 ft.) in length by Santonian time (Everhart and Everhart, 1997). They were the "heavyweights" of the first wave of mosasaurs, with heavy, conical teeth capable of capturing a variety of prey, including other mosasaurs (Martin and Bjork, 1987) and plesiosaurs (Sternberg, 1922; Everhart, 2004a). *Platecarpus* was probably the most common in terms of the sheer number of individuals in the first wave and reached lengths of about 7 m (23 ft.). Judging by its slender, backward-curving teeth (Massare, 1987; Russell, 1970), it is likely that *Platecarpus* fed on smaller, soft-bodied prey, including squid and other cephalopods that were abundant at the time. *Clidastes*, the smallest and probably the most primitive of the three major groups, was generally less than 5 m (16 ft.) in length when full grown and fed on fish and other small marine creatures. Recently, several specimens of an unusually large

Figure 9.6. Tylosaurus proriger *was the largest of the mosasaurs in the Western Interior Sea during the deposition of the Smoky Hill Chalk, reaching 10 m or more in length. The larger mosasaurs, such as* Tylosaurus, *preyed upon a variety of marine vertebrates, including fish, plesiosaurs, birds, and even other, smaller mosasaurs. Drawing by Russell Hawley.*

(6–7 m) *Clidastes* sp. have been discovered in the Smoky Hill Chalk and also noted to be in the collection of the Yale Peabody Museum from the 1870s (Ott et al., 2002). They have yet to be described.

By the early Campanian, *Tylosaurus* had reached lengths of 13–14 m (42–45 ft.) as evidenced by the "Bunker Mosasaur" specimen on exhibit in the Natural History Museum at the University of Kansas. The remains of this huge tylosaur (KUVP 5033) were discovered just above the contact of the Smoky Hill Chalk and the Pierre Shale near Wallace, Kansas, in 1911 and represents what is probably the largest mounted mosasaur skeleton in the United States. The Sternberg Museum has a similar-sized but incomplete *Tylosaurus* specimen (FHSM VP-2496) in its collection, which is also from the Sharon Springs Member of the Pierre Shale. The bases of the teeth of this specimen are almost as large as a man's clenched fist (Schumacher, 1993). Similar, very large *Tylosaurus* remains have also been found in Texas and South Dakota.

Later in the Campanian, a "second wave" of new species began making their appearance as mosasaurs diversified and dispersed into the Cretaceous oceans around the world (Table 9.1). *Prognathodon* and *Globidens* evolved more robust skulls and heavy, crushing teeth for eating hard-shelled prey. *Plioplatecarpus* replaced *Platecarpus*, and *Mosasaurus* replaced *Clidastes*. Tylosaurine mosasaurs such as *Hainosaurus* (U.S. and Europe), *Taniwhasaurus oweni* (New Zealand), and a new species (*Lakumasaurus antarcticus*) from Antarctica (Novas et al., 2002) are evidence of the worldwide distribution of this group. Highly derived mosasaurs such as the ichthyosaur-like *Plotosaurus* and *Plesiotylosaurus* in California (Camp, 1942), and a number of poorly known species from Africa suggest that mosasaurs were also evolving in other directions as isolated populations in these areas. Within the space of a few more million years, by Maastrichtian time (71 mya), some mosasaur species were truly huge, with several lineages (*Mosasaurus* and *Hainosaurus,* a close relative of *Tylosaurus*) reaching more than 15 m (50 ft.) in length. One giant specimen of *Mosasaurus hoffmanni* from Europe was 17 m (almost 55 ft.) in length (Lingham-Soliar, 1999b). Recently described specimens of *Prognathodon* from Utah (Kass, 1999), the Netherlands (Dortangs et al., 2002), and Israel (Christiansen and Bonde, 2002) suggest that this genus was also becoming larger and more heavily built. During this age of the last great marine reptiles, there was no doubt which were the biggest and "baddest" predators in the oceans of the Late Cretaceous.

A complete listing of mosasaur species in the Smoky Hill Chalk is still being worked out. Even though they have been collected in Kansas for more than 130 years, it has only been recently that much attention has been paid to their biostratigraphy. Williston (1897) briefly discussed the stratigraphic occurrence of mosasaurs in the *Pteranodon* Beds (Smoky Hill Chalk) for the first time. He also made the observation that *Clidastes* does not occur in the *Rudistes* Beds (lower chalk), indicating that other genera probably occurred within 100 feet of the contact of the chalk with the underlying Fort Hays Limestone.

In the years that followed, S. W. Williston, Charles H. Sternberg and his sons, H. T. Martin, and others continued to collect spectacular examples of mosasaurs and other marine species from the Smoky Hill Chalk without adding significantly to the knowledge of the stratigraphic record of these creatures. Williston (1898a) published the first comprehensive description of the systematics and comparative anatomy of mosasaurs from the Smoky Hill Chalk, and discussed their range and distribution in comparison with specimens discovered earlier in New Zealand and Europe. He commented that *Tylosaurus*, "so far as was known, begins near the lower part of the Niobrara [Smoky Hill Chalk] and terminates at its close or in the beginning of the Fort Pierre [Pierre Shale]." Of *Platecarpus*, he stated that the species on which the genus is based are "known nowhere outside of Kansas and Colorado, and are here restricted exclusively to the Niobrara." He again concluded that the lowest horizon of *Clidastes* "is the upper part of the Niobrara in Kansas."

Russell (1967) reviewed the specimens collected by the Yale scientific expeditions and postulated that the Smoky Hill Chalk could be divided into a lower, *Clidastes liodontus–Platecarpus coryphaeus–Tylosaurus nepaeolicus* zone and an upper, *Clidastes–Platecarpus ictericus–Tylosaurus proriger* zone. He also suggested that the increased abundance of *Clidastes* specimens in the upper portion of the chalk was an indication of a gradual change from a midocean to a near-shore environment. Russell (1970) noted significant differences between the distribution of mosasaur species in the Smoky Hill Chalk compared to Gulf Coast species occurring in the Selma Formation of Alabama. In his initial work concerning the biostratigraphy of the Smoky Hill Chalk, Stewart (1988, p. 81) stated that he was aware of several exceptions to Russell's 1967 stratigraphic distribution of mosasaurs in the Smoky Hill Chalk that caused him to regard it with "a degree of skepticism."

It was not until Hattin (1982) published his composite measured section of the Smoky Hill Chalk that significant progress could be made in documenting the vertebrate biostratigraphy of this formation. Hattin used bentonites (layers of volcanic ash) and other physical features to delineate his twenty-three lithologic marker units, and divided the chalk into five biostratigraphic zones based on the occurrence of invertebrate species. In doing so, he provided field workers with the first dependable method of determining their stratigraphic location in the section.

Stewart (1990) incorporated Hattin's marker units as upper and lower boundaries for his six proposed biostratigraphic zones (Table 2.1). He also provided the first comprehensive description of the distribution of known invertebrate and vertebrate species in the Niobrara Formation and made the first attempt to assign specific stratigraphic ranges for mosasaur species within the Smoky Hill Chalk.

Schumacher (1993) and Sheldon (1996) reviewed existing collections of mosasaur material in the Sternberg Museum of Natural History, the Yale Peabody Museum, and other institutions and further refined the occurrence of mosasaurs in the Smoky Hill Chalk by building upon the stratigraphic methodology provided by Hattin (1982) and Stewart (1990). Since that time, additional discoveries (Everhart, 2001; Everhart and Johnson, 2001; Everhart, 2002a) of mosasaur remains with associated stratigraphic data have aided in the refinement of the known temporal ranges of *Tylosaurus proriger, Platecarpus planifrons,* and *Clidastes liodontus.*

Based on the latest discoveries, we can now say that the genus *Tylosaurus* is found throughout the chalk, with *T. nepaeolicus* and an as-yet undescribed species living during the late Coniacian and being replaced by (or evolving into) *T. proriger* by the early Santonian. Tylosaurs appear to continue well into the Campanian, and possibly the Maastrichtian, with a closely related genus, *Hainosaurus,* evolving during the early Campanian. *Platecarpus tympaniticus* first appears in the lower chalk (late Coniacian) and continues essentially unchanged through the entire deposition of the Smoky Hill Chalk. Note that this species is the senior synonym of *P. coryphaeus* and *P. ictericus* (Russell, 1967). Sometime in the early Campanian (Pierre Shale), *P. tympaniticus* is replaced by *Plioplatecarpus.* Another mosasaur, *P. planifrons,* appears to be limited to the low chalk near the Coniacian-Santonian boundary, with a single upper chalk occurrence reported by Sheldon (1996). The genus *Clidastes* may appear during the deposition of the Fairport Chalk during middle Turonian time (Martin and Stewart, 1977), with *C. liodontus* being recognized from the late Coniacian (Everhart et al., 1997) and *C. propython* arriving in the late Santonian. *Clidastes* is then replaced by an explosion of *Mosasaurus* species during the early Campanian. About the same time, *Globidens* first appears in Kansas (Everhart and Everhart, 1996), apparently as the genus moved northward from the Gulf Coast. As yet, there are not enough specimens of either *Ectenosaurus* or *Halisaurus* to say more than that they appear in the fossil record during the late Santonian/early Campanian. While this scenario is a considerable improvement over that envisioned by Williston (1897) or Russell (1967), it can certainly be improved by future discoveries. Enlarging a bit on Williston's 1897 plea, I would reiterate that additional collecting is required to improve the accuracy of these ranges, and it is essential that good stratigraphic information be obtained for all specimens from the Smoky Hill Chalk.

As mosasaurs increased in size and diversity during the last 25 million years before the end of the Cretaceous, they also spread around the world. While they have been collected most often in Kansas and South Dakota, their remains are known from every continent and have even been recovered from Late Cretaceous rocks exposed on islands off the coast of Antarctica (Chatterjee and Zinsmeister, 1982; Case et al., 2000; Novas et al., 2002). Recent discoveries in Israel (Christiansen and Bonde, 2002) and Turkey (Bardet and Tunoglu, 2002) further document a worldwide distribution during the Late Cretaceous.

The word "mosasaur" means Meuse River lizard, a tribute to the locality where the first and one of the largest known specimens of a mosasaur was discovered in an underground mine near Maastricht in the Netherlands in about 1780 (Mulder, 2003). Another, less complete specimen (No. 7424) in the Teylers Museum (Haarlem, the Netherlands) had been found several years earlier but did not generate the same interest (Mulder, 2003). A military doctor, J. L. Hoffman, realized the scientific value of the new specimen and brought the Meuse River specimen to the attention of several noted scientists of the day. At the time, nearly fifty years before the discovery of dinosaurs, the identity of this strange creature was a genuine mystery, and it was believed to be everything from a large fish to a giant crocodile. In 1808, Baron Cuvier wrote about "le grand animal fossile de Maastricht" (Bardet and Jagt, 1996) and agreed with Adrian Camper (see Williston, 1898a) that the animal was closely related to modern monitor lizards. Fourteen years later, the name "*Mosasaurus*" was coined by W. D. Conybeare, and mosasaurs began to be recognized as a group in their own right. Mosasaur remains were soon discovered in England, the east coast of North America, and as far away as New Zealand. It was not until 1829, however, that Gideon Mantell authored the species name "*hoffmanni*" in honor of the man who was responsible for bringing it to the attention of the scientists of the day.

Mitchell (1818) published the first North American record of mosasaur remains (a tooth and part of a jaw) from the Navesink Formation of New Jersey. It is possible, however, that the first mention of the discovery of mosasaurs in the American West came from notes kept by the Lewis and Clark Expedition of 1804–1806. Mitchell (ibid.) recounts the discovery of a "petrified skeleton of a very large fish, seen in Sioux country, on the Big Bend of the Missouri River," noting that "Sergeant Gass, who had found the skeleton in 1804, wrote in his journal that it was forty-five feet long, and lay on the top of a high cliff." Harlan (1834) wrote, "It is not improbable that Lewis and Clark, in their Expedition up the Missouri, allude to the remains of a similar animal in the following extracts: 'Monday, September 10th, 1804, we reached an island (not far from the grand detour, between Shannon creek and Poncarrar River), extending for two miles in the middle of the river, covered with red cedar, from which it takes the name of *Cedar Island*; just below this island, on a hill, to the south, is the back-bone of a fish forty-five feet long, tapering towards the tail, and in a perfect state of petrifation, fragments of which were collected and sent to Washington.' " This point on the river is several miles below the confluence with the White River and is probably in the northwest corner of present-day Gregory County, South Dakota.

Four members of the expedition wrote slightly different accounts of the discovery in their journals. (See Moulton,

1983–1997, for entries of September 10, 1804). Supposedly, a portion of the specimen was collected and sent back to Washington, D.C., but was subsequently lost (Simpson, 1942). While the remains were probably the articulated vertebrae of a large mosasaur, we will probably never know for certain. Almost thirty years later, another mosasaur fossil was discovered in the same general area along the Missouri River in what is now South Dakota. A nearly complete skull with associated vertebrae and limb material was collected and brought back to St. Louis by an Indian Agent named Major Benjamin O'Fallon (Goldfuss, 1845), where it was displayed in the formal garden of his home. The bones eventually came to the attention of Prince Maximilian zu Wied during his travels in the American West and were acquired by him. The material was shipped to Bonn, Germany, where it was examined by Dr. August Goldfuss (1782–1848), a well-known naturalist. Dr. Goldfuss prepared the specimen and named it *Mosasaurus Maximiliana* in honor of the man who brought it back to Germany. His paper, "Der Schädelbau des *Mosasaurus*" (The Structure of the Skull of *Mosasaurus*) was originally presented at a scientific meeting in Mainz, Germany, in the fall of 1842 and later published in 1845. This historic specimen is still in the collection of the Goldfuss Museum (University of Bonn), Bonn, Germany, along with other remains described by Dr. Goldfuss.

It is worth noting here that the end of the snout and the tips of the lower jaws were missing from the skull described by Goldfuss. The rest of the story may be just as interesting. A paper published in 1834 by Dr. Richard Harlan (1796–1843), a Philadelphia surgeon and naturalist, described and named a new species of "Ichthyosaur" from a fragment which later turned out to be the anterior end (premaxilla) of a mosasaur snout. Harlan reported that his specimen was found by a beaver trapper from "Missouri" (probably from along the Missouri River in South Dakota) in about 1832. He gave it the name of *Ichthyosaurus missouriensis*. Soon it became clear that the fragment was part of a mosasaur and not an ichthyosaur as suggested initially by Harlan. In fact, it came from the same species of mosasaur as the one described by Goldfuss more than ten years later. In spite of Harlan's erroneous identification, the name of *Mosasaurus Maximiliana* Goldfuss 1845 was short-lived and eventually became a junior synonym of *Mosasaurus missouriensis* (Harlan 1834).

It appeared likely that Harlan's "ichthyosaur" fragment was the actually the missing snout of the Goldfuss mosasaur skull. The "mystery" of the missing piece was noticed soon after the Goldfuss paper was published. Although Russell (1967) indicated that the missing piece could not be located, he also credits an 1845 letter from Hermann von Meyer to Professor Bronn, published in the *Neues Jahrbuch für Mineralogie, Geognosie, Geologie und Petrefaktenkunde* (pp. 308–313) with the first mention that Harlan's "ichthyosaur" fragment might be a part of the Goldfuss mosasaur. Camp (1942) provides a more detailed explanation of the circumstances surrounding this mosasaur mix-up. However, the story does have a happy ending. In May 2004, just before the "First Mosasaur Meeting" at the Natural History Museum of Maastricht in the Netherlands, Gordon Bell and Mike Caldwell were examining the mosasaur collection in the National Museum of Natural History in Paris and came across a fragment of a mosasaur snout (No. 1314). When they looked closer, they could see that Harlan's name had been written in ink on the bone. Photographs taken the following week of this specimen and the rest of the skull in the Goldfuss Museum in Bonn by my friend Takehito Ikejiri confirmed that Harlan's "*Ichthyosaurus*" was the missing piece of the skull of "*Mosasaurus Maximiliana.*"

Since 1845, mosasaur remains have been discovered in many parts of the United States, from New Jersey to California, and from Texas and the Gulf Coast to North Dakota. In fact, they are known from just about everywhere in the world where a Late Cretaceous marine fossil record is preserved. They are, however, most numerous and best preserved from the chalks and shales of the Western Interior Sea and especially in western Kansas. According to Williston (1898a), "Kansas, *par excellence,* has been the great collecting ground of the world for these reptiles."

Mosasaur fossils occur in Late Cretaceous rocks deposited as near-shore or shallow marine sediments in many places around the world. Although mosasaurs probably preferred the shallower coastal waters where their prey was most abundant, they apparently also were excellent open-water swimmers. Mosasaurs were certainly capable of traveling across large bodies of water as evidenced by the numerous examples of their remains recovered from the Smoky Hill Chalk. These rocks were deposited near the middle of the Western Interior Sea, hundreds of miles from the nearest land. From this evidence, it appears likely that mosasaurs were living continuously in midocean and not just migrating through it. Fossils discovered in Canada and in Africa also suggest that near the end of the Cretaceous at least two mosasaur species, *Plioplatecarpus* (Holmes et al., 1999) and *Goronyosaurus* (Lingham-Soliar, 1999a), were also adapting to estuarine and freshwater habitats that had been the primary domain of alligators and crocodiles for millions of years. It would have been interesting to see how mosasaurs might have evolved had they survived the extinction at the end of the Cretaceous. The fact that primitive whales (e.g., *Basilosaurus*) first adapted a "mosasauroid" body plan suggests that the long, sinuous form and swimming style of mosasaurs continued to be successful for a time, even in mammals.

As the pioneers and the railroads moved into the Midwest after the Civil War, railroad survey teams, geologists, and other explorers came across new sources of fossils. One of the first and most famous Kansas fossils (*Elasmosaurus platyurus*) was discovered in 1867 by an Army doctor named Theophilus H. Turner in the Pierre Shale near McAllaster Butte in Logan County (Chapter 7). The news of the discovery of these fossil remains (Cope, 1868a) created a sensation in the East. *Elasmosaurus* and the other fossils that had been shipped from Kansas by Turner and two other military doctors, and by Prof. B. F. Mudge, sparked a "fossil rush" by Marsh, Cope, and other paleontologists of the day. The first Kansas mosasaur, "*Macrosaurus*" *(Tylosaurus) proriger* (MCZ 4374), was named by Cope (1869a) from a specimen obtained by Dr. Louis Agassiz from near the "overland stage station" at Monument Rocks in Gove County. Marsh followed quickly, and in 1870 he led the first of the Yale College scientific expeditions, collecting fossils in many places in the West, including Kansas.

Late in November of that year, Marsh and his students made a brief visit to the area south of Fort Wallace. As might be expected for Kansas, the weather was already turning cold. Accompanied by a military escort, they were able to collect for several days in the Smoky Hill Chalk along the Smoky Hill River. One of the students later wrote (Betts, 1871, p. 671) that they spent four days digging out the nearly complete skeleton of a mosasaur "allied with the genus *Mosasaurus*." His note regarding the length of the specimen ("not been less than sixty feet") was certainly an exaggeration. The find was important, however, because it provided the first evidence that mosasaurs had hind limbs, something that was unknown to Cope and others at the time.

Marsh (1871a) reported in a brief note shortly after their return to Yale that "some interesting reptilian and fish remains" were collected during their visit to Kansas. Later, Marsh (1871b) published a more detailed account of the mosasaurs found by the 1870 expedition, including the nearly complete *Clidastes* specimen mentioned by Betts (1871).

For the next several years, Cope and Marsh competed to see who could collect and name the most new animals from Kansas and elsewhere. While Cope and Marsh were both in the Kansas chalk during 1871, they usually employed others, including professional collectors, to find and secure most of the fossils they studied. Over a four-year period (1870–1873), the Yale expeditions collected more than a thousand mosasaur specimens from Kansas for the Peabody Museum. They also collected fragmentary remains of the first fossil bird (*Hesperornis*) and the first *Pteranodon* in the chalk. After B. F. Mudge found the first bird with teeth (*Ichthyornis*) in 1872, Marsh became preoccupied with the pursuit of the remains of toothed birds from the chalk and hired Mudge to locate as many specimens as possible (Shor, 1971). While this strategy established Marsh as the expert on these primitive birds, it also allowed Cope to collect and describe many new species of fish and marine reptiles.

Cope hired a young Charles Sternberg and others to collect for him, and he seemed to always be one step ahead of Marsh in naming new species of mosasaurs from Kansas (Bell, 1997a). In retrospect, Marsh may simply have been overwhelmed by the sheer volume of new and undescribed material that was being returned to Yale by his employees. Cope was certainly the more agile of the two, producing many more papers over the course of his career than Marsh. One adverse result of the competition between Cope and Marsh, however, was a large and unrealistic number of new species of mosasaurs from the Niobrara of Kansas, cited as seventeen by Cope (1872) early in the exploration of the chalk.

Although other workers of the time (Baur, 1892; Merriam, 1894) published papers on mosasaurs, S. W. Williston (1898a) was the first to describe the occurrence of mosasaurs as a group. He suggested that many of the differences observed by Cope and Marsh were due to distortion of the bones during preservation and that fully "four-fifths of all the described species must be abandoned." Russell (1967) published an extensive review of the systematics and morphology of American mosasaurs and significantly reduced the number of valid taxa. Bell (1997b) applied cladistic analysis to mosasaurs as a group for the first time in his phylogenetic revision of North American and Adriatic Mosasauridae. Since then, however, new species of mosasaurs continue to be found and described. We are some years from being able to say that we fully understand their evolution and diversity on a worldwide basis, or even in Kansas.

We are still trying to understand their adaptations and their success in the relatively short time (25 million years or so) they existed at the end of the Late Cretaceous (Williston, 1914). Even the way they swam was a departure from the proven methods used by the previously successful ichthyosaurs and plesiosaurs. The long, muscular tails of mosasaurs were flattened from side to side, and they used an undulating motion (like a snake's) of their tails to propel their bodies through the water. Because this method of swimming is inefficient compared to that used by more streamlined fish and other marine reptiles, like the ichthyosaurs and the short-necked pliosaurs and polycotylids, they were probably not capable of traveling long distances at high rates of speed (Massare, 1987). Rather, they probably hunted from "ambush," surprising and outrunning their prey over a short distance. Even though mosasaurs were far less streamlined than the "fish-lizard" ichthyosaurs, their method of hunting conserved much of the energy ichthyosaurs were required to expend to swim constantly in pursuit of prey. This adaptation, probably more than anything else, enabled mosasaurs to complete successfully with the bony fish (*Xiphactinus* and *Ichthyodectes*) and sharks (*Cretoxyrhina*) that were also becoming much larger and common during the Late Cretaceous.

The tail accounts for half or more of the body length in most mosasaurs and is composed of about 100 relatively short (compared to their diameter) vertebrae. Osborn (1899, p. 178) was the first to report a "tail bend" in the specimen of a *Tylosaurus proriger* (AMNH FR 221) that is now on exhibit in the American Museum of Natural History, noting that measurements "tend to show that the vertebral centra were slightly longer above than below and thus produced the [downward] curve." Wiman (1920) noted a "tail curve" in specimens of *Platecarpus tympaniticus* and *Halisaurus sternbergi* he had received from Kansas, and noted a similarity to the modifications of the caudal vertebrae of ichthyosaurs. In mosasaurs, as in ichthyosaurs, this adaptation would have most likely served to deepen the end the tail from top to bottom and provide a greater area for producing thrust, as well as produce a slight downward thrust to assist in overcoming buoyancy. The work by Osborn and Wiman, however, has been largely ignored in reconstructions of mosasaurs over the years. Schumacher and Varner (1996) reported the same tail bend in *Clidastes, Platecarpus,* and *Tylosaurus,* and observed a slight wedge shape in each of the vertebrae in area of the tail bend. They also noted that the transverse processes of the caudal vertebrae are also reduced in size while the neural spines become taller (Schumacher, pers. comm., 2004). This means the tail becomes laterally compressed at the same time as it gets taller, forming a more efficient paddle. Some reconstructions, including those of Wiman (1920, figs. 2–3) depict a fleshy lobe projecting from the lengthened neural spines above the bend. We may never know what the exact shape of the tail was in mosasaurs, but it was likely much broader dorsally than is shown in many illustrations.

The tail of *Clidastes* is shorter in proportion to its body (about 2/5 of total length) than that of other mosasaurs. *Clidastes* also has elongated dorsal processes on the caudal vertebrae that would have supported a proportionately deeper tail. This may indicate that *Clidastes* was faster than other mosasaurs. Swimming faster would have been a useful adaptation when your larger cousin (*Tylosaurus*) saw you as his next meal. However, speed wasn't always enough, as demonstrated by preserved stomach contents of a large *Tylosaurus proriger* discovered in South Dakota which included the remains of a smaller *Clidastes* (Martin and Bjork, 1987).

In addition to differences in size and proportion between the three most common genera, mosasaurs also differed in the number of vertebrae that made up their extremely long bodies (Russell, 1967). *Tylosaurus* had the most vertebrae: 7 cervicals, 23 dorsals, and 95–119 caudals, including the pygals, for an average of about 140. *Platecarpus* had between 125 and 130 vertebrae, and *Clidastes* had the fewest, with 108. Mosasaur vertebrae were procoelous in form; that is, the anterior face of the centrum (cotyle) was concave and the posterior face (condyle) was convex. The round end of one vertebra was set securely in the cupped centrum of the following vertebra with little or no space for cushioning between the vertebrae, forming a long column that had little vertical flexibility. Lateral movement of the vertebral column was also limited in the anterior dorsal vertebrae but increased toward the hips and tail. While swimming at cruising speeds, the head and body of a mosasaur would have been held fairly stiff, with the paddles probably folded against the body, while the tail undulated from side to side. In most cases, the paddles were probably used only for steering, although the limbs of some mosasaurs, including *Plio-*

platecarpus, Mosasaurus, and *Plotosaurus,* were much more heavily constructed than those of *Tylosaurus* and *Platecarpus.* There has been a fairly recent discussion regarding the possible use of such paddles by *Plioplatecarpus* as an adaptation to "subaqueous flying" similar to that used by plesiosaurs (Lingham-Soliar, 1992; Nicholls and Godfrey, 1994). While anything was possible as mosasaurs evolved and spread into a variety of different environments, it is likely that most of them continued to use their tails as their primary means of movement.

Because mosasaurs had adapted so well to living in the ocean, it is improbable that they ever voluntarily returned to land. As they adapted to life in the ocean, their upper limb bones became shortened and their feet became modified into broad, flat paddles. In contrast to the extreme hyperphalangy (multiple finger bones) seen in plesiosaurs and ichthyosaurs, the number of bones in each of the fingers of a mosasaur was fairly small, ranging from three to five in *Clidastes* to five to eleven in *Tylosaurus.* The five digits of each paddle were widely separated and loosely webbed together in most species. Mosasaurs probably also retained more flexibility in their limb joints compared to the rigid, bony flippers of the plesiosaurs and ichthyosaurs.

The shoulder and hip girdles of mosasaurs were no longer solidly connected to their rib cages and backbones. Over time, much of what had been bone in their limbs was replaced by cartilage since the excess strength and rigidity of bone was not needed in the weightless environment of the ocean. This meant, however, that mosasaur limbs were basically useless out of water and could not support the weight of their bodies on land. While mosasaurs may have been able to slither through shallow water much like a snake, they would have been nearly helpless on dry land. Once beached, they would have died either of suffocation due to the collapse of their lungs or from overheating, much like a beached whale. The larger the mosasaurs got, the more their weight made them helpless on land.

Ichthyosaurs and plesiosaurs had a much longer history than mosasaurs and appear to have been scaleless. It is possible that a smooth skin was much better adapted for a life in the ocean than was one with scales because it offered less resistance in the water, or because it provided fewer opportunities for the attachment of parasites. Mosasaurs, on the other hand, were covered with small scales, the shape and arrangement of which resembled those of modern monitor lizards (Williston, 1898a). The scales of *Tylosaurus* (Fig. 9.7) were first described by Snow (1878) from impressions preserved with remains (KUVP 1075) found on Hackberry Creek in Gove County, Kansas. They are rhomboidal and quite small, roughly 3.3 mm in length by 2.5 mm in width, and have a raised keel along the long axis. The scales of *Platecarpus* (private collection) and *Ectenosaurus* (FHSM VP-401) appear to be about the same size but are completely smooth. All of the skin impressions that I am aware of are from relatively early mosasaurs.

Figure 9.7. Diamond-shaped scale impressions from the Tylosaurus proriger *(KUVP 1075) specimen described by Snow (1878). Several square feet of scale impressions are preserved with this specimen.*

We may find that Maastrichtian mosasaurs were smooth-skinned, like the ichthyosaurs and plesiosaurs. An early misinterpretation of mosasaur remains should be mentioned here. The large "dermal scutes" described by Marsh (1872a) were re-examined by Williston (1891) and found to be fragments of the thin sclerotic ring that covered much of the outer surface of the eye of the mosasaur (Fig. 9.8). Mosasaurs do not have "dermal scutes."

Williston (1898, 1899) made a mistake in the interpretation of soft tissues (tracheal cartilages) that led to the erroneous belief that mosasaurs had a dorsal fringe. Charles Knight, a well-known artist of the day who specialized in recreations of prehistoric animals, then painted a mosasaur with a frilly mane along its back, and the mistake has been repeated time and again ever since. In his report regarding a wonderfully preserved specimen of *Platecarpus tympaniticus* (KUVP 1001—found by Albin Stewart in 1898), Williston noted that it included scales, preserved sternal cartilage and "a row of dermal processes." The specimen was first reported by Williston (1898b) in a brief editorial note in the *Kansas University Quarterly* and then followed by a short article (Williston, 1899). About the same time, H. F. Osborn (1898) reported the same structures preserved in a large *Tylosaurus proriger* from western Kansas and interpreted them correctly as tracheal rings. However, he also credited Williston for his discovery of the "nuchal fringe" and published a figure by Charles Knight that showed the fringe.

Russell (1967, p. 88) briefly discussed "cartilaginous structures in the thoracic region of mosasaurs" and diplomatically stated that both Williston and Osborn had seen similar "tracheal structures" in these "two excellent skeletons from the Niobrara chalk." To

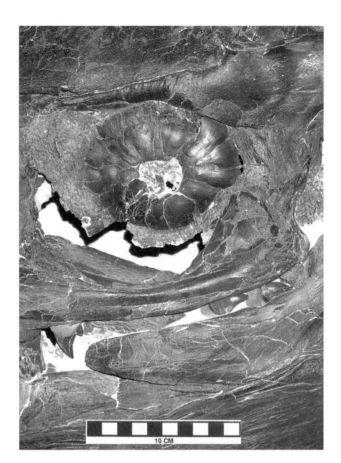

Figure 9.8. The scleral ring in the left orbit of a large Tylosaurus proriger *(FFHM 1997-10) skull collected from Smoky Hill Chalk (middle Santonian) of Gove County in 1996.*

Williston's credit, he (1902) did recognize his error and made the following statement: "In conclusion, I wish to correct an error made by myself. That which I considered to be the nuchal fringe in the mosasaurs is evidently only the slender cartilaginous rings of the trachea, first described and figured by Professor Osborn. I have no excuses to make for the mistake, which I recognized when too late to correct." Indeed, it was too late to correct, and for more than a century images of mosasaurs have continued to be plagued with an ugly accessory which they do not deserve.

The structure of their skeletons shows that mosasaurs were certainly reptiles, but their bodies had changed in many other ways as they adapted to life in the ocean. The discovery of a *Plioplate-carpus* "mother mosasaur" in South Dakota with the remains of several unborn young preserved inside her abdomen demonstrates conclusively that these animals bore their babies alive (Bell et al., 1996). While similar specimens have not been found in the Smoky Hill Chalk of Kansas, some remains, like the fragmentary skull of a very small mosasaur (FHSM VP-14845) in the Sternberg collection, suggest a near-term fetus or newly born individuals. Like the ichthyosaurs, mosasaurs probably only gave birth to a maximum of four or five fairly large babies at a time. This number is small compared to other reptiles such as crocodiles, alligators, and turtles that lay dozens of small eggs and depend on sheer numbers for the survival of their young.

Reproduction has been a controversial topic among mosasaur scientists for almost as long as mosasaurs have been studied. While acknowledging that mosasaurs would have been "practically helpless" on land, early paleontologists, such as Williston (1898a), firmly believed that mosasaurs must have somehow crawled up on the shore to lay their eggs. They also concluded that young mosasaurs lived in estuaries or other near-shore, protected environments because their remains had not been located along with those of adult mosasaurs. Whether or not those remains were simply overlooked or ignored by Williston, Cope, Marsh, Sternberg, and others in the rush for bigger, more spectacular specimens, we now know that mosasaurs of all ages lived and died in the middle of the Western Interior Sea (Sheldon, 1996; Everhart, 2002b).

The fairly common presence of the remains of young individuals in the Smoky Hill Chalk of Kansas suggests to me that mosasaurs were giving birth in midocean, 200 miles or more from the nearest land. While there is no evidence for parental care or involvement, the fact that very young animals are found in open water suggests to me that female mosasaurs, out of necessity, may have lived in groups for the protection of their young. There were simply too many other hungry predators around, including six-meter sharks (*Cretoxyrhina mantelli*), giant predatory fish (*Xiphactinus*), and other species of mosasaurs for young mosasaurs to have survived long without some kind of parental care or protection. While modern monitor lizards do not care for their young, the female American alligator (*Alligator mississippiensis*) is known to move newly hatched babies from their nest to the water, and to protect them to some extent from predators. Unlike alligators and monitor lizards, mosasaurs were not egg-layers, and they were well adapted in many other ways for their life in the ocean. Like most other animals in the wild, the mortality rate among young mosasaurs must have been high. However, from the number of adult specimens that have been found and their success in spreading around the world, it appears that many of them survived and successfully reproduced.

While it is possible that the poorly circulated Western Interior Sea had masses of floating seaweed that young animals could hide in, there is no fossil evidence to support this idea. The presence of many fast-swimming predators, such as the ichthyodectid fish *Xiphactinus audax* and short-necked polycotylids like *Polycotylus* and *Dolichorhynchops,* seems to argue for large areas of unobstructed, open water. One variety of primitive swordfish (*Protosphyraena perniciosa*) had long pectoral fins that extended two to three feet outward on either side of its body, hardly a good design for efficient hunting in a kelp forest. The other marine reptiles, including the giant marine turtles (*Protostega* and *Archelon*) and the elasmosaurs, also moved through the water with long, outstretched limbs. The snake-like mosasaurs, on the other hand, might have been well suited for such an environment if it existed.

Mosasaurs probably fed primarily on fish and cephalopods such as squid and belemnites, but the fossil record shows that they would have eaten just about anything they could swallow. The preserved stomach contents of a large *Tylosaurus proriger* (SDSMT 10439) on exhibit in the Museum of Geology at the South Dakota School of Mines and Technology contains the bones of a smaller mosasaur (*Clidastes*), a toothed, swimming bird (*Hesperornis*), a fish (*Bananogmius*), and several partially digested shark teeth (Martin and Bjork, 1987). Fish remains were observed in a specimen of *Plotosaurus* from California (Camp, 1942) and a *Tylosaurus proriger* from Kansas (Everhart, pers. obs.). The remains of a turtle are preserved in a fifteen-meter specimen of *Hainosaurus* from Belgium (Dollo, 1887). One of the most recently "rediscovered" items in the mosasaur diet was plesiosaurs (see Chapter 8). A large *Tylosaurus* specimen discovered in Logan County in 1918 included the remains of a juvenile polycotylid plesiosaur as stomach contents (Sternberg, 1922; Everhart, 2004a). While the *Tylosaurus* specimen was acquired and quickly exhibited by the USNM (Gilmore, 1921), the plesiosaur remains were largely ignored. Sternberg's brief report in the Transactions of the Kansas Academy of Science had been overlooked by mosasaur workers until 2001, when I came across it while searching for something else. It appears that mosasaurs, especially the larger individuals, were generalists when it came to suitable prey and ate about anything they could get in their mouths.

A *Globidens* specimen (SDSMT in prep.) from the Pierre Shale of South Dakota appears to contain fragments of bivalve shells in the abdominal area (Martin and Fox, 2004). Ammonite shells have been discovered with what appear to be mosasaur bite marks (Kauffman and Kesling, 1960), but there is some question as to whether or not these holes were caused by mosasaur teeth or by shell-boring invertebrates (Kase et al., 1998). To me, the question now is not so much whether or not mosasaurs fed on ammonites, but rather if their bites were responsible for the marks left on ammonite shells. As far as I am concerned, it is likely they did prey on ammonites, but I think such predatory attacks would have resulted in the shells being crushed, and not simply punctured.

Tooth count (Table 9.3) is one method of distinguishing among the various taxa of mosasaurs, and it suggests a general trend toward a reduction in the number of jaw and pterygoid teeth over time. Most mosasaur teeth are cone-shaped and slightly recurved (Fig. 9.9). They may be smooth or distinctly striated, slender or robust, but usually do not have the sharp, flesh-cutting edges typical of shark or theropod dinosaur teeth. They were also replaced continuously by new, slightly larger teeth throughout the life of the mosasaur, with the new tooth dissolving away the root of the older tooth and pushing it out as it grew (Caldwell et al., 2003).

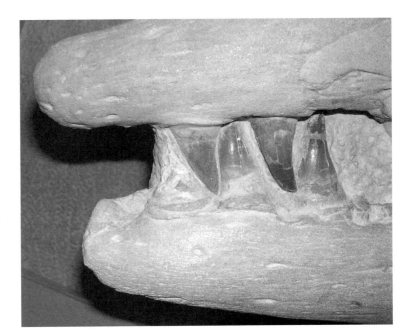

Figure 9.9. The anteriormost upper and lower teeth of Tylosaurus proriger *(FHSM VP-3). Note the extensions of both the premaxilla and the dentary in front of the teeth. Tylosaurs used their teeth for seizing and killing their prey, not for tearing flesh. Mosasaurs were able to swallow large prey (juvenile plesiosaurs, other mosasaurs) because of special adaptations in their skull and lower jaws.*

With the exception of a few specialized, late-evolving species, mosasaurs used their jaws and teeth for seizing, killing, and holding their prey until it could be swallowed, and not for tearing it apart. After capturing large prey in open, sometimes deep water, it is unlikely that a mosasaur would have had the luxury of tearing the meal apart into manageable pieces without risking its sinking to the bottom or losing it to other predators. Short of dismembering its prey by the sheer force of its bite, a mosasaur essentially had to swallow whatever it caught whole and headfirst, much like a modern snake.

Mosasaurs and snakes are also similar in the way their skulls are constructed (Caldwell, 1999; Lee et al., 1999), including a flexible joint between the tooth-bearing dentary and the surangular (and other bones) that make up the back half of the lower jaw. In addition, the dentaries were not fused together at the anterior end of the lower jaw, allowing several degrees of movement at that point. This flexibility allowed the mosasaur to "ratchet" prey into the throat with the aid of two rows of sharp "holding" teeth on the pterygoid bones in the roof of the mouth.

A simple demonstration devised by Cope (1872) can help to visualize the mechanics of how the mosasaur's lower jaw worked. Hold your arms straight out in front of you, hands extended, with your fingertips touching. Now try to pull your hands toward your face. Notice how your elbows have to flex outward and that the angle between your fingers widens. This is the same process that occurred repeatedly as a mosasaur swallowed large prey. Each time the prey was pulled further into the mouth, the pterygoid teeth on the roof of the mouth caught the prey again and prevented it from escaping. Such feeding adaptations were necessary for survival in the Cretaceous oceans, and mosasaurs were certainly winners in that regard. Russell (1975) noted that *Globidens dakotensis* is the only mosasaur species known that does not have pterygoid teeth, but Bell (pers. comm., 2002) suggested the teeth of that specimen had been removed accidentally during the collection or preparation of the specimen.

Only a few of the later genera (Campanian–Maastrichtian) of mosasaurs developed specialized teeth for cutting or tearing flesh (*Leiodon*) or for crushing clams and other hard-shelled invertebrates (*Globidens*). The only specimen of *Globidens* (FHSM VP-13828) known from Kansas was discovered in 1995 by Pete Bussen near the top of the Sharon Springs Member (early Campanian) of the Pierre Shale in Logan County (Fig. 9.10). Several other specimens, including the type specimen of *G. dakotensis*, are known from the Pierre Shale in South Dakota. An unusual and poorly known species from West Africa, *Pluridens walkeri*, appears to have reversed a trend toward shorter, heavier skulls and fewer teeth and became more "ichthyosaur-like" with thirty small, closely set teeth in its long narrow jaws (Lingham-Soliar, 1998). *Goronyosaurus*

Figure 9.10. The right dentary (lower jaw) of the only Globidens *specimen (FHSM VP-13828) known from Kansas. The specimen was discovered in 1995 by Pete Bussen in the Sharon Springs Member of the Pierre Shale Formation, Logan County.*

Figure 9.11. Globidens *was a late arrival in the Western Interior Sea, moving north from the warmer waters of the Gulf Coast during the Campanian. The ball-shaped teeth in its jaws were well adapted for crushing the hard shells of clams and other shellfish. Drawing by Russell Hawley.*

nigeriensis, another African species, had unusually large "canine-like" teeth that interlocked and fitted into deep sockets in the upper and lower jaws, very similar to the appearance and function of the dentition of crocodiles (Lingham-Soliar 1999a; 2002).

Life in the oceans of the Cretaceous was dangerous at best and mosasaurs didn't always have things their way. Early workers in the chalk, Mudge (1876) and Williston (1898a), noted signs of sharks scavenging on mosasaur carcasses. The serrated teeth marks of *Squalicorax falcatus* and *S. kaupi* are clearly visible on the scavenged bones of many fish and marine reptile specimens from the oceans of the Late Cretaceous (Bardet et al., 1998; Schwimmer et al., 1997). Recent evidence (Shimada, 1997; Everhart, 1999), including shark teeth embedded in partially digested mosasaur bones (FHSM VP-13283) from the late Coniacian (86 mya) Smoky Hill Chalk of Kansas, indicates that mosasaurs were frequently fed upon by the large ginsu sharks (*Cretoxyrhina mantelli*). While it is not possible to determine whether this feeding activity was the result of an attack on a live mosasaur or the scavenging of a carcass, it is probable that a 6-m (20-ft.) ginsu shark would have behaved much like a modern, similarly sized great white shark. The painting on the cover of this book by Dan Varner depicts a shark attack on a living mosasaur that could have produced the fossil evidence described from the specimen mentioned above. Most likely, ginsu sharks probably attacked small, injured, or sickly prey, including mosasaurs, because they were vulnerable, but they probably would not have passed up a free meal on a dead mosasaur. The fossil record, however, also indicates that these giant *Cretoxyrhina* sharks became extinct a few million years later (early Campanian) at roughly the same time that mosasaurs were becoming larger, more numerous, and more widespread. Did the success of mosasaurs cause the extinction of these large sharks? No one knows for sure, but we do know that modern sharks suffer greatly from overfishing by humans because they do not reproduce rapidly. It seems reasonable that bite-sized, juvenile sharks were on the menu for the expanding population of large, hungry mosasaurs.

Mosasaurs ruled the oceans of the Late Cretaceous and were apparently beginning to invade freshwater environments such as estuaries, swamps, and rivers when the Age of Dinosaurs ended. At that point, there were more than forty-five species of mosasaurs, many of which are only known from the Maastrichtian (Lingham-Soliar, 1999a). Like the dinosaurs, it is unlikely that any mosasaur species survived much past the Cretaceous-Tertiary boundary. Did they all die suddenly due to the catastrophic effects of an asteroid impact in the Yucatan, or was their extinction more gradual, following a general collapse of the marine ecosystem? We may never know.

Addendum

The study of mosasaurs appears to be moving along at a more rapid pace, as more specimens are found worldwide and more researchers become interested in these fascinating animals. In May of 2004, I attended the First Mosasaur Meeting in the Natural History Museum of Maastricht in the Netherlands, along with about twenty other paleontologists from around the world (the Netherlands, Sweden, Denmark, Germany, France, Bulgaria, Japan, Canada, and the United States). It was the first time that such a meeting had been held on mosasaurs only. We were able to tour the historic limestone quarries around Maastricht where the first mosasaur remains had been found and to visit the Teylers Museum in Haarlem where specimens that were discovered before the type specimen of *Mosasaurus hoffmanni* have been on exhibit for more than 200 years. In between, we spent two days listening to papers and sharing information on mosasaurs. The meeting was an overwhelming success, thanks in large part to the work done by the staff at the museum (Anne Schulp, John Jagt, Eric Mulder, and others) and the enthusiastic support of the museum director, Douwe Th. De Graaf. As a result, one of the questions raised during the meeting was "When and where is the next one?" The answer is 2007, probably at the Sternberg Museum of Natural History in Hays, Kansas. Time for me to get busy and make sure the second gathering lives up to the standards set by the First Mosasaur Meeting.

Ten

Pteranodons: Rulers of the Air

Soaring above the seemingly endless stretch of bright blue water on kitelike wings, the young male pteranodon effortlessly rode the sea breeze upward as he searched for signs of a school of fish feeding near the surface. From his vantage point, he could look down on a large area of the water's surface, seeking the shifting dark shadows made by huge schools of small fish as they migrated across the shallow sea. Making swift movements of his head, he also kept track of other members of his flock, watching them for a sudden dive toward the surface that indicated prey had been sighted.

It was hatching time and there were many new babies to be fed. This flock of males had left the rookery late the previous afternoon, catching the rising air currents and prevailing winds that carried them far out to sea during the hours of darkness. Navigating long distances over water was both instinctive and necessary in order for the pteranodons to reach the most productive feeding grounds where there was the least competition.

Now the sun was again rising in the sky and his sharp eyes were searching as he soared in large, counter-clockwise circles. Off to one side, he saw something dark moving slowly in the water. Adjusting his wings slightly, he dipped and turned to the right, spiraling downward toward the object. He quickly recognized the long dark object as it flew slowly through the water on wings of its own. It was a plesiosaur, a huge, long-necked marine reptile that fed on the same small fish the pteranodon was seeking. Still, it was only one predator and it was no threat to the pteranodons. Having other predators in the area sometimes meant that schools of fish might be frightened and forced to the surface, creating a feeding opportunity for the flock. Keeping an eye on the plesiosaur, the pteranodon turned into the wind, climbed higher, and resumed his aerial search pattern.

The hours wore on without sighting any schools of migrating fish. If necessary, the pteranodons were capable of staying in the air for days at a time, but that would not feed their hungry young or the females waiting back at the rookery. About noon, he saw one of his flock mates turn and rapidly lose altitude, followed almost immediately by several others. As they descended, he could see a large shadow forming near the surface. He was the furthest away from what was developing and would be the last to arrive. As he turned and descended quickly toward what was certainly a school of fish, he saw other dark, sinuous forms circling the shoal. A pod of snakelike mosasaurs had surrounded the fish and were moving around them in an ever-tightening circle. In the center of the trapped school, the closely crowded fish began to jump out of the water, causing the surface to erupt into a foaming melee.

The first group of pteranodons skimmed quickly across the surface, dipping their beaks in to the water and scooping up mouthfuls of small fish. They were quickly joined by a flock of small white birds with toothed beaks that dove into the water and took their prey one at time from the school. The first group of pteranodons wheeled about and started another run across the now boiling school of fish as a second group completed its pass.

Gauging his approach and watching out for his flock mates, the young male pteranodon adjusted his wings to slow his forward speed as much as possible as he descended in a shallow dive. Flying low across the school of fish, just inches from the surface, he dropped the sharp tip of his lower jaw into the water and immediately felt the impact of the small prey as they were scooped up and forced back into his mouth and throat. Lifting his beak out of the water, he began flapping his wings to gain altitude as he swallowed the wriggling fish. Rising slowly at first, then more rapidly, he turned to make another pass at the school. This time he saw that the mosasaurs were now actively feeding along the outer edges of the bait ball, but too far away to endanger the feeding pteranodons.

Again and again, he skimmed the water's surface, each time adding a number of small fish to his catch. Soon his gullet was full and it took much more effort to gain altitude at the end of a pass. Finally, as he turned into the wind and labored upward, he saw that the rest of the flock was gathering above him. Below, the school of small fish had been dispersed by the attacking mosasaurs.

He could see that a number of larger fish, including sharks, had joined the feeding frenzy. Small groups of prey fish scattered and then reformed into larger groups, only to be scattered again. Even as the well-fed pteranodons gathered to leave the area, another flock arrived and began to skim the surface. They were a smaller species, although still much larger than the birds that still wheeled and screeched over the dwindling bounty.

Once the pteranodon flock was back together, an older male took the lead and headed back toward the distant shore where the rookery had been established. The flock followed the leader, flying one behind the other in a relatively close V-formation that took advantage of "slipstreaming" to reduce wind resistance and energy use. They would have to fly for many hours to reach the rookery on the shore. Ahead, however, a fast-moving cool front was building storms across their flight path. Within an hour, they began to run into turbulence. For the lightly built pteranodons, this was a dangerous situation. The flight leader instinctively moved between the rising thunderheads, seeking calmer air. Still the flock was buffeted about, breaking up into smaller and smaller groups as flight conditions became more difficult. As darkness fell, flashes of lightning allowed the young male to adjust and maintain his position near the center of a group of five. The other four pteranodons were older and larger than he was, but they were still having trouble with the winds. The effort of flying in this weather was wearing him down more quickly, however, and he was having a difficult time keeping up. Then a sudden downdraft scattered the small group and he found himself alone, fighting wind gusts near the wave crests of the storm-swept sea. Moments later his wing tip caught the crest of a wave and he crashed heavily into the water. Dazed by the impact, he tried to untangle his wings, but they only wrapped more tightly around him. Struggling to breathe, he was submerged and trapped beneath the dark water. Mercifully, blackness soon enveloped him . . .

The story above, especially the long flight far out to sea to feed, is fiction. We don't know very much about these giant pterosaurs even though an estimated two thousand specimens have been found in the Smoky Hill Chalk since the 1870s. We know they ate fish, but we don't know if they fed while on the wing or while sitting in the water. For that matter, we don't know if they soared for long periods over the Western Interior Sea like a modern albatross, or only periodically migrated across the sea. We have never found evidence of where or how they nested, and we certainly know nothing about their social behavior. We do not know why so many of their remains are found hundreds of miles from what would have been the nearest shore, or how they died, or, for that matter, how they lived.

The Smoky Hill Chalk has produced a wealth of *Pteranodon* (meaning "wing without tooth") specimens. Note here that the word "pteranodon" is both a scientific name that should be capitalized and italicized (*Pteranodon*) when referring to the genus, and a common name that refers generally to the winged and toothless flying reptiles of the Late Cretaceous (i.e., *Quetzalcoatlus northropi*, the largest known flying reptile, is also a pteranodon). Years after his initial discovery of the first *Pteranodon* wing bones in 1870, Marsh (1884, p. 424) reported that the Yale Peabody Museum had the remains of about six hundred individuals in its collection. From a survey of the Yale collection and others, Russell (1988) calculated that a total of 878 specimens were known, or roughly 12 percent of all vertebrates in museum collections from the Smoky Hill Chalk, placing them a distant third behind fish (58 percent) and mosasaurs (25 percent). Four years later, Bennett (1992) examined more than 1,100 specimens in his study of size differences between male and female specimens of *Pteranodon*.

In the years I have spent collecting in the chalk, I have only found half a dozen sets of remains, most of which were very fragmentary. I found my first one in 1990 when I came across the ends of several wing bones eroding from the chalk in northeastern Lane County. Once uncovered, it was easy to see that the thin-walled bones had been crushed. Working with them was like trying to handle two layers of eggshell. Even so, I was able to collect the lower jaw and most of one wing of a medium-sized *Pteranodon* (wingspan 4.2 m [14 ft.]). All of the bones were severely damaged by "root rot" (a situation where the bones are literally enclosed by a web of plant roots that eventually destroy the bone in a search for nutrients) and there was nothing left of the large, shield-shaped sternum except an empty space filled with roots. The nearly complete lower jaw of the specimen was quite interesting. The narrow, V-shaped jaw was 63 cm (25 in.) long and narrowed down to a point that was as small as a pencil lead (about 2 mm). I still have to wonder how something that small and delicate could survive flying low over the water and grabbing fish near the surface (Brower, 1983, p. 118). If that was how *Pteranodons* fed, I suspect they weren't taking very big fish.

Our next discovery was a bit more spectacular. It happened late in the afternoon of June 1, 1996. We had been collecting all day and were about ready to call it quits and head for home. As I was walking back to the van, my wife Pam called me over to where she had found the ends of two bones eroding out of the chalk. It was clear from their flattened appearance that they were from a *Pteranodon*. They were about halfway up the east side of an eight-foot-high, nearly vertical wall of a narrow gully in the middle chalk (early Santonian) of southern Gove County. The bones had apparently been exposed fairly recently but were already bleached to a light blue-gray color that contrasted well against the pale tan of the chalk. After digging back into the soft chalk about six inches, I uncovered most of the still articulated "elbow" (the joint between the humerus and the radius/ulna) of a fairly large *Pteranodon*. The "fresh," unweathered material was a reddish brown color and was extremely fragile (again, I'll emphasize that it is comparable to working with two layers of broken eggshell held together with a soft, chalky matrix). The bones were oriented in the shape of a "V," with the open end of the "V" pointing into the side of the gully. That they were from a large individual (humerus about 24 cm in length), and were still articulated, indicated to me that more of the *Pteranodon* might still be in the chalk. Finding them certainly made it easy for us to plan another trip to the site. The humerus was completely exposed and removed. The two other bones (radius and ulna) were already broken near the middle, so I carefully removed what I could and covered the rest. Since the length of the humerus is roughly 4.5 percent of the length of the wingspread (Bennett, 1992, p. 431), I figured we had found the remains of a fairly large *Pteranodon* (wingspread about 5.3 m or 17.4 ft.).

On June 29, we returned to the site about 8 A.M. to take advantage of as much of the cool morning temperature as possible. The first order of business was to start taking the overburden off where the rest of the fossil was expected to be. There was about a meter of hard tan chalk interbedded with calcite seams, covering an area about two meters on a side that needed to be removed. Part of this was done right away in order to get a minimum working area and to expand the initial entry point to locate additional remains. Additional chalk was removed as needed over the next day and a half. All the initial overburden removal was done with a heavy pick and flat-bladed shovel (and lots of sweat!). The chalk was fairly hard (for chalk) and partially cross-bedded with calcite seams, causing it to break up in mostly small, uneven pieces. This was hot work and not a lot of fun, so it was done in small increments.

When I stopped digging with the pick and shovel, there was about 10 cm (4 in.) of chalk left on top of the fossil. This layer was removed slowly, with a sharp ice pick and brush and occasional, careful use of a rock hammer. As soon as I started working at the level where the original material was found, I quickly encountered additional bone, including the distal portions of the radius/ulna that I had been unable to take out initially. This led to additional bones that were scattered randomly to the right (southwest) of the initial find. These included the other humerus, a scapulo-coracoid, two cervical vertebrae, and additional wing bones, including most of the number IV metacarpal (the longest wing finger bone) that was 56 cm in length. These were exposed, coated with a preservative, photographed, and then removed. The preservative hardens the bone but can be removed with acetone if necessary during final preparation.

By early afternoon, another ridge of bone was encountered at the back (east side) of the excavation. As this was uncovered, it became apparent by the size and shape that we had found the lower jaws, lying together with the right jaw on top of the left. Further excavation at the posterior end of the jaw showed additional bone above the upper surface of the jaw and eventually the entire skull was found to be in place, and still articulated with the lower jaw.

The skull was lying on its left side, with the lower jaw closed. Working around the skull, I found several other wing bones, including one that went directly under where the crest should have been.

By the time that I had the skull exposed, it was almost five o'clock and I knew we were not going to get the remains out that day. I covered up the specimen with a tarp and we checked back into the motel for the night (a twenty-five-mile drive to the nearest small town along I-70).

Sunday was supposed to be even hotter, so we were back in the field by 7:45 A.M. The *Pteranodon* skull was cleaned further in preparation for the application of more preservative and jacketing. At this point, the last of the overburden to the east and south of the skull had to be removed in order to make room for jacketing of the skull and turning of the jacket. Several of the wing bones around the skull were removed individually in order to cut down on the size (and weight) of the final jacket. This was important to me because I was going to have to carry the skull and jacket back to our van. In the process, several small pieces of bone were isolated at the back of the skull. Field examination showed one wing claw mixed in with other material. Two more unarticulated cervical vertebrae (third and fourth?) were also found at the back of the skull and removed. From what I could see without too much digging around, it appeared that the atlas/axis vertebrae were still connected to the base of the skull.

At this point, the block containing the skull was isolated by digging a four-inch-deep trench completely around it. Once this was done, the block was undercut slightly to allow the jacket material to fill in and support as much of the chalk as possible and to find a seam in the chalk that could be used to separate the block from the underlying matrix. Once the undercut was completed, the exposed bones of the skull were again painted with the preservative and allowed to dry.

When the preservative had dried, aluminum foil was placed over the block to prevent the jacketing foam from contacting the specimen. A narrow piece of thin plywood was added along the length of the skull to reinforce the jacket. Then a temporary form for the jacket was made of cardboard, placed around the block containing the skull, and supported with pieces of broken chalk. A two-part mixture of isocyanate foam was prepared and poured into the form. The foam expands and cures in about half an hour, producing a rigid yet light jacket for the specimen. A plaster and burlap jacket could have been used, but I was concerned about being able to carry the specimen once I had it safely jacketed. A comparable plaster jacket would have weighed nearly as much as the piece of chalk containing the skull. In this case, high-tech, modern chemistry won the day.

When the jacket had cured and the form was removed, thin chisels and metal blades were driven under the block of chalk to free it from the matrix. Once it was loosened, the block was carefully but quickly turned over so that it was now upside down and resting safely inside the "top" of the foam jacket. This is always the "moment of truth," when you have something inside a jacket that needs to be turned over. Occasionally, if you don't do it right, the matrix and the fossil can fall out of the jacket as it is being rolled. In this case, everything stayed in place. I removed some of the excess chalk and then carried it about fifty yards to our van for the trip home. We left the field about 1 P.M. on Sunday, feeling pretty good about our efforts.

In May 1998, almost two years after a long-delayed preparation of the skull, Chris Bennett, then at the University of Kansas, identified our discovery as a remains of a young male *Pteranodon sternbergi*. Interestingly, the only bones from below the shoulders that have been found so far are several tarsals and a toe claw that Chris noted had been lying under the beak of the animal. It's possible that other remains were still at the site, or had eroded out and been destroyed earlier, or that the lower body of the *Pteranodon* had been severed by scavengers before burial. As noted in the fictional story above, the wings were literally wrapped around the skull. It is possible that they helped protect and preserve the skull itself.

In June 1998, we made a final trip to the site, hoping to collect additional post-cranial material. After removing a lot of overburden, however, we were only able to find one more bone, the other scapulo-coracoid. In 1999, the specimen was donated to the Cincinnati Museum Center, where the preparation was completed. The skull is currently on exhibit (CMC VP-7203). It turned out that, according to Bennett (1992), our specimen was almost exactly the average size he calculated for a male *Pteranodon* from the Smoky Hill Chalk (wingspan = 5.6 m [18 ft.]).

Pterosaurs were flying reptiles. They were superbly adapted for flight, with hollow, air-filled bones, a relatively large, birdlike brain (Seeley, 1871; Wellnhofer, 1991), and membranous wings that were supported by the elongated fourth finger of each hand (Fig. 10.1). Their upper bodies were stiffened by rigidly binding the fused dorsal vertebrae, ribs, and sternum together into a solid structure that supported the large muscles needed to power their wings. Some pterosaurs were covered with fur and it is likely that they were "warm-blooded" to some extent. They had three clawed fingers on each hand and four clawed toes on each foot. The smallest (*Pterodactylus*) was about the size of an American robin, and the largest (*Quetzalcoatlus*) had a wingspread as large (11–12 m [36–39 ft.]) as a light airplane. The largest species found in Kansas had a wingspread of about 8 m (26 ft.) but would probably have weighed no more than 11 kg (25 lb.). Compared to their heads and their wings, their bodies are almost absurdly small. One comparison provided by Hankin and Watson (1914) was that "with a body little larger than that of a cat, they had a span of wing asserted in some cases to have reached 21 feet or more!"

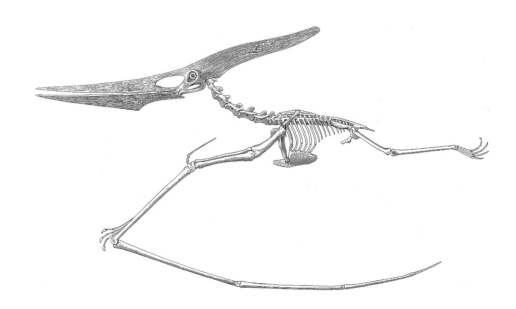

Figure 10.1. A drawing of the skeleton of a male Pteranodon longiceps from Eaton (1910, pl. XXI). Note the size of the skull compared to the rest of the body.

Although pterosaurs (wing-lizards) had been known from Europe since the early 1800s, they were not discovered in North America prior to 1870. The European specimens from the Jurassic period were generally small with long tails and had jaws filled with sharp teeth. They looked much like a reptilian version of a bat and were often depicted as something similar. The whole concept of these flying lizards was about to be radically changed by a chance discovery in western Kansas.

During the summer and fall of 1870, O. C. Marsh and his Yale scientific expedition collected fossils from as far west as Wyoming and Utah. Late in November, they made a brief visit to western Kansas in the vicinity of Fort Wallace. The weather was cold but they were still able to collect for several days along the Smoky Hill River in what are now Wallace and Logan Counties. Charles W. Betts (1871), one of Marsh's students from Yale, wrote a detailed and well-illustrated popular account of the expedition's adventures. Their time in Kansas, however, was only a small part of their journey and was allocated only a half page of the eight-page essay. Betts (ibid., p. 671) did indicate that they spent four days digging out the nearly complete skeleton of a "sea-serpent . . . allied with the genus *Mosasaurus.*" The story was published in *Harper's New Monthly Magazine* in October of the following year.

In a short note published after returning to Yale in mid-December, Marsh (1871a) noted only that "some interesting reptilian and fish remains" were collected during their two-week stay in Kansas. A few months later, Marsh (1871b) published a more detailed account of the mosasaurs found by the 1870 expedition in Kansas, including the nearly complete *Clidastes* specimen mentioned by Betts (see Chapter 9).

While this has been a rather long introduction to the discovery of the first *Pteranodon* in Kansas, I think it is important to understand that it wasn't readily apparent or expected, even by Marsh. At the time, larger, more complete specimens of strange fish and marine reptiles were being collected, and mosasaurs were among the most popular of the "new" discoveries. In the first mention of the *Pteranodon,* Marsh (1871c) reported that the distal ends of two long bones (two metacarpals from the wings of two individuals, YPM 1160 and 1161) had been found, and he noted that they were not unlike those figured by Richard Owen in 1851. He (ibid., p. 472) also noted that the bones were thin-walled and hollow. From these few fragments, he named the new species *Pterodactylus Oweni,* "in honor of Professor Richard Owen of London." Marsh also estimated the size of the creature from the fragments and noted that the outstretched wings would have measured "not less than twenty feet!"

Surprisingly, and as a complete fabrication since no skull material was reported to be present, Marsh (ibid.) also indicated that "the teeth are smooth, and compressed." In his much more complete description of additional remains of this and two other species collected by the 1871 expedition, Marsh (1872, p. 244) again noted the presence of teeth in *Pteranodon:* "The teeth found with remains of this species, and supposed to belong to them, are very similar to the teeth of Pterodactyls from the Cretaceous of England. They are smooth, compressed, elliptical in transverse outline, pointed at the apex and somewhat curved." Of the teeth of a second species (*Pterodactylus ingens*), Marsh wrote (ibid., p. 247) that "the dental characters of this species are at present only known from a single crown of a tooth; found with one series of the specimens and from two larger and very perfect teeth found by themselves, which agree so closely with the former that they deserve notice in this connection. These specimens are less curved and less compressed than the teeth referred to *Pt. occidentalis,* but in other respects they are nearly identical." Apparently Marsh assumed that the American "Pterodactyle" would have teeth like its European cousins and was hedging his bet on their discovery when the first skull was found. In that regard, he was not the only one who would be playing "fast and loose" with this conclusion. His rival E. D. Cope (1872a, p. 337) also indicated, more conservatively, that *Pteranodon* skulls "were slender and the teeth indicated carnivorous habits." As shown in an illustration (See Fig. 13.1) in a book called *Buffalo Land* (Webb, 1872), it is clear that Cope believed that at least one of the species he had named (*Ornithochirus umbrosus*) had teeth.

In the summer of 1871, Marsh and the second Yale scientific expedition returned to Kansas. Marsh was able to return to the spot where he had found the first remains of a *Pteranodon,* and he located additional pieces of the same bone. By the time Marsh (1872) published more complete descriptions, however, the giant *Pteranodons* were already taking a back seat to the recent discovery of toothed birds in western Kansas (Chapter 11). Marsh (ibid., p. 241) noted that the name given to the first specimens (*Pterodactylus Oweni*) was preoccupied by a specimen described by Seeley and replaced it with the name *Pterodactylus occidentalis*. Besides this species, Marsh (ibid., pp. 246–247) collected several specimens of "the most gigantic of Pterosaurs," which he called *P. ingens* (YPM 1160 and 1172) and estimated had a wingspread "of nearly 22 feet!" Marsh also named another, smaller species (*P. velox;* YPM 1176) found in 1871 on the basis of what he believed were differences he saw in the wing bones. Bennett (1994) examined the Marsh collection at the Yale Peabody Museum and indicated that because of the stratigraphic level in which the pterosaur remains collected by Marsh in 1871 and 1872 occurred, they were all probably *Pteranodon longiceps*. In that regard, Bennett (ibid., p. 14) considered all the early names given by Marsh and Cope (below) to be *nomina dubia* because the material does not exhibit any species-specific characters and was too fragmentary to be accurately identified beyond the genus *Pteranodon*.

Cope apparently found at least two sets of *Pteranodon* remains during his trip to Kansas in late 1871. In a short note, Cope (1872a, p. 337) named two species, *Ornithochirus umbrosus* (AMNH 1571) and *O. harpyia* (AMNH 1572), that were apparently distinguished from one another only on the basis of size. In doing so, he accepted Seeley's name for the genus and apparently ignored Marsh's *Pterodactylus* without further comment. In a narrative that preceded the listing of the two new species, Cope (ibid., p. 323) described them in their natural habitat: "The flying saurians are pretty well known from the descriptions of European authors. Our Mesozoic periods had been thought to have lacked these singular forms until Professor Marsh and the writer discovered remains of species in the Kansas chalk. Though these are not numerous, their size was formidable. One of them, *Ornithochirus harpyia* Cope, spread eighteen feet between the tips of its wings, while the *O. umbrosus,* Cope, covered nearly twenty-five feet with his expanse. These strange creatures flapped their leathery wings over the waves, and often plunging, seized many an unsuspecting fish; or, soaring, at a safe distance, viewed the sports and combats of the more powerful saurians of the sea. At night-fall, we may imagine them trooping to the shore, and suspending themselves to the cliffs by the claw-bearing fingers of their wing-limbs." While this image of pteranodons hanging from rocks along the seashore has been shown in numerous recreations over the years, it is probably just a fantasy. As noted by Stein (1975), it would have been impossible for *Pteranodon* to land on all fours on level ground without collapsing the wings first and thus losing lift. Performing such a landing against the vertical wall of a cliff would seem to be a death-defying act. Furthermore, none of the cores of the wing claws I have collected or examined show any damage on the tips as might be expected from a daily routine of hanging from one's fingertips.

In a follow-up paper read before the American Philosophical

Society, Cope (1872b) provided a more complete description of the specimens (both consisting of the fragments of wing bones). Cope (ibid., pp. 420–421) noted that some of the phalanges (finger bones) bore large claws that were similar to Seeley's genus *Ornithochirus*. He also stated that "as it is not likely on other grounds that the species of the Niobrara cretaceous strata belong to the genus *Pterodactylus* of Cuvier, which is chiefly known from the Jurassic period, I place the Kansas species for the present in *Ornithochirus*, as established by Seeley." In addition, Cope suggested that "this species is the largest Pterodactyle as yet known from our continent, the end of the wing metacarpal exceeding the diameter of that of the species described by Professor Marsh."

Clearly the challenge had been given, and a response was not long in coming. In a report regarding Cope's paper published in the American Journal of Science, an "anonymous reviewer" (1872, pp. 374–375) noted that "*Ornithochirus umbrosus* Cope = *Pterodactylus ingens* Marsh, and *Ornithochirus harpyia* Cope = *Pterodactylus occidentalis* Marsh. As separate copies of Prof. Marsh's article were distributed March 7th, while the paper of Prof. Cope was not issued before March 12th, the names given by Prof. Marsh have priority." Cope had lost the race by five days. Not one to give up easily, Cope (1875, p. 67) noted that the species had been first "described by the writer in 1872 under the name of *P. harpyia;* but a fire occurring in the establishment printing the paper, its publication was delayed until two days after Professor Marsh had republished his species as *P. occidentalis.*"

Cope (1874, p. 26) included a list of Pterosauria from the Niobrara Cretaceous and conceded *Pterodactylus ingens*, *P. occidentalis*, and *P. velox* to Marsh. However, he kept *Pterodactylus umbrosus* Cope and placed it at the top of the list. Keeping things a bit confused, Cope (1875, p. 65 and pl. VII) refers to the winged reptiles as Ornithosauria (Seeley), while conceding them to the Pterosauria only on page 249. Cope (ibid., pl. VII) was the first, however, to illustrate *Pteranodon* wing bones, in this case both the AMNH 1571 and 1572 specimens.

In 1874, Marsh had the definite advantage of having B. F. Mudge, Harry Brous, and S. W. Williston collecting for him in Kansas. Cope, on the other hand, had hired a relatively young and inexperienced Charles H. Sternberg to work for him in 1875 (Sternberg, 1909, pp. 32–34). Sternberg was twenty-five years old that year. While Sternberg was still learning about collecting fossils in general, Marsh's better-organized, more experienced field crew was collecting birds and pteranodons. These fossil-collecting crews watched each other's activities closely. In 1876, Williston wrote (Shor, 1971, p. 77), "I do not think they will have any success in pterodactyls or small birds. One or two excellent ones they had struck carelessly with a pick and abandoned." Later he (ibid., p. 78) noted that Sternberg was "down on the Smoky. . . . He doesn't know what a pterodactyl looks like & hardly what a saurian is. He has directions from Cope to collect all vertebrates—and we will take pains to leave him plenty of fishes." Williston was correct. Sternberg would collect many fishes, but the outcome was probably not quite what he or Marsh expected. Cope would eventually name most of the fish that have been described from the Smoky Hill Chalk. Marsh would name none.

In the case of the American pterodactyls, however, Marsh had the final words in on the subject between the two men. Apparently believing that American species were different enough to deserve their own order, Marsh (1876a, p. 507) created a new order (Pteranodontia) based on "the absence of teeth," and a new genus (*Pteranodon*). He also described a new species, *Pteranodon longiceps,* which was the first to include a complete skull (YPM 1177). The specimen had been collected by S. W. Williston in May 1876. Marsh noted that "there are no teeth, or sockets for teeth, in any part of the upper jaws, and the premaxillary shows some indications of having been encased in a horny covering. The lower jaws, also, are long and pointed in front, and entirely edentulous [toothless]." It should be recognized here that H. G. Seeley had reached the same conclusion about a new genus of flying reptiles discovered in the Cambridge Greensand several years earlier. Seeley (1871, p. 35) indicated that "they have the ordinary dagger-shaped snout, but appear to be entirely destitute of teeth. I provisionally name the genus *Ornithostoma.*" Seeley originated the term "ornithosaur" which means "bird-reptile."

In a later note entitled "Principal Characters of American Pterodactyls," Marsh (1876b) reported that the remains from the Upper Cretaceous of Kansas "differ widely from the Pterodactyls of the old world, especially in the *absence of teeth* [italics by Marsh]." It had been six years since the first discovery of a *Pteranodon* and Marsh had finally understood that the fish teeth that had been collected with the earlier specimens did not belong to the specimens. Bennett (1994, p. 4) reported that the remains found in 1870 "consisted of the distal ends of two right metacarpals and a tooth of a *Xiphactinus* [a large fish] that Marsh believed came from the pterosaur." When I asked Chris Bennett (pers. comm., 2004) for more information, he provided the following comment: "I found the actual teeth referred to by Marsh in the YPM collections. They were not kept with the *Pteranodon* specimens, but rather were separated out after it was realized that they did not belong with the *Pteranodon* specimens. However, they still had the appropriate YPM numbers on them. The teeth are teeth of *Xiphactinus* and perhaps *Ichthyodectes* as well, and so they were hidden away in the YPM fish collection." There are three teeth currently catalogued in the Yale Peabody collection that were collected by Marsh in 1870 and later identified as *Xiphactinus audax* (two teeth as YPM 1163A and one as YPM 1171A).

Two other new and noticeably smaller *Pteranodons* were briefly mentioned for the first time and named by Marsh (1876a). One of those, "*Pteranodon*" *gracilis* Marsh, actually represented a new genus (*Nyctosaurus*), which would be recognized as such in a subsequent paper. The second new species, *Pteranodon comptus* (YPM 2335), was found by Mudge in 1875, but is considered to be a *nomen dubium* by Bennett (1994, p. 48). It was Eaton (1910, p. 3) who first noted that Marsh had misidentified the lower leg bones (tibiae) of a *Nyctosaurus* as wing bones of the new species.

Marsh (1876b, p. 480) renamed "*Pteranodon*" *gracilis* as *Nyctosaurus gracilis* (*Nyctosaurus* means "night-lizard"). Although he still considered it to be closely related to *Pteranodon,* he noted (ibid.) that it could be distinguished from the larger species because the coracoid was not "coossified with the scapula." In the larger and heavier *Pteranodon,* the coracoid and the scapula are fused together into a single, U-shaped element. The ends of this bone are connected to the sternum and the fused dorsal vertebrae (notarium) and form a strong support for the attachment of the wing. Marsh believed *Nyctosaurus* would have had a wingspread of about eight to ten feet. The humerus of *Nyctosaurus* also has a distinctive "hatchet" shape that is readily distinguishable from that of *Pteranodon.* Adding a bit of confusion to the name, Marsh (1881, p. 343) noted later that "the name *Nyctosaurus* . . . appears to have been preoccupied, and hence may be replaced by *Nyctodactylus.*" However, Williston (1903, p. 125) stated that "the name has never been used otherwise for a genus of animals. Doubtless he [Marsh] thought the term conflicted with Nyctisauria, used for a group of sauria. It does not, however, according to the accepted canons of nomenclature, and the original name should not be displaced."

The collection of *Pteranodon* specimens from the Smoky Hill Chalk continued at a steady pace during the 1870s and 1880s. Marsh (1884, p. 423) noted that "the remains of more than six hundred individuals of these reptiles have been secured . . . and now are in the museum of Yale College." Marsh (ibid.) further described the skull of a specimen of *Pteranodon longiceps* that he first reported in 1876 (YPM 1177). In addition, he figured the skull (ibid., pl. 15) and noted that it was about 760 mm (30 in.) long from the tip of the upper jaw to the back of the sagittal crest (Fig. 10.2). The illustration was the first ever to show four views of a three-dimensional reconstruction of a *Pteranodon* skull and was quite well done, although Williston (1892) would question the length of the crest.

In Marsh's last paper on *Pteranodon* (1884, p. 425), he indicated that "there was apparently no ring of bony sclerotic plates, since in the best preserved specimens no traces of this have been found." Several years later, Williston (1891, p. 1124; 1892, p. 4) reported that the remains of a sclerotic ring had been discovered in a new specimen of *Pteranodon longiceps* (KUVP 2212). In two brief papers written largely in defense of Marsh's uncompleted work on *Pteranodon,* Eaton (1903; 1904) attacked Williston's conclusions repeatedly, sometimes without justification. In one instance, Eaton (1904, p. 318) erred in citing a reference to the sclerotic ring of *Nyctosaurus* (Williston, 1902a, p. 528) instead of Williston's decade-earlier original comments (Williston, 1891, p. 4)

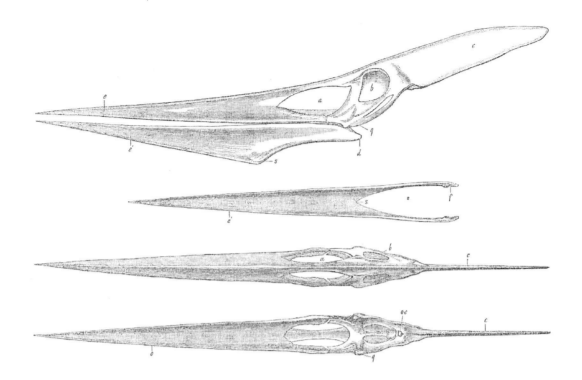

on *P. longiceps,* stating that "oddly enough, in this revision, it is *Pteranodon* that is credited with a sclerotic circle, and not *Nyctosaurus* in which Williston observed the structure." Eaton apparently missed the point that Williston had found the sclerotic ring in both *Pteranodon* and *Nyctosaurus.*

After Marsh, Williston was the next to concentrate on the study of *Pteranodons* of the Smoky Hill Chalk. He had worked for Marsh for eleven years, from 1874 to 1885 (Shor, 1971), and had himself collected many of the Yale Peabody specimens from Kansas. However, Williston and others were not allowed to publish papers on them so long as they were working for Marsh. Soon after he joined the faculty at the University of Kansas in 1890, he began collecting specimens from the chalk, including *Pteranodon,* for the KU Museum of Natural History. A year or so later, Williston (1891, p. 1126) provided probably the first complete description of a *Pteranodon:* "About five or six species are known, varying in size, when alive, from four feet to not over twenty feet in expanse of wing.[4] The head (in all the larger species, at least) was elongate and slender, with a well-developed occipital crest, and without teeth. The jaws may have been encased in horn, but I have never seen any evidence whatever that such was the case. . . . The body was short, the pelvis of moderate size, the hind legs comparatively small, with great freedom of movement, the tail short, and the feet with out much, if any, prehensile power. Their food probably consisted of fishes.[5]" In the footnotes, Williston added (ibid., note 4) that although the wingspread of *Pteranodon* had earlier been re-

Figure 10.2. The first published drawing of the skull of Pteranodon longiceps (YPM 1177) by Marsh (1884, pl. 15). Top to bottom: left lateral view of the complete skull; dorsal view of the lower jaw; dorsal view of skull, and; ventral view of skull, lower jaw removed.

ported as being larger, he had measured the largest specimen and found the distance could not have exceeded twenty feet. He also indicated (ibid., note 5) that "several coprolites found within the above described pelvis, ellipsoidal in shape, and about the size of an almond, showed bones so finely comminuted that their precise character could not be made out." Barnum Brown (1943, p. 106) mentioned a specimen at the American Museum of Natural History that included the "backbones of two species of fish and the joint of a crustacean, lying in the position of the throat pouch when death overtook the animal." The nearly complete lower jaws (AMNH 5098) were found by Charles H. Sternberg in 1877 in Lane County (Mehling, pers. comm., 2004). A small mass (3 cm) of fish vertebrae located behind the posterior margin of the symphysis of the jaw was interpreted by Bennett (pers. comm., 2004) as a bolus of food regurgitated at the time of death. Carpenter (1996, pp. 44–46) reported that "fish bones were found in two small coprolites associated with a *Pteranodon* sp. skeleton" from the Pierre Shale.

Williston (1892) redescribed the skull of *Pteranodon longiceps* from a new, more complete specimen (KUVP 2212) that had been discovered by E. G. Case during the previous field season. He also criticized Marsh's reconstruction of a specimen he himself had collected in 1876 for Marsh (*P. longiceps,* YPM 1177; above). Williston (ibid., p. 1124) noted that the type specimen was essentially complete when collected, except for the crest. Then he noted that Marsh had restored the crest from "indications presented by the basal portion, but without indicating that such a conjectural restoration had been made. The result is unfortunate." According to Williston, the new specimen indicated that the crest was half as long as Marsh's reconstruction, and with a different shape. The specimen cited by Williston (KUVP 2212) is still on display as a part of a composite specimen in the University of Kansas Museum of Natural History. When Marsh's figure (1884, pl. XV) is compared to Williston's (1892, pl. I), it is readily apparent that the reconstructed crest is much larger in comparison to the skull (Fig. 10.3). As measured from the back of the orbit, the crest on the YPM 1177 skull repre-

Figure 10.3. The skull of Pteranodon longiceps *(KUVP 2212) from Williston (1892, pl. I), with a much shorter crest than suggested by Marsh's 1884 reconstruction.*

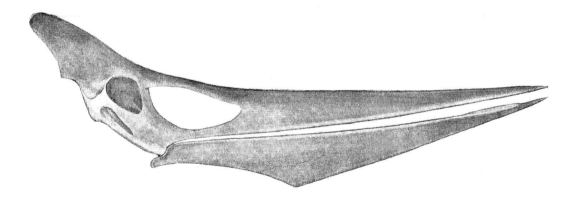

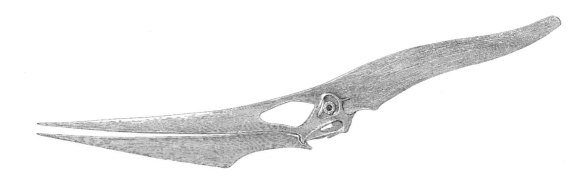

sents about 32 percent of the total length compared to 26 percent of the slightly larger KUVP 2212 specimen.

However, the argument did not end there. Eaton (1903, pl. VI) reported and figured the crest of another specimen (YPM 2473, collected by Brous and Williston in 1876) that was being prepared at the Yale Peabody Museum. When the outline of the YPM 2473 crest was overlaid on Marsh's drawing of the YPM 1177 skull, the new crest was substantially longer (43 percent of the skull length) than what Marsh had reconstructed (Fig. 10.4). However, all three men apparently failed to consider the differences in the size and age of the animal, or sexual dimorphism, as reasons why the crests might be of different sizes. In Eaton's case, he probably had a few other things on his mind at the time. The new Yale Peabody restoration of *Pteranodon longiceps* was being prepared under Eaton's (1904, p. 318) direction "as the contribution of the Department of Vertebrate Paleontology of the Yale Museum to the University's exhibit at the Louisiana Purchase Exposition" at the 1904 World's Fair in St. Louis, Missouri. That same plaster reconstruction of a *Pteranodon* with a ten- to twelve-foot wingspan is currently on display in the Great Hall of the Yale Peabody Museum (C. Bennett, pers. comm., 2004).

Returning to Williston's work in the early 1890s, it is apparent that Williston had trouble initially with the toothless character of *Pteranodon*. His 1892 paper was devoted mostly to the description of a new specimen of *Nyctodactylus (Nyctosaurus) gracilis* Marsh that was essentially complete except for the skull. The specimen had been collected during the previous field season near Monument Rocks (Williston, 1892, p. 6). Williston noted (ibid., p. 11) that "not a single character has been given to distinguish this genus from *Pterodactylus*, and it is not at all impossible that it may prove to be the same; its location in *Pteranodon* rests solely on the assumed absence of teeth, and that is a character that is yet wholly unknown. . . . It seems very probable that the genus *Nyctodactylus* has no teeth in the jaws; it agrees in *every other respect* with the genus *Pterodactylus*, so far as is known. If the genus has teeth it must be united with *Pterodactylus*." Until the skull of *Nyctosaurus*

Figure 10.4. Composite drawing of the skull of Pteranodon longiceps *(YPM 1177) overlaid with the crest of a partial skull (YPM 2473) in Eaton (1910, pl. IV). This crest is substantially longer than the one reconstructed by Marsh in 1884. Note the presence of the scleral ring around the eye in this figure.*

was found, however, Williston was unable to distinguish its remains from those of *Pterodactylus*.

In his conclusion, Williston (ibid., p. 12) suggested that *Pteranodons* would also be found "in Europe, and if so, it is probable that the name *Pteranodon* must eventually be given up." In this case, he was referring to toothless remains of pterodactyls that had been described by Seeley in 1871 and named *Ornithostoma*. This was the same genus to which Cope had originally referred his discoveries from Kansas. Williston, however, had not seen Seeley's material and was careful not to make a decision at that time. By the following year, however, Williston (1893, p. 79) had received copies of Seeley's papers and was convinced that "there can no longer be any reasonable doubt of the congenerousness [membership in the same genus] of our species with those included in the genus *Ornithostoma*." He also noted, and credited Seeley for recognizing, that it was Cope "whose acumen led him to refer his fragmentary material to the genus *Ornithocheirus;* an acumen all the more noteworthy in contrast with the total incomprehension of their affinities displayed by Marsh, notwithstanding his wealth of material upon which to base an opinion." The comparison here is that Marsh had hundreds of specimens to work with while Cope had only a few. It seems likely that Williston was expressing some anger regarding the scientific and management methods of his former employer. In a later note, Williston (1897, p. 35) quoted Seeley (1871) as saying, "There is, so far as I can discern, no evidence of generic difference between *Ornithostoma* and *Pteranodon*."

Williston wrote a number of papers on *Pteranodon* between 1891 and 1904. He described the lower jaw (Williston, 1895) from a specimen found by H. T. Martin. Previously, the remains of the skull and lower jaw had only been found crushed from side to side. In this specimen, the lower jaw was preserved more or less upright, allowing the true shape and other features to be seen for the first time. Williston (1896) described a new skull of *Ornithostoma* (*Pteranodon*) found by C. H. Sternberg and continued to criticize the faults in Marsh's reconstructions. The following year, Williston (1897) published a "Restoration of *Ornithostoma* (*Pteranodon*)" and again indicated his belief that the American pterodactyls probably should be included in the European genus since there were no discernible differences between the two groups.

Following the discovery of an unusually complete specimen of *Nyctodactylus* (*Nyctosaurus*) by H. T. Martin, Williston (1902a) provided a further description of this smaller pterosaur. The general measurements he gives for the specimen (FMNH 25026) are of interest because they provide a more accurate look at these strange animals. Williston (ibid., p. 300) noted that "while the wings gave a spread of very nearly eight feet, the body proper was less than four inches in diameter and not more than six inches in length, exclusive of the small tail. . . . One wonders where sufficient surface was presented for attachment of the strong muscles necessary for control of the wings. When it is remembered, however, that even the largest bones of the skeleton had walls less than a millimeter in thickness, and that many of the smaller ones were almost like cylinders of writing paper, he will perceive that, notwithstanding the extraordinary development of the anterior extremities and head, the creature, when alive, must have weighed but little. I very much doubt whether the living animal attained a weight of five pounds." In conclusion, Williston (ibid., p. 305) stated again, "I still believe that the genus *Pteranodon* is identical to *Ornithostoma,* and that the former term must be abandoned." In making a point that will be discussed later in this chapter, Williston (ibid.) noted at the end of the paper that when the skull of *Nyctodactylus* was fully prepared, it had no occipital crest. This statement is of current interest in a more modern controversy regarding the crest of *Nyctosaurus.*

A follow-up paper on the same specimen (Williston, 1902b) provided a detailed description of the skull of *Nyctodactylus* and included one of the first known photographs of a pterosaur skull. The picture (ibid., pl. I) shows a ventral view of the posterior of the skull. Two drawings (ibid., pl. II), showing the first dorsal and ventral views of the skull of *Nyctodactylus,* are also included. Williston (1902c) wrote "Winged Reptiles" for *Popular Science Monthly* magazine in which he discussed the *Pteranodon* in more general terms. It is interesting to note at this point that he had largely given up on the idea of changing the genus back to *Ornithostoma,* although he still includes the older term in parentheses.

As mentioned earlier, George Eaton (1903) wrote a short note on a newly discovered crest of *Pteranodon longiceps* in which he criticized Williston's work. Eaton was a graduate student at Yale at the time and was in the process of completing his doctorate. Eaton's critical comments did not go unnoticed by Williston, who replied to them in his 1903 paper on the osteology of *Nyctosaurus* and American pterosaurs. In 1902, Williston accepted a new job as professor of paleontology at the University of Chicago and had been able to secure a complete *Nyctosaurus* specimen (FMNH 25026) that had been found by H. T. Martin for the Field Museum. Williston (1903, p. 162) also noted, with some venom in his comments, that "had Mr. Eaton done me the honor to have read more attentively the article which he quotes; or had he examined the extended article on the skull of *Nyctodactylus* with plates, published in the Journal of Geology for August, 1902; or even had he examined the figure of the skull in Zittel's text-book of Paleontology, published last autumn, of all which he seems strangely ignorant, he would have learned that *Nyctodactylus has no crest whatsoever,* not even the vestige of one." Ouch! Eaton (1904) followed up with a second article in which he again improperly criticized Williston on the sclerotic ring in *Pteranodon* (see above) and in the count of vertebrae in the neck. So far as I am aware, Williston did not reply in publication to these criticisms.

Williston's 1903 paper also included considerably more detail in the descriptions of other specimens. One item that was briefly mentioned (ibid., p. 160) was the discovery of the upper end of a femur from the Kiowa Shale (Early Cretaceous) of Clark County that Williston believed was from a pterosaur. He named it *Apatomerus mirus* (KUVP 1198). Years earlier, the same bone fragment had been figured by Williston (1894, pl. I; 1898, fig. 3) when he believed it might be part of a crocodile. If it was in fact a pterosaur femur, then it would have been an early record for a North American species that would have been substantially larger than *Pteranodon*. Brown (1943, p. 108) accepted Williston's later identification of the bone as a pterosaur but did not indicate that he had actually examined the material. Schultze et al. (1985, p. 58), however, disagreed that it belonged to a pterosaur, but he was unable to identify it. Bennett (pers. comm., 2004) also indicated that he did not believe it was from a pterosaur. In May 2004, when I examined the KUVP 1198 specimen with Larry Martin, it was apparent that the limb bone was much too heavily constructed to be from a *Pteranodon*. In July 2004, I located another specimen (KUVP 16216) of the upper end of a "femur" in the KU collection that was similar to the fragment described by Williston. The second specimen was collected in 1969 by Orville Bonner in Clark County, Kansas, in association with two plesiosaur vertebrae. The shaft of both specimens is hollow but the bone is much too thick for a pterosaur limb. Both specimens were collected from the Kiowa Shale, and I have concluded that both represent the weathered upper end of a plesiosaur limb (propodial). The hollow shaft is apparently an artifact of preservation, similar to a number of hollowed plesiosaur vertebral centra that Larry Martin pointed out to me.

Williston (1904) published a short note on "The Fingers of Pterodactyls" in which he discussed the clawed fingers and elongated wing finger of *Pteranodon*. In his final paper on pterosaurs, Williston (1911) described his restoration of *Nyctosaurus*, including a discussion of the hand and wing finger as they relate to their reptilian origins. Williston makes the point that the wing finger is the fourth finger in the hand (ring finger) and not the fifth or little finger as is sometimes assumed.

In 1910, Eaton published his doctoral dissertation on the "Osteology of *Pteranodon*." Besides updating the description of *Pteranodon*, the paper provides excellent photographs and line drawings of the catalogued specimens in the Yale Peabody Museum collection. In the text, Eaton is not as apologetic for Marsh's mistakes as he was in his 1903 and 1904 criticisms of Williston's work. He wrote pointedly (ibid., p. 2) that "to the critical reviewer of Professor Marsh's early work on this subject, it is evident that the two skulls referred by him to *P. ingens* [YPM 2594] and *P. occidentalis* were specifically identified by their size alone." Because he believed that many of the specimens were so incomplete as to have little value for describing the differences between species, Eaton limited his discussion to the skeleton of *Pteranodon*. In doing so, he quotes repeatedly from Williston's earlier work and credits him his descriptions. It appears that at least some of what Williston had to say in his 1903 rebuke had made a point with Eaton.

On one issue, however, Eaton followed Williston too closely. On the subject of whether or not *Pteranodon* had a fibula (the small bone paired with the tibia in the lower leg of most animals, including *Pteranodon*), Eaton (1910, p. 35) quotes from Williston (1903): "There is no trace of any fibula, in either of the preserved remains, or of any tibial articulation." Eaton (1910) then goes on to say, "This is also true of the *Pteranodon* material contained in the Marsh Collection." That same year, however, in a short article regarding a recently mounted specimen of *Platecarpus*, Williston (1910, p. 540) uses the opportunity to state "incidentally I may mention that in *Pteranodon* among pterodactyls the fibula is supposed to be absolutely wanting, yet in a specimen in our collection I find distinct remains of it fused with the tibia." Wiman (1920, pp. 9–10) noted the exchange between Eaton and Williston in his description of specimens purchased from C. H. Sternberg for the Palaeontological Museum in Uppsala, Sweden. One of the specimens he figured (ibid., fig. 1a) was a *Pteranodon* tibia that included the remnant of the fibula. Bennett (2001; pers. comm., 2004) indicated that "the fibula is readily apparent in well preserved specimens" of *Pteranodon*.

In a brief review published in the *Journal of Geology*, Williston (1912, p. 288) took the opportunity to comment on Eaton's work, noting that "the writer [Williston], whose acquaintance with vertebrate paleontology began with the collection of a specimen of *Pteranodon*, takes especial pleasure in the expression of his appreciation for the present memoir by Dr. Eaton. The rich material of this genus in the Yale collections is unsurpassed, and it has been well utilized in the present paper, with its large number of excellent illustrations." In noting that the occipital crest published by Marsh was well documented by the photograph of the specimen itself, Williston admits his 1892 criticism of the reconstruction was probably unjustified. Ending on a note suggesting that more could have been done, Williston (ibid.) wrote that "one could wish that Dr. Eaton had entered more fully into some of the disputed points about the relationships and characters of the genus, but the omissions are immaterial in comparison with what he has given."

In his review of the fossil vertebrates of Kansas, Lane (1946) had relatively little to say about the *Pteranodon* collection at the University of Kansas. This reflected the fact that few additional remains had been collected since Williston had left Kansas in 1902 and that no one had been actively working on *Pteranodon* since Eaton's major publication.

In 1952, however, a large and unusual *Pteranodon* skull was found by George F. Sternberg in Graham County. Instead of the usually long, tapering crest that projected behind the skull, the new specimen had a short, wide crest that sat more or less on top of the skull. The strange-looking skull (FHSM VP-339) was briefly described as a new species by Harksen (1966) and named *Pteranodon sternbergi* (Fig. 10.5). Although the upper and lower jaws were fragmentary and incomplete, Harksen's (ibid., p. 76) illustration of

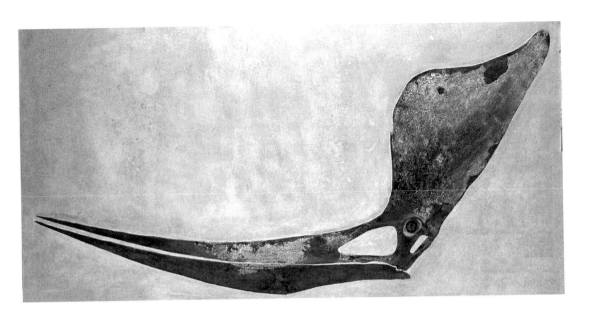

the skull was impressive. The length of the lower jaw (121 cm) was more than twice that of the type specimen of *P. longiceps* (58 cm) and suggested a truly huge flying reptile. Bennett (1994, p. 33), however, suggested that "the mandible was probably not as long as reconstructed."

Following his discovery of the type specimen of *Pteranodon sternbergi*, Sternberg found a smaller, headless specimen in September 1956 near WaKeeney in Ellis County in the lower chalk (Sternberg and Walker, 1958). The remains (FHSM VP-184) included both wings with the wing fingers and claws, the pelvis and both legs and feet, along with the sternum and fragments of vertebrae. The wings and legs were still articulated. From the measurements given (ibid., pp. 84–85), the wingspread would have been about 3.8 m (12.5 ft.). This specimen is currently on display at the Sternberg Museum of Natural History. In 2003, a complete and very large (1.8 m [6 ft.]) skull of *Pteranodon sternbergi* was found in Ellis County by a private collector but little else is known about it.

In 1962, George Sternberg found a nearly complete specimen of *Nyctosaurus* (FHSM VP-2148) near Elkader in Logan County (Fig. 10.6). The specimen was small enough (skull length 30 cm) to be contained with within a single slab of chalk roughly 14.5 by 27.5 inches (Bonner, 1964, p. 11). The wingspread (ibid., table 6) was a little over 2.6 m (8 ft.). By comparing the measurements of it and other known specimens, Bonner recognized that the specimen was a new species and named it *Nyctosaurus sternbergi*. However, the species name was preoccupied and Miller (1971a, p. 13) changed the name to *N. bonneri* in honor of Bonner's work.

Miller (1971a) reviewed the literature regarding *Pteranodon*, provided brief descriptions of major specimens, including those in

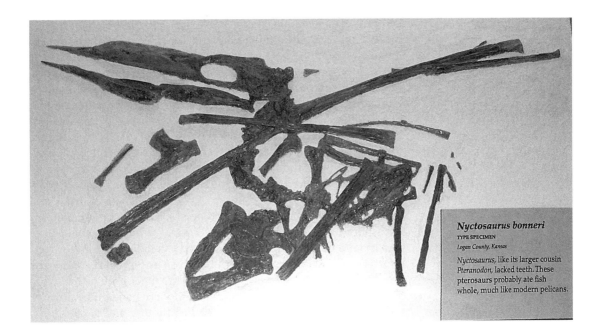

Nyctosaurus bonneri
TYPE SPECIMEN
Logan County, Kansas

Nyctosaurus, like its larger cousin
Pteranodon, lacked teeth. These
pterosaurs probably ate fish
whole, much like modern pelicans.

the collection of the Sternberg Museum, and suggested several changes in the systematics of the genus. Miller's proposed revision was faulty, however, and has been disregarded by subsequent workers (Bennett, 1994, p. 8). In the same journal, Miller (1971b) also described a new specimen of *Pteranodon longiceps* (FHSM VP-2183) that had been also had been found near Elkader in Logan County by George Sternberg. Unlike many specimens, it included a skull and post-cranial material (1971b, p. 21).

In the first hundred years after the discovery of *Pteranodon,* little work had been done to understand how such strange-looking animals were able to fly. Early reconstructions of pteranodons that showed their wing membranes attached to their legs like bats (Williston, 1902c; Eaton, 1910) were probably influenced by the drawings made of Jurassic pterodactyls by earlier European workers, including Seeley (1901). Padian (1983) notes that there "is no evidence that the hindlimb was attached to the wing in any pterosaur." Essentially, this means that pteranodon wings were long and narrow, like those of modern marine soaring species such as albatrosses and frigate birds, and this factor is important in determining how they flew.

At a time when humans were just beginning to learn about heavier-than-air flight, Hankin and Watson (1914, p. 11) provided the first aerodynamic analysis of pterosaur flying capabilities, suggesting that it was "highly probable that their habitual mode of flight was soaring rather than flapping." The dynamics of flight in *Pteranodon ingens* (*longiceps*) were re-examined by Stein (1975) by conducting wind tunnel tests using models of pterosaur wings. Stein concluded (ibid., p. 547) that *Pteranodon* was a highly

Figure 10.6. The type specimen of Nyctosaurus bonneri (FHSM VP-2148). Originally described as new species ("N. sternbergi") by Orville Bonner in 1964, it was renamed by Miller (1971a). Bennett (1994) considered N. bonneri to be a junior synonym of N. gracilis. The skull is 29 cm (11.5 in.) long.

evolved, maneuverable, and unstable flyer." In this case, "unstable" means that when the animal's flight is disturbed, it has to be able to make changes to restore itself to its course. This requires a highly evolved sensory and nervous system. In contrast, outside forces acting on the wings of a stable flyer will tend to restore stability automatically. The critical issue here is that "high stability and high maneuverability are mutually incompatible" (ibid.). The analysis of the data indicated that *Pteranodon* was a slow-speed, long-distance flyer and suggested (ibid., p. 541) that 4.5 m/s (about 10 miles/hr.) would have been about the minimum flight speed of a large (15 kg [33 lb.]) *Pteranodon*. Stein also discussed the means by which *Pteranodon* would have been able to take off, maneuver, feed, and land, all of which he considered to be more important adaptations than pure flying. He also concluded that *Pteranodon* would have had to land on its back feet (as opposed to landing on all four feet) because in order to get the front feet on the ground, the wings would have to be collapsed and would no longer provide lift.

Using the flight characteristics of a modern hang-glider, Brower (1983) modeled the performance of *Pteranodon* and *Nyctosaurus* and noted that both were probably limited to slow horizontal and vertical speeds. While both were accomplished at long-duration soaring, Brower (ibid.) suggested that unlike *Nyctosaurus*, *Pteranodon* was probably not capable of continuous flapping flight. This certainly raises the question of why so many *Pteranodon* specimens have been found hundreds of miles from the nearest shoreline. It is certainly possible that the remains we find were those animals lost from long-distance migrations, but judging from the relative abundance of the remains we find, it seems more likely that they maintained a relatively constant presence in the Western Interior Sea. To do so, in my opinion, means they had to have been strong flyers.

One feature that distinguishes *Pteranodon* and *Nyctosaurus* from other pterosaurs is that they are considered to be tailless. Eaton (1910, p. 23) noted that "the number of caudal vertebrae in *Pteranodon* is not definitely known, but the opinion of Professor Marsh and Professor Williston expressed in various papers, that the tail was short, is well supported by the material preserved in the Marsh collection." However, a study of three specimens in the Marsh collection in the Yale Peabody Museum (YPM 2489, 2546, and 2462) and another in the American Museum of Natural History (AMNH 6158) by Bennett (1987) indicated that *Pteranodon* had a somewhat longer tail that was complete in form and may have functioned to support a rearward extension of the wing membrane. Bennett (ibid., p. 23) suggested that the caudal vertebrae and membrane would have formed a control surface which would have provided pitch control when moved up or down. The caudal vertebrae would have been at least 19 cm long on a *Pteranodon* with a 7.5 m wingspread (ibid.). Since that time, however, Bennett (pers. comm., 2004) has concluded the there was no connection between the wing membrane and the caudal vertebrae.

One of the problems in the study of pteranodons from the very start has been the determination of exactly where they were found and more specifically their stratigraphic occurrence. Locality information, especially on the early discoveries, is generally lacking or inaccurate. During the 1870s, western Kansas was largely unsettled and localities were often described as being "near the Smoky Hill River." Maps from the 1870s show an elongated Wallace County that included present-day Logan County. However, Logan County was not incorporated until 1885 (Bennett, 1990, p. 44) and specimens found before 1885 with localities recorded as Logan County are therefore suspect. Later specimens had better locality information, but their stratigraphic occurrence was often limited to "upper" or "lower" chalk. Hattin (1982) provided detailed descriptions of twenty-three "marker units" within the Smoky Hill Chalk and enabled workers to make accurate determinations of the stratigraphic occurrence of specimens for the first time. Stewart (1990) based his six biostratigraphic zones on Hattin's work and noted that *Pteranodon* and *Nyctosaurus* made their first appearances near the middle of the chalk (Santonian). In that regard, I have collected *Pteranodon* remains as low as Hattin's marker unit 4 (late Coniacian) in Stewart's (1990) zone of *Protosphyraena perniciosa*. The distal end of a *Pteranodon* femur reported from the basal Greenhorn Formation (Upper Cenomanian) of Russell County (Liggett et al., 1997) suggests that pterosaurs were at least occasional inhabitants of the Western Interior Sea before the deposition of the Niobrara Chalk.

Bennett (1990) inferred the stratigraphic occurrence of several specimens from various sources of historical and geographical information, noting that the type specimen of *Pteranodon sternbergi* (FHSM VP-339) came from the low chalk (near Hattin's marker unit 4). He (ibid.) also suggested that the type specimen of *Pteranodon longiceps* (YPM 1177) came from the upper chalk between marker units 15 and 16. The stratigraphic separation of these two species was an important discovery that was not readily evident until the right stratigraphic tools became available. Bennett (ibid., p. 44) noted that "the post-cranial skeleton [of *Pteranodon*] is of no taxonomic value at the species level." This means that, without certain portions of the skull, there is no way to determine the species of *Pteranodon* remains except by their stratigraphic occurrence.

After a study involving the measurement of more than 400 sets of *Pteranodon* remains from the Smoky Hill Chalk, Bennett (1992) discovered that there were two "size-classes" represented among the specimens, with the average large size-class measurements being about 50 percent larger than those of the average small size-class specimen. In addition, it was the larger size-class that was found with crests, while the pelvis of the smaller size-class was proportionally larger. Bennett (ibid.) interpreted these findings to indicate that individuals in the larger size-class were males while those in the smaller size-class were females. He also found that the number of specimens of females outnumbered those of males by about two to one.

A recent discovery (Wang and Zhou, 2004) in China of an Early Cretaceous (Aptian, about 121 mya) pterosaur embryo preserved inside an egg appears to have answered the question of whether or not they laid eggs. The 53 mm by 41 mm fossilized egg contained a nearly-ready-to-hatch embryo that would have had a wingspread of about 27 cm (11 in.). While the question of egg laying in the much larger *Pteranodon* is still open to discussion, it seems likely that they would have reproduced in a similar fashion.

In a discussion of possible functions of the crest, Bennett (1992, p. 430) concluded that it was most likely used as a display since the larger ones were only associated with the larger size-class. Regarding stratigraphic occurrence, while noting that *Pteranodon sternbergi* is found in the lower chalk and *P. longiceps* is found in the upper chalk, Bennett (ibid., 422) suggested that they are part of "a chronospecies forming parts of a single anagenetic lineage." In other words, it is likely that only a single species of *Pteranodon* is found throughout the chalk, rather than two, and that that species would have to be called *P. longiceps*. The shape of the crest and other small differences would be evolutionary changes that occurred gradually over the five-million-year deposition of the chalk. The taxonomy and systematics of *Pteranodon* were revised by Bennett who noted again (1994, p. 1) that "*Pteranodon sternbergi* seems to be ancestral to *P. longiceps*," and (ibid., p. 18) that "*N[yctosaurus] bonneri* is a junior synonym of *N. gracilis*."

At the Society of Vertebrate Paleontology annual meeting in Mexico City, Bennett (2000) reported the discovery of three specimens of *Nyctosaurus* from western Kansas that were significantly different from previous known remains. One of the new specimens had a wingspan of 4.5 m (14.8 ft.) and was thus the largest *Nyctosaurus* known. The other two specimens had been found near WaKeeney (Trego County) in the middle chalk (Santonian) and both had crests that were large and rather bizarre for such small individuals.

Since *Nyctosaurus* had been considered to be "crestless" almost since Williston's original description of the skull, this discovery obviously and abruptly changed the status quo of this genus. Bennett (2003) provided a complete description, measurements, and detailed photographs of the specimens, and he suggested that the hyper-attenuated crest occurred only in mature males. According to Bennett (ibid., p. 62), the tall, slender, and branched crest of the smaller but more complete specimen is roughly three times the length of the skull (71.7 cm vs. 24.5 cm). The skull of the larger specimen is about 1/3 longer (31.6 cm), but the crest is incomplete. Viewed another way, the crest is almost as long as each of the wings (43 percent of the total wingspan; ibid., p. 73). There is no evidence of a membrane between the branches of either crest (ibid.). It is important to note, however, that Bennett did not see any reason to indicate that these animals represented a new species of *Nyctosaurus*.

While Bennett (ibid.) noted that "the aerodynamic effect of the crest could have been rather minor [since] the base and rami of the crest are streamlined and would have caused only a small amount of drag, I find it difficult to believe that a small, winged reptile (wingspread 1.7 m) could fly hundreds of miles from the nearest land with what was essentially a long pole sticking out of its skull. However, I have no other explanation for these bizarre specimens.

Pteranodons were often portrayed in the old drawings of life in the oceans of Kansas as hanging by their wing claws from tall cliffs. This raises a couple of questions. For me, the first is, "Where are these cliffs?" I don't have a good answer for that because most of the shorelines, especially on the eastern side of the interior sea, were eroded away millions of years ago. One thing that can be said, however, is that the cliffs were never in western Kansas as was suggested by many of the early art works. While there were certainly cliffs (or large trees) someplace along the east and west coasts of the Western Interior Sea, the closest land of any sort that we can be reasonably sure about was probably in the southwestern corner of present-day Arkansas, more than 200 miles away from the chalk of western Kansas. The other coasts at the time were much farther away. Did they fly that great distance just to feed in the middle of the ocean? I don't think so. During the season when the young were born or hatched, at least one of the parents would have foraged close to where they nested since hungry, fast-growing young usually require frequent feedings. In addition, the shallower coastal areas, swamps, and estuaries provided a much richer bounty of prey in the small sizes that pteranodons probably fed upon. It is possible, however, that adult pteranodons were solitary animals and lived a lifestyle much like a modern albatross or frigate bird, flying for long periods and staying at sea most of their lives. We simply do not know at this point.

Second, wherever they nested, I don't believe they hung from anything as a general practice. Their wing claws may have been useful for climbing or clinging but it is likely that they nested near the coast in large, reasonably level rookeries like most modern sea birds. They apparently were able to walk fairly well (after folding up those huge wings, of course). Whether they were bipedal or quadrupedal is still the subject of some debate, although the analysis of pterosaur skeletal structure by Padian (1983) appears to indicate that they were primarily bipedal. There are pterosaur tracks along a shore in Jurassic rocks in Wyoming that preserve marks that are apparently from the front limb touching the substrate, but it is not clear that the front limbs were used necessarily to support the animal's walking. We are also relatively certain that some pterosaurs in other places and times, like *Pterodaustro*, fed while standing in shallow water, much like modern flamingos. While they may not have been fast runners, they were certainly able to walk around while on the ground.

The question of why we find their remains hundreds of miles from the nearest land is still unanswered. The story at the beginning of this chapter has a flock of pteranodons flying for hours to find a school of fish to feed upon because that seems to be the generally accepted idea. That being said, however, I really don't believe they were flying clear out to the middle of the Western Interior Sea just to feed and then fly ten or fifteen hours back to their nests. One direction or the other would have been against the wind and, as lightly as *Pteranodons* were built, I don't see them as being able to fly against a headwind for long distances. If they were feeding far out to sea, they would be using more energy getting back and forth than they gained from what they were able to catch. At this point, I believe that what we are seeing in the many fossil remains from the Smoky Hill Chalk are the *Pteranodons* that died while migrating across the seaway. These would be the sick, old, and infirm flyers that dropped out of massive seasonal migrations . . . or, quite possibly, the results of migratory flights that ran into bad weather. A kite in a thunderstorm is probably a fairly good representation of a *Pteranodon*'s chances in bad weather. In any case, while I do not believe that these flying reptiles were routinely feeding in midocean, I don't have a better explanation for why we find so many of their remains in the Smoky Hill Chalk.

Eleven

Feathers and Teeth

It was a bright, cloudless day, and the flock of small white birds flew in wide, lazy circles over the nearly calm surface of the Western Interior Sea. They were searching for their favored prey, a school of small fish feeding near the surface. From their vantage point high in the air, such schools appeared as dark, ever-changing shapes within the lighter blue background of the water. More often, however, the flock was attracted to the feeding activities of other, larger predators. From experience, they had learned that watching for the skin-winged pteranodons skimming the surface, or for marine reptiles or large fish as they swarmed around the shifting masses of tiny fish, was often a more successful strategy. It was also more dangerous.

Today they found a group of sharks swimming around a school of tiny fish. Occasionally, the sharks would dart through the mass of trapped prey with their mouths open, scooping in large numbers of the much smaller silver fish. Almost as one, the birds wheeled and flew as quickly as they could toward the feeding activity. As the birds approached, the circle of sharks tightened and the surface of the water erupted in a spray of water and fish as the prey tried in vain to escape upwards, the only direction left to them. Their escape was only momentary.

Just as the birds reached the surface, they folded their wings at the last moment and dove headfirst into the water, creating a barrage of small impacts. Once underwater, they opened their strong wings again and used them to propel themselves in pursuit of the little fish, much like some modern sea birds. They didn't have to swim very far to catch the panicked prey in their long jaws filled with sharp teeth. Soon, the birds popped back to the surface, each with a small wiggling fish in its mouth. With a jerk of the head, each fish was quickly flipped up into the air, caught headfirst, and then swallowed. After bobbing their heads up and down several times to help move the prey into their gullets, the birds again plunged into the roiling mass of small fish. This frenzied feeding activity continued for several minutes.

Then, without warning, a giant, six-meter long ginsu shark flashed into the tightly packed school of fish. As he passed through the melee, taking hundreds of the small fish into his open mouth, he also unknowingly swallowed two of the white, toothed birds. The momentum of his sudden attack scattered the ring of smaller sharks and allowed the school of fish to briefly disperse. Frightened by his appearance, the remaining birds took wing and regrouped in the air, and began their search for another feeding opportunity.

The story above, of course, is mostly fiction. We don't know very much about the daily activities of *Ichthyornis* ("fish bird") or the other toothed flying or swimming birds of the Late Cretaceous. Most paleontologists who work with these fossil birds believe they were most like modern terns or gulls, although Clarke (2004) has recently provided a more accurate reconstruction based on her study of *Ichthyornis* specimens in the collection at the Yale Peabody Museum. Given the variety of ways that modern sea birds dive, paddle, or even "fly" underwater, however, I tend to believe that *Ichthyornis* may have pursued their prey in the water, feeding directly on the bounty of small fish that was available to them near the surface. This might help explain the relative "wealth" of bird remains in the middle of the Western Interior Sea. Further study is certainly needed in this area. Like modern migratory birds, *Ichthyornis* and other toothed flying birds found in the chalk were apparently capable of remaining in the air for long periods of time and flying hundreds of miles.

The much larger *Hesperornis* ("western bird") might be considered the "only marine dinosaur" and is certainly the only feathered "dinosaur" for which we have a fossil record in the Western Interior Sea. It apparently returned to the oceans of the Late Cretaceous about the same time as the demise of the ichthyosaurs, and successfully competed for prey with plesiosaurs, mosasaurs, and large fish in the more northern (cooler?) oceans. While hesperornids were much smaller than the adults of most marine reptiles, it is important to remember that most of the marine reptile population at any one time consisted of subadult or juvenile animals that would have been feeding on the same size prey as the adult birds. From the fossil record, we know that *Hesperornis* arrived in what is now Kansas during the early Campanian. However, it was never as common in the Smoky Hill Chalk as in Late Cretaceous deposits further north. Its remains indicate that it was very much a bird, with feathers, but it still had teeth and solid (not hollow) bones. Its upper limbs were atrophied to the point of almost disappearing and were most likely useless. Its lower limbs were so specialized for swimming that *Hesperornis* probably could not walk very well on land, if at all. It is likely that these birds built their nests close to the water's edge and, outside of nesting and raising their chicks, spent very little time on dry land. In that regard, they may well be considered as a modern analog to penguins. *Hesperornis* and its kin were apparently very much at home in the ocean, since all of the remains found in the Kansas chalk would have been more than 200 miles from the nearest coast. They apparently fed on fish and other small prey, and in turn were occasionally eaten by mosasaurs (Martin and Bjork, 1987).

Professor Benjamin F. Mudge (1866a; 1866b), then a professor at the Kansas State Agricultural College (KSAC, now Kansas State University, Manhattan, Kansas), first reported the discovery of four three-toed "Ornithichnites," or footprints, in a piece of what was probably Dakota Sandstone from a bluff above the Republican River, several miles east of Concordia (Cloud County), Kansas. From differences observed in their size, Mudge believed that the tracks were made by at least two different individuals, both "long-legged waders." Although he went to great pains to measure and otherwise document the tracks, the slab containing them was lost when his horse-drawn wagon was overturned the next day while fording a stream. About twenty years later, Snow (1887) reported a single bird track recovered from the spoils of a well dug into the Dakota Sandstone of Ellsworth County. That specimen is in the collection of the University of Kansas. Williston (1898c) reprinted Snow's note along with a photograph of the footprint.

While it was O. C. Marsh who found the first remains of a bird (*Hesperornis regalis*) in the Smoky Hill Chalk of Trego County on July 25, 1871 (YPM 1200), the specimen was without a skull and Marsh had no way of knowing the full significance of his discovery. Marsh (1872a) wrote, "One of the treasures secured during our explorations this year [1871] was the greater portion of the skeleton of a large fossil bird, at least five feet in height, which I was fortunate enough to discover in the Upper Cretaceous of Western Kansas. This interesting specimen, although a true bird—as is clearly shown by the vertebrae and some other parts of the skeleton—differs widely from any known recent or extinct forms of that class, and affords a fine example of a comprehensive type. The bones are all well preserved. The femur is very short, but the other portions of the legs are quite elongated. The metatarsal bones appear to have been separated. On my return, I shall fully describe

Figure 11.1. The reconstruction of the skeleton of Hesperornis regalis *in Marsh (1883, fig. 18). The legs of* Hesperornis *were greatly modified for swimming, however, and it is unlikely that it could have stood up in such a pose.*

this unique fossil under the name *Hesperornis regalis.*" The locality given in the Yale Peabody Museum records (Trego County?) is suspect because there is little or no Campanian-age chalk in that part of Kansas and no other remains of *Hesperornis* have since been found that far east (Fig. 11.1).

Actually, Marsh had found *Hesperornis* about a year earlier but didn't realize it. Williston (1898a) noted that "late in the season of 1870, Professor Marsh, with an escort of United States soldiers, spent a short time on the upper part of the Smoky Hill River collecting vertebrate fossils. The material then collected served for the description of a number of interesting types by Marsh. It included the first known specimen of 'Odontornithes,' a foot bone brought in with other material, but which was not discovered in the material until after other specimens had been obtained later. In June of the following year Marsh again visited the same region, with a larger party and a stronger escort of United States troops, and was rewarded by the discovery of the skeleton which forms the type of *Hesperornis regalis* Marsh, together with other material."

The first recognized remains of a toothed bird from the Smoky Hill Chalk were collected the following year by B. F. Mudge. Williston (1898a) noted "that Professor Mudge found the remarkable specimen of *Ichthyornis*, from the North Fork of the Solomon, which furnished to the world the discovery of the then startling fact of birds with genuine teeth." It was a small, very delicate specimen, contained in a slab of chalk, and, while it eventually created considerable excitement, it also led to a bit of confusion.

As was the custom at the time, Mudge had been sending his better specimens to well-known paleontologists on the East Coast for examination and identification. In previous years, he had sent most of his material to E. D. Cope. In an 1870 letter to Mudge (Williston, 1898a, pp. 29–30), Cope complimented him on the scientific value of the material he had sent and named a "new species" of mosasaur (*Liodon mudgei*) after him. In fact, many of the new species of fish and mosasaurs described by Cope between 1869 and 1872 were from specimens sent to him by Mudge (Everhart, 2002). Cope (1872) also visited Mudge in late 1871 and examined his collection in Manhattan, Kansas. However, it was a letter from Marsh a year later that initiated a chain of events that would deny a spectacular specimen to Cope and give his rival, O. C. Marsh, the honor of describing the first toothed bird from the chalk.

According to Williston (1898a), on September 2, 1872, Marsh wrote to Mudge "inquiring about his summer collections in the Cretaceous, with the offer to 'determine any reptilian or bird remains without expense,' and stating that he would give him 'full credit' for their discovery." At the time, Mudge had a shipment of specimens already prepared to send to Cope. Apparently, he changed his mind at the last moment and instead sent the box to O. C. Marsh for examination.

Williston (1898a) related the following story of what had occurred: "An incident related to me by Professor Mudge in connection with this specimen is of interest. He had been sending his vertebrate fossils previously to Professor Cope for determination. Learning through Professor Dana that Professor Marsh, who as a boy had been an acquaintance of Professor Mudge, was interested in these fossils, he changed the address upon the box containing the bird specimen after he had made it ready to send to Professor Cope, and sent it instead to Professor Marsh. Had Professor Cope received the box, he would have been the first to make known to the world the discovery of 'Birds with Teeth.' "

Upon seeing the unusual specimen, Marsh quickly realized that he was looking at the skeleton of a small bird that was very different from his newly discovered *Hesperornis*. The skull, however, was not visible, and Marsh recognized what he thought were the lower jaws of a small reptile in the same block of chalk. Williston (1898a) noted that Marsh wrote to Mudge again on September 25, 1872, acknowledging the receipt of a box of fossils, and stating that the "hollow bones are part of a bird, and the two jaws belong to a small saurian. The latter is peculiar, and I wish I had some of the vertebrae for comparison with other Kansas species." Having no additional remains available at the time (or at least not recognizing them), Marsh concluded that Mudge's specimen actually represented the remains of two new species, and he briefly described the new bird in the *American Journal of Science* (1872b, p. 334):

"One of the most interesting of recent discoveries in Paleontology is the skeleton of a fossil bird, found, during the past summer, in the upper Cretaceous shale of Kansas, by Prof. B. F. Mudge, who has kindly sent the specimen to me for examination. The remains indicate an aquatic bird, about as large as a pigeon, and differing widely from all known birds in having *biconcave vertebrae*. The cervical, dorsal, and caudal vertebrae preserved all show this character, the ends of the centra resembling those in *Plesiosaurus*. The rest of the skeleton presents no marked deviation from the ordinary avian type. The wings were large in proportion to the posterior extremities. The humerus is 58.6 mm. in length, and has the radial crest strongly developed. The femur is small, and has the proximal end compressed transversely. The tibia is slender, and 44.5 mm. long. Its distal end is incurved, as in swimming birds, has no supratendinal ridge. This species may be called *Ichthyornis dispar*. A complete description will appear in an early number of this journal." While the records of the Yale Peabody collection indicate that the remains (YPM 1450) came from the Niobrara Formation of Scott County (which didn't even exist as an incorporated county in 1872—the boundaries were not defined until 1873), it is more likely that Mudge collected them much farther to the east (Williston, 1898a; Peterson 1987, p. 231). Walker (1967, p. 61) noted that the slab containing the bird remains came from an exposure "on Bow Creek in northwestern Rooks County," a tributary of the Solomon River. This locality is about 75 miles (120 km) from the closest chalk in what is now northeast Scott County.

Later in the same year, Marsh (1872c) followed with another brief note describing and naming a new species of marine reptile: "An interesting addition to the reptilian fauna of the Cretaceous shale of Kansas is a very small Saurian, which differs widely from any hereto discovered. The only remains at present known are two lower jaws, nearly perfect, and with many of the teeth in good preservation [Fig. 11.2]. The jaws resemble in general form those of the Mosasauroid reptiles, but, aside from their very diminutive size, present several features which no species of that group has been observed to possess. The teeth are implanted in distinct sockets, and are directed obliquely backward. There were apparently twenty teeth in each jaw, all compressed, and with very acute summits. The rami were united in front only by cartilage. There is no distinct groove on their inner surface, as in all known Mosasauroids. The dentigerous portion of the jaw is 41 mm in length, its depth below the last tooth is 5 mm and below the first tooth in front 3 mm. The specimen clearly indicates a new genus which may be called *Colonosaurus*, and the species may be named

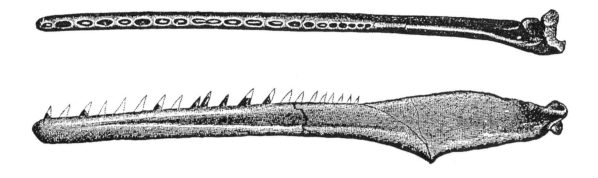

Figure 11.2. The lower jaw of Ichthyornis dispar, *discovered by B. F. Mudge in Rooks County and identified initially by Marsh (1872c) as the remains of a new marine reptile which he named* Colonosaurus Mudgei. *Adapted from Marsh (1875, pl. 2).*

Colonosaurus Mudgei for the discover Prof. B. F. Mudge, who found the remains in the upper Cretaceous shale of Western Kansas. [Yale College, Oct. 7th, 1872]."

By the following January, however, Marsh (1873a, p. 161) had apparently recognized his mistake and noted that the specimen proved "on further investigation to possess some additional characters, which separate them still more widely from all known recent and fossil forms. The type species of this group, *Ichthyornis dispar* Marsh, has well developed *teeth in both jaws* [italics by Marsh]." In February of that year, Marsh (1873b, p. 230) redescribed *Ichthyornis* as "a bird, about as large as a Pigeon, and differing from all known birds in having *teeth* and *biconcave vertebrae* [italics by Marsh]. The known remains were found in the Upper Cretaceous shale of Kansas, and are preserved in the collection of Yale College." *Hesperornis* was also mentioned in the same note but without any indication that Marsh was then aware that it also had teeth. *Colonosaurus*, the new genus based on the jaws of the little reptile that wasn't, was left to die a quiet death. Williston (1898a) recounted the story briefly in Volume 4 of the *Geological Survey of Kansas*.

In 1875, Marsh published a major article, "On the Odontornithes, or Birds with Teeth," in which he established a new subclass and orders for the recently discovered toothed birds. In the article (Marsh, 1875, p. 403), the history of the discovery of birds in western Kansas appears to have been rewritten slightly when Marsh noted that "[t]he first species in which teeth were detected was *Ichthyornis dispar* Marsh, described in 1872. Fortunately the type specimen of this remarkable species was in excellent condition, and the more important portions both of the skull and skeleton were secured." No mention was made of the discoverer (Mudge) or the resulting confusion over the lower jaws. Years later, Marsh (1883, fig. 27) provided a drawing of the skeleton of *Ichthyornis dispar* (Fig. 11.3).

Marsh (1872, p. 404) noted that the "most interesting bird with teeth yet discovered is perhaps *Hesperornis regalis*, a gigantic diver, also from the Cretaceous of Kansas, and discovered by the writer in 1870." Marsh neglected to mention that the original dis-

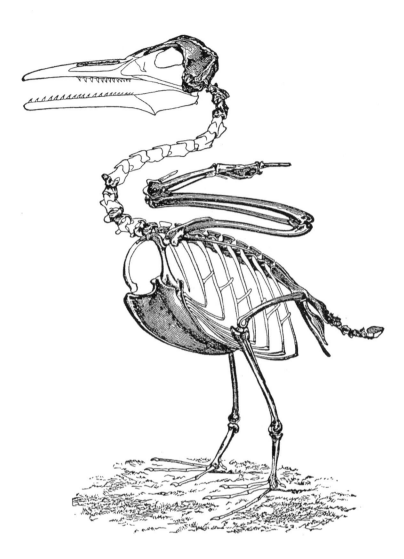

Figure 11.3. The restoration of
Ichthyornis dispar *in Marsh*
*(1883, fig. 27). A more recent
reconstruction by Clarke (2004)
indicates that the head is smaller
in relation to the size of the body
than was shown by Marsh.*

covery did not include a skull or teeth, but adds that "a nearly perfect skeleton was obtained in Western Kansas by Mr. T. H. Russell
and the writer in November, 1872." The reader is left to assume
that this was the first specimen of *Hesperornis* (YPM 1206) found
with teeth.

Lane (1946, p. 393) noted that *Hesperornis* had fourteen teeth
in each maxilla and thirty-three teeth in each side of the lower jaw.
There are no teeth in the premaxilla. Lane repeated (ibid., p. 293)
Marsh's explanation of the mode by which the teeth were replaced,
noting that the new tooth bud forms on the lingual (inner) side of
the older tooth, and dissolves the root of the older tooth as it
grows. Eventually the older tooth falls out and the new tooth fills
the vacated space. Lane (ibid., p. 394) also indicated that the pelvis
of *Hesperornis* is more reptilian in form than that of any modern
bird. The tail of *Hesperornis* is also unusual because it is composed
of twelve vertebrae (more than any modern bird), some of which

have expanded transverse processes (Fig. 11.4) "and clearly indicate that the tail was moved mostly up-and-down, evidently an adaptation to diving" (ibid., p. 395).

In a brief report on Kansas geology, Mudge (1877) noted with more than a bit of humor intended that "it may not be amiss to add that birds with teeth have been found only in the United States, and one genus, the Ichthyornis, only in Kansas. Thus, our fossil pterodactyls have no teeth, and our birds have, in direct contrast with those found in other parts of the world."

Marsh published *Odontornithes: A Monograph on the Extinct Toothed Birds of North America* in 1880. In the appendix (pp. 191–201) entitled "Synopsis of American Cretaceous Birds," he listed and briefly described nine genera and twenty species, most of which were in the Yale College collection. Shortly before Marsh's death in 1899, Williston (1898b) published an article on "Birds" in Volume IV of the *University Geological Survey of Kansas*, crediting Mudge with the "most important" discovery (*Ichthyornis*), giving a brief history of the specimens known at the time, and describing additional specimens in the University of Kansas collection. In a short paragraph, Williston also demolished Marsh's Odontornithes: "The systematic position of the toothed birds from Kansas is by no means yet settled. All ornithologists are, however, agreed that they do not form a separate group, and the name Odontornithes is in consequence generally abandoned. The value of the teeth is subordinate; they do not in themselves justify a separate subclass." Like Cope's Pythonomorpha and Streptosauria years before, Marsh's Odontornithes vanished from the language of paleontology.

Marsh (1883, p. 51) reported that the total number of bird specimens in the Yale museum, made up mostly of those collected from the Smoky Hill Chalk, represented "about one hundred and fifty different individuals." Clarke (2004) reported that 78 of these specimens were *Ichthyornis*. An earlier tally by Russell (1993) estimated a total of 225 bird specimens (or about 3 percent of the total) were found in museum collections from the Smoky Hill Chalk, similar to the number of turtle specimens (210), and only about 1/4 of the number of *Pteranodon* specimens (878).

Until very recently (Clarke, 2004), little new information had been added to our knowledge of the toothed birds from Kansas. Williston's (1898b) summary of Late Cretaceous bird specimens in Museum of Natural History at the University of Kansas was the previous major work on the subject. At that time, Williston reported that a few new specimens had been collected, but nowhere near the number found when Marsh made them a priority for his field workers. Williston (ibid.) noted that there were six species of *Ichthyornis* known at the time, all of which were named by Marsh between 1872 and 1880: *I. dispar, I. agilis, I. anceps, I. tener, I. validus,* and *I. victor.* Clarke's (2004) study of the same specimens in the Yale Peabody collection indicated that only one species of *Ichthyornis* was valid—*I. dispar,* with "*I. tener*" becoming *Guil-*

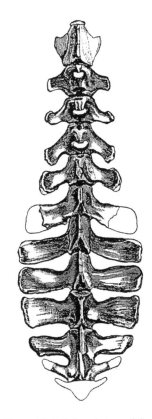

Figure 11.4. A dorsal view of the caudal vertebrae of Hesperornis regalis *(Marsh, 1883, fig. 17). These vertebrae supported a broad, flat tail could be moved up or down, and was probably used by* Hesperornis *for directional control while swimming underwater.*

davis tener and the other names becoming junior synonyms of *I. dispar*. Clarke also named a new species (*Iaceornis marshi*) from one of the other specimens in the collection.

Partially because of their small size and fragility, bird fossils remain rare and elusive. Putting things into a personal perspective, I've collected fossils in the Smoky Hill Chalk since 1969 and the only *Ichthyornis* remains I have ever found is a half-inch (11 mm) fragment of a shoulder bone (coracoid; FHSM VP-15574) from Gove County in 2002. Worse yet, the only reason that I found it was that I was down on my hands and knees looking for shark teeth. In her review of the Yale Peabody collection, Clarke (2004; pers. comm., 2003) further humbled my little discovery when she noted that the bones of the wing and shoulder, including the humerus and coracoid, are the most commonly collected bones of *Ichthyornis*. Her research indicated that limb bones (wings and legs) were among the first pieces to drop off a rotting carcass, and she concluded that many of the specimens in museum collections were derived from the decomposing bodies of dead birds that may have floated some distance from the location where they died. While I have trouble understanding how the body of a small bird could float on the surface for any period of time without attracting scavengers, we do know that several much larger dinosaur carcasses floated hundreds of miles into the Western Interior Sea during the same time period before settling to the bottom (Chapter 12).

There are some bright spots in the research on these marine birds since Williston's 1898 report. A "headless" but otherwise nearly perfect articulated specimen of *Hesperornis* (AMNH FR 5100) was acquired by the American Museum of Natural History from Charles H. Sternberg in 1907. It had been found in Logan County, Kansas, probably very close to Elkader, where the Sternberg family made many of their discoveries. Sternberg (1909) credits his second son, Charles M. Sternberg, with the discovery. The specimen is interesting in that the remains were found with the limbs positioned on either side of the body as they were in life. Martin and Tate (1966) noted that the exhibit specimen of *Hesperornis* in the University of Nebraska State Museum was a composite of two specimens collected from the Smoky Hill Chalk in the 1940s by George Sternberg.

One of the hesperornid fossils (KUVP 2287) in the University of Kansas collection described by Williston (1898b) was a nearly complete specimen of *Hesperornis gracilis* collected by H. T. Martin in 1894 from Graham County, Kansas. Williston (ibid., p. 46) noted that this species is smaller than *H. regalis*, "but has never been adequately described." Martin (1984) re-examined the KU specimen and concluded that it represented both a new genus and species of an even more primitive swimming bird from the Late Cretaceous: *Parahesperornis alexi* ("*para*" = near). Additional specimens of another genus of swimming bird, *Baptornis advenus*, were described by Martin and Bonner (1977). Martin and Tate

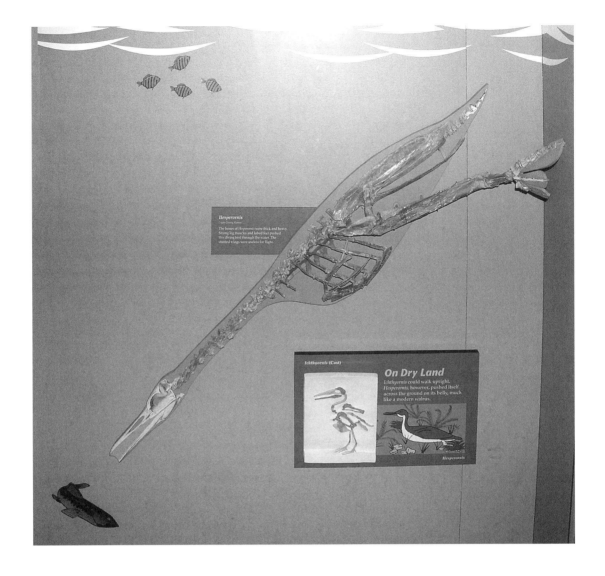

In the image, the following labels are visible:

Hesperornis
Logan County, Kansas

The bones of Hesperornis were thick and heavy. Strong leg muscles and lobed feet pushed this diving bird through the water. The stunted wings were useless for flight.

Ichthyornis (Cast)

On Dry Land
Ichthyornis could walk upright. Hesperornis, however, pushed itself across the ground on its belly, much like a modern walrus.

Hesperornis

(1976) noted the presence of small coprolites in association with a *Baptornis* specimen and were able to identify the jaw of a small fish (*Enchodus* sp.) in one of them.

In September 1958, M. C. Bonner collected the fairly complete, but headless, remains of a *Hesperornis regalis* from south of Russell Springs in Logan County. The specimen (FHSM VP-2069) is currently on exhibit in the Sternberg Museum of Natural History (Fig. 11.5).

An excellent *Ichthyornis* specimen (Fig. 11.6), including portions of the skull and lower jaws, was collected by J. D. Stewart in eastern Graham County in 1970 (Martin and Stewart, 1977). The locality is probably within a few miles of the original site where Mudge collected the first specimen in 1872. The specimen (FHSM VP-2503) is currently on exhibit in the Sternberg Museum of Nat-

Figure 11.5. A partial skeleton of Hesperornis regalis (FHSM VP-2069) on exhibit at the Sternberg Museum of Natural History. This specimen was collected by M. C. Bonner in 1958 from the early Campanian chalk of Logan County.

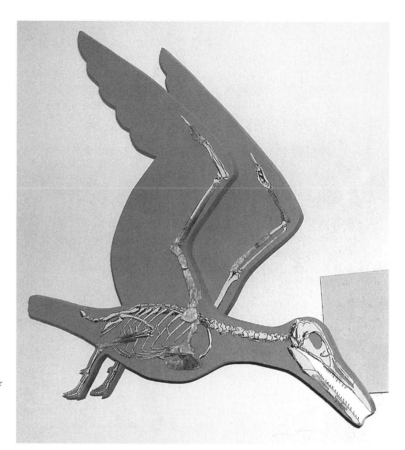

Figure 11.6. A reconstruction of Ichthyornis dispar *(FHSM VP-2503) at the Sternberg Museum of Natural History. This specimen was collected by J. D. Stewart in 1970 in the lower chalk of Graham County.*

ural History. Another *Ichthyornis* specimen in the University of Kansas collection was found by Greg Winkler and Pete Bussen in northern Lane County in the early 1990s and is currently under study.

Ichthyornis is also known from somewhat older rocks in Kansas and Canada. Walker (1967) reported the proximal end of a tarso-metatarsus (FHSM VP-2139) from the lower Fairport Chalk or upper Greenhorn Limestone (middle Turonian) of Ellis County. Another as-yet unreported specimen (FHSM VP-6318; tarso-metatarsus) was found in the Carlile Shale (Turonian) of Russell County. The known geographical range of *Ichthyornis* was extended some 2300 km (1400 mi.) to the northwest of Kansas when a left humerus was discovered in the early Turonian of Alberta, Canada (Fox, 1984). More recently, a diverse bird fauna was reported by Tokaryk et al. (1997) from the Upper Cenomanian of Saskatchewan, Canada, and fragments of small, birdlike bones have been found the basal Greenhorn (Upper Cenomanian) of Russell County, Kansas (unreported data; pers. obs.).

In the more than one hundred years since Williston's (1898b) note on Kansas toothed birds, it has become apparent that

Kansas was near the southern range of *Hesperornis*. Remains of these birds are extremely rare to the south of Kansas and have only recently been found in the Gulf region (David Schwimmer, pers. comm., 2003). As you go further north, however, *Hesperornis* becomes much more common. According to Russell (1967), almost one-third of the early Campanian vertebrate remains collected from the Anderson River (Northwest Territories, Canada) in 1965 were identified as *Hesperornis*. Nicholls (1988) reported that bird remains are "one of the dominant components" of collections made from the Pembina Shale of that region (172 of 476 specimens counted, or 36 percent). *Hesperornis* may have been a Cretaceous analog to modern penguins, preferring cooler waters or feeding on prey that was more abundant in cooler waters. During the Cretaceous, however, these birds were limited to the Northern Hemisphere, just the opposite of where penguins are found today.

As Marsh (1883) noted, "the remains of birds are among the rarest of fossils." Complete bird skeletons are extremely rare in the Western Interior Sea. Because of their smaller size and relatively light construction, most of the material in museum collections consists of a few parts or just a single bone. A survey of the collection at the Sternberg Museum in 2002 indicated five partial specimens of *Hesperornis,* including the exhibit specimen (FHSM VP-2069), and eight specimens of *Ichthyornis,* including the exhibit specimen (FHSM VP-2503). In most cases, the remains consisted of elements of the lower limbs. It is difficult from fragmentary remains such as these to figure out what these birds were like and how they lived.

Since it was without wings and had strong, well-developed legs, it is fairly obvious that *Hesperornis* was a swimming bird. *Ichthyornis,* on the other hand, had strong wings and a well-developed keel (breastbone) for anchoring flight muscles. Chinsamy et al. (1998) noted that it was "generally accepted that *Ichthyornis* used its long jaws and recurved teeth for scooping fish from surface waters (as gulls and terns scoop fish from surface waters today)." While that is certainly a strong possibility, another method of feeding that would be well suited for jaws with teeth might be diving and pursuing prey underwater, as is done by modern sea birds like shearwaters, guillemots, and puffins. Such a scenario might also better explain the occurrence of the remains of these relatively small birds in the chalk.

Sharing the ocean with other large predators sometimes made life short for the unwary. The remains of a *Hesperornis* were discovered in 1979 as gut contents of a large mosasaur, *Tylosaurus proriger* (SDSM 10439), along with another mosasaur, a large fish, and possibly a shark (Martin and Bjork, 1987). Hanks and Shimada (2002) reported bite marks of a shark (*Squalicorax*) on a hesperornithiform bird bone. The fragmentary remains that make up most of the collections of the Yale Museum and other institutions are indicative of the "leftovers" when these birds were partially consumed by other predators.

Both ichthyornids and hesperornids are now regarded as unsuccessful side branches in avian evolution. In spite of their superficial similarities to extant birds, they have no modern relatives. While the number of specimens collected since the first remains were found by Marsh and Mudge increases but slowly, research into these fascinating toothed birds continues.

Twelve
Dinosaurs?

The small herd of plant-eating dinosaurs moved noisily through thick underbrush as they fed on the lush vegetation growing along the bank of a flood-swollen river. It was late in the rainy season and the new growth near ground level was easy picking for the blade-like teeth of the squat, heavily built nodosaurs. In order to get the nourishment they needed to feed their large bodies, these animals fed almost continuously on low-growing, woody shrubs. The floor of the forest was still muddy from the last rain and the slow-moving animals were churning it into a sticky muck that clung to their stubby legs.

One of the nodosaurs was a four-year-old female. She was one of only seven surviving nodosaurs of her age group in the herd. At just over 4 m (15 ft.) in length, she was not yet full-grown. However, she was old enough and large enough to feed on the outer edge of the herd with the adults and to help provide a ring of protection for the juveniles and the much smaller hatchlings in the center. While the older animals were heavily armored and built low to the ground for protection against attack by predators, the younger ones were at risk of becoming prey if left unattended for long. The large carnivores that roamed the same scrub forest were always looking for an opportunity to make a meal out of a young no-dosaur.

The young female worked her way slowly through the thick brush, her sharp teeth clipping efficiently at the tender new growth on the plants around her. Although she couldn't see it through the undergrowth, she was on the side of the herd nearest the river. The roaring sound of the river at flood stage seemed to be getting louder, drowning out the feeding noises made by the herd. Then, as she moved forward across a small clearing, the ground began to tremble under her. Suddenly, the edge of the riverbank shifted and gave way beneath her feet. Startled, she raised her head and saw a crack open in the muddy soil between her and the rest of the herd. Before she could move, the earth dropped from under her with a loud splash as it crashed into the rain-swollen river below. As she fell, her heavy body and all the vegetation around her were immediately immersed by muddy water. She was quickly pulled under the surface by the current and tumbled along with the flow. Unable to raise her head out of the water, she soon drowned. Her body was carried quickly downstream along with many uprooted trees and other flood debris.

After few miles, the river reached the edge of a shallow ocean and the current slowed. The body of the dead nodosaur was too heavy for the diminished current to carry it further, and it soon lodged under the edge of a tangle of brush and tree limbs in a brackish estuary along the coast. Micro-organisms living in the gut of the dead nodosaur, however, continued to break down the plants the animal had eaten that day but also began working to decompose her internal organs. In doing so, they generated gases that bloated the nodosaur's body and made it steadily more buoyant. After a few days in the warm water, the bloated carcass tugged free of the last of the branches that held it and drifted slowly up to the surface. The weight of the heavy dermal scutes that made up the protective armor of the dinosaur turned the carcass over so that it now floated upside down. Once the current from the diminishing floodwaters caught the body, it began to move seaward. Slowly but relentlessly, the swollen body of the nodosaur moved away from the land and into the reaches of the vast shallow sea.

After drifting at sea for many days, the bloated carcass was rotting from the inside out and beginning to come apart. The thick skin of the dinosaur still held the body together and the bony armor protected most of it from scavenging by all but the largest sharks, but it was only a matter of time before the abdominal cavity ruptured. Eventually, the layer of skin and muscle covering the belly ripped open, releasing the gases trapped inside, and the dinosaur began to sink toward the muddy floor of the chalk sea. Still largely intact after its long journey, the nodosaur's large bones and dermal scutes made it heavy enough to drop rapidly toward the bottom some 400 feet below. Landing hard on its right side amidst a sparsely populated community of inoceramid clams, some of the remains were driven several inches into the soft mud. At last at rest in the pitch-black darkness of the sea bottom, the carcass would be slowly buried without much further damage by scavengers.

The preceding story is fiction, except for the fact that the reasonably complete remains of a nodosaur now called *Niobrarasaurus coleii* were actually found in the chalk of Gove County in 1930. We will never know for certain how this nodosaur died and was carried out to sea. Even more of a mystery is how it traveled hundreds of miles from the nearest shore to the middle of the Western Interior Sea, where the carcass sank and was eventually covered by millions of years of accumulated marine sediments. Although it is a rare occurrence, this specimen is not the only dinosaur to be discovered in the Smoky Hill Chalk. Five sets of dinosaur remains that had been discovered since 1871 were reported by Carpenter et al. (1995), and another, more recent one by Hamm and Everhart (2001). Of these, one is a hadrosaur and the other five are nodosaurs. As I will explain below, the first two nodosaur specimens may actually be from the same dinosaur.

In the summer of 1871, Professor O. C. Marsh and his scientific expedition of students from Yale College were the first to collect the remains of a dinosaur in the Niobrara chalk. In Marsh's (1872) words, the specimen "was the greater part of a skeleton of a small *Hadrosaurus,* discovered by the writer in the blue Cretaceous shale near the Smoky Hill River, in Western Kansas." Although the exact locality is unknown, the dinosaur remains (YPM 1190) were probably found near the Smoky Hill River in Logan County according to Carpenter et al. (1995). Marsh (1872) initially called the new dinosaur *Hadrosaurus agilis* and indicated that it "would be fully described in this Journal [*American Journal of Science*] at an early day." The specimen was mounted as an exhibit in the Yale Peabody Museum, but Marsh never made good on his promise to describe the remains. Some years later, Marsh (1890) did reassign the remains to a new genus, renaming it *Claosaurus agilis.* According to Carpenter et al. (ibid., p. 290), however, "*C. agilis* has never been adequately described and figured."

It was more than forty years before the next dinosaur was discovered in the Smoky Hill Chalk. In 1905, Charles Sternberg (1909) found the remains of what he thought was "a large new sea-tortoise with an ossified carapace." By then Sternberg had collected a number of large marine turtles from the Smoky Hill Chalk (see Chapter 6) and should have been an expert on them. The locality information he provided ("five miles south of Castle Rock, and 3 miles south of Hackberry Creek"; Wieland, 1909, p. 250) is unclear, but it is likely the remains were found in the low chalk of southeastern Gove County. At the time, Sternberg thought the bones were too weathered to bother collecting, but his son, George F., later discovered the specimen on his own and collected it anyway. At some point, Charles Sternberg sent two of the "peculiar" pieces he believed to the part of the shell (neurals) of the "new turtle" to the Yale Museum for study (Fig. 12.1). There, Dr. G. R.

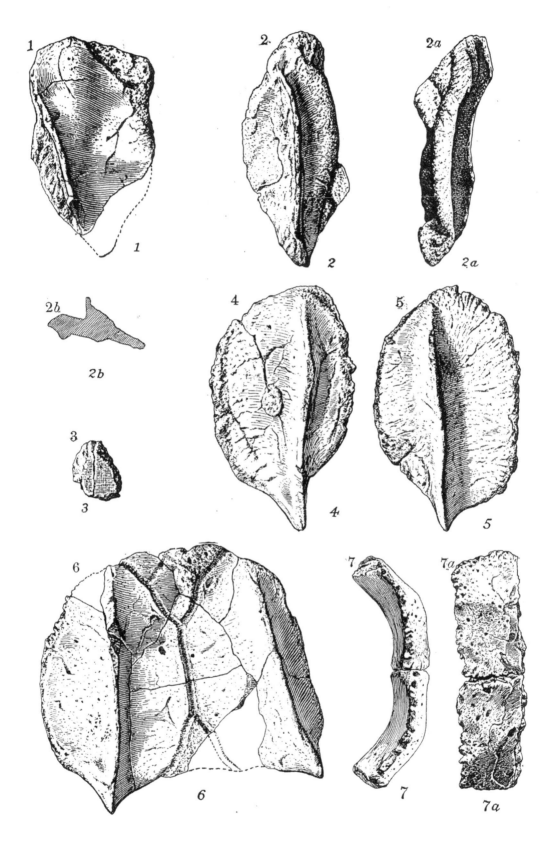

Wieland, who was working with fossil turtles at the time, saw the unusual remains included in Sternberg's turtle material. Wieland told Sternberg they were dermal scutes of an armored dinosaur. The Sternbergs collected the rest of the remains and sent at least some of them to Yale, where they were later described by Wieland (1909, p. 250) as *Hierosaurus Sternbergii* (now *H. sternbergi* per Carpenter et al., 1995). Although Sternberg (1909) suggested that they had found a large portion of the skeleton and dermal armor, it is unclear what happened to a substantial portion of the specimen. According to Carpenter et al. (1995, p. 276), about "34 scutes, skull and rib fragments" are all that remain of the skeleton (YPM 1847). To make things even more difficult, we are unsure if Sternberg collected one dinosaur from the chalk or two. Wieland (1909) figured two dermal elements from the original material provided by Charles Sternberg in 1907, then acknowledged the receipt of "six dermal scutes" from the "last season" (1908). The Yale Peabody Museum collection has two specimens donated by Charles Sternberg. YPM 1847 is the original specimen and YPM 55419, curated later, apparently contains the dermal scutes donated the following year. It is unclear if they represent the same set of remains (Carpenter et al., 1995, p. 275). Given the relative rarity of dinosaurs in the Smoky Hill Chalk, however, I think they are most likely from the same individual.

Years passed, and in February 1930, Virgil B. Cole, a geologist doing surface mapping in western Kansas for Gulf Oil, found the remains of what he called a "baby dinosaur" in southeastern Gove County (roughly twenty miles south of Quinter). A handwritten letter by Cole to Dr. M. G. Mehl at the University of Missouri at Columbia described the discovery and indicated that he was shipping it (300 pounds of bones and plaster) to Missouri. Cole had been a student of Mehl and received his master's degree from the University of Missouri at Columbia in 1923. The letter and some of Cole's field notes were retained with the specimen.

From his letter, it is apparent that Cole initially thought the remains were those of a plesiosaur. However, after excavating the fairly complete articulated right hind limb, he realized that it was a dinosaur. From his letter, he seemed disappointed at first that it wasn't a plesiosaur, but was justifiably proud of his discovery for the rest of his life (Walters, 1986).

After receiving the specimen from Cole and removing it from the plaster that had been poured directly on the bones, Mehl briefly reported on the discovery in an abstract for the Geological Society of America (GSA) meeting in 1931. He then followed with a more complete description (Mehl, 1936), naming the new dinosaur *Hi-*

Figure 12.1. *(facing page) Dermal scutes (YPM 1847 and 55419) from the first nodosaur remains found in the Smoky Hill Chalk (Wieland, 1909, figs. 1–7). The specimen was discovered by Charles H. Sternberg in 1905 (probably in Gove or Trego counties) and named* Hierosaurus sternbergi *in his honor by Wieland.*

erosaurus coleii in honor of Virgil Cole. In the title of his paper, Mehl (1936) indicated his belief that *Hierosaurus* was an aquatic dinosaur, possibly to explain why it had been found in the middle of the Western Interior Sea, so far from the nearest land.

After being described and named, the specimen more or less sat unnoticed in the collection of the University of Missouri at Columbia until Carpenter et al. (1995) re-examined the type material and published a revised description in their "Dinosaurs of the Niobrara Chalk" paper. Because of the differences the authors noted between it and the type of *Hierosaurus,* they renamed it as *"Niobrarasaurus coleii,"* a separate genus from *Hierosaurus* (now a *nomen dubium,* ibid.).

In 1973, J. D. Stewart, who was then a student at the University of Kansas, found the badly weathered remains (rib and limb fragments and dermal scutes) of another Niobrara nodosaur in the middle chalk (Santonian) from Rooks County. This specimen (KUVP 25150) is currently on exhibit in the Museum of Natural History at the University of Kansas. It was described by Carpenter et al. (1995) but is so damaged and incomplete that it could be identified only as Nodosauridae *incertae sedis.* Stewart was only the fourth person to find dinosaur remains in the chalk, and it would be almost thirty years before it happened again.

In October 2000, Shawn Hamm, a geology student at Wichita State University in Wichita, Kansas, became the fifth person to find dinosaur remains in the Smoky Hill Chalk. He was collecting fossils in northeast Lane County when he came across two unusual bones lying on the surface of the chalk near Hattin's marker unit 8 (early Santonian). They were quite dense and already bleached gray-white by exposure to the sun. Unable to identify them, he showed them to me. I was also puzzled because they were like nothing I had ever seen from the chalk. My first impression was that they were limb bones of a large marine turtle, like *Protostega,* but they didn't look like the figures in the references that I had available. The larger one also appeared to have been crushed, suggesting that it had originally been hollow, something that is not characteristic of turtle bone. I told him that was my best guess, but that I would continue to try to get them properly identified. A few days later, when I was going through Carpenter et al. (1995) while looking for information on another subject, I realized that Shawn's strange bones were from a dinosaur. The two bones (Fig. 12.2) appeared to be the right radius and ulna from a young nodosaur (Carpenter, ibid., Fig. 10 and 12), apparently very close to *Niobrarasaurus.* When I called Shawn and told him the good news, we decided the discovery was worth reporting at the upcoming Society of Vertebrate Paleontology (SVP) meeting.

In early 2001, after Shawn Hamm and I submitted our abstract (Hamm and Everhart, 2001) for the SVP annual meeting in Bozeman, Montana, I decided we needed to compare the two bones Shawn had found (FHSM VP-13985) with those of the type specimen (MU 650 VP). So, after tracking down the type material at the

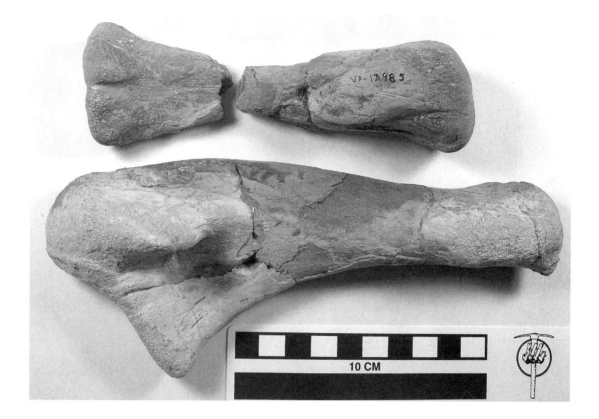

Figure 12.2. The right radius (top) and ulna of a juvenile nodosaur (FHSM VP-13985) collected from the Smoky Hill Chalk in Lane County in 2000 by Shawn Hamm of Wichita, Kansas. There are two parallel bite marks on the distal end of the radius (top right, under number). These bones appear to have been partially digested by a large shark, probably Cretoxyrhina mantelli.

University of Missouri at Columbia, I contacted the geology department in March 2001 and requested the right radius and ulna be sent on loan to me in care of the Sternberg Museum. My request was granted and the right lower limb of *Niobrarasaurus coleii* arrived for study within a month.

Even though VP-13985 represents a much younger individual (about half the length of the comparable elements in the type specimen, their overall characteristics appeared to be very similar to those of the holotype of *Niobrarasaurus*. From the bite marks on the radius and the corroded appearance of both bones, it appeared that the lower portion of the front limb had been ripped off a floating carcass, swallowed and partially digested by a large shark. In August 2002, when I returned the type material to the University of Missouri, I took a chance and asked if it might be possible that the entire specimen could be transferred to the Sternberg Museum for use in a "Kansas Dinosaur" exhibit. The answer turned out to be yes, and in December 2002, I stopped by the University of Missouri at Columbia and picked up the whole specimen.

While I was packing the specimen for shipment, I found an unpublished note from Virgil Cole written on a scrap of an old cardboard box that carried an additional bit of information about Cole's discovery, some of which was confusing: "*Hierosaurus coleii*

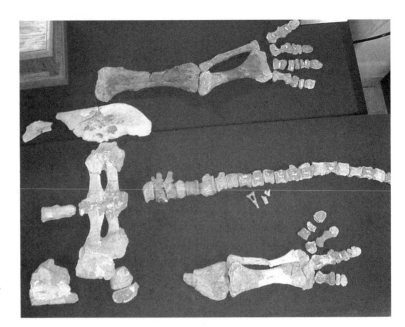

Figure 12.3. The rear limbs, pelvis, and caudal vertebrae of Niobrarasaurus coleii *(FHSM VP-14855)* as they were displayed in May of 2003.

more pieces picked up after the rains June 8, 1935, Trego Co. Kan. Horizon 122' up in Niobrara Chalk V. B. Cole."

This bit of stratigraphic information was important to me because Cole's 1930 letter to Mehl had originally placed the specimen some 190 feet above the base of the chalk. Cole had also provided information on the section, range, and township where the remains had been found, and I was very familiar with the area. However, his original determination of the horizon of the specimen had me worried because the whole area was much closer to the base of the chalk than he indicated. This could mean his locality data was wrong or that his estimate of the horizon was wrong, or both. His second determination of the horizon was much closer to the actual distance above the Fort Hays Limestone in his designated section. I was also a little concerned about the "Trego Co." notation since the original locality given by Cole is in Gove County. However, the given locality was also just a few miles to the east of the county line between Gove and Trego Counties. I hoped his change in the name of the county was just an honest mistake.

On May 1, 2003, a media conference was arranged at the Sternberg Museum to announce the acquisition of the type specimen of *Niobrarasaurus coleii* (now FHSM VP-14855). I spent the morning arranging a display of the bones of *Niobrarasaurus* on a piece of dark blue cloth and talking to reporters and photographers who had arrived early. It was possibly the first time the specimen had been laid out since it was first prepared (Fig. 12.3) and certainly the first time it had been photographed so extensively. Then, on May 3, after most of the news stories had played out, I took a group of amateur paleontologists into the field. By coincidence, we were collecting fossils in the same section (square mile) that had supposedly produced the bones of *Niobrarasaurus* in 1930. The locality also happens to be one of my favorite areas of the chalk and one for which we had access from the property owner. And there begins the rest of this strange and wonderful story . . .

Niobrarasaurus coleii and the Rest of the Dinosaur

The morning promised a nice day for fossil hunting. It was cool and cloudy, with just a bit too much wind at times. I was leading a group of my friends from the New Jersey Paleontological Society on a field trip in the chalk. When we arrived at the site in southeastern Gove County, just south of the Smoky Hill River, the five of us scattered quickly across the shallow valley. We were all carrying radios, but, surprisingly, there was little of the chatter we usually had when we were having good luck finding fossils. My day started pretty well, finding a nice, though partially digested, mosasaur axis vertebra (shark food), and it ended with the discovery of my first *Martinichthys* rostrum (a rare plethodid fish—see Chapter 5) since 1996. About 5:00 P.M., just as I began to walk back to the van to leave for the day, Tom Caggiano called me on the radio and told me he had found some unusual bones.

After making a joke about the "*Bovinasaurus*" (cow bones) in the pasture I could see him walking through, I asked him what they were. Tom said he didn't know. When I met up with him, he showed me four bones (three metacarpals and a terminal phalanx), bleached gray-white by exposure to the Kansas sun. I recognized them immediately as dinosaur bones, just like the ones I had been arranging for the *Niobrarasaurus* news conference at the Sternberg two days earlier. "Great!" I thought to myself. Tom had discovered another Smoky Hill Chalk dinosaur . . . only the sixth person to ever do so (after O. C. Marsh, Charles Sternberg, Virgil Cole, J. D. Stewart, and Shawn Hamm). We were both excited about the discovery as we walked back to the site where he had found the bones.

Tom had seen the terminal phalanx (end of the toe) lying on the side wall of a seven-foot-deep gully that he was walking in, at about eye level. When he had climbed up to see where it came from, he found the three metacarpals lying close together near the edge. Then, as I examined the bones again, I noticed that all three of the metacarpals still had bits of white plaster clinging to them. I suddenly realized that this wasn't a "new" dinosaur discovery at all, but rather the lingering remains of Cole's original *Niobrarasaurus* specimen. By sheer chance, Tom had found the site where Cole discovered the holotype. To me, this was even more important than a new dinosaur because it provided accurate stratigraphic information on the original specimen. When I looked around, I determined that the remains occurred below Hattin's (1982) marker unit 3. It was located about eighty-five feet above the contact with the underlying Fort Hays Limestone and was definitely late Coniacian in age. Tom's discovery also verified what I

Figure 12.4. Tom Caggiano holds four bones from the right front foot of the type specimen of Niobrarasaurus coleii *(FHSM VP-14855) that he discovered in May 2003 in the lower (late Coniacian) chalk of Gove County. The bones still have remnants of the plaster that had been poured on them by Virgil Cole in February 1930.*

had concluded from their localities about how early (late Coniacian) some of the nodosaurs had been found were.

Of course, it also added four of the missing bones (Fig. 12.4) of the right front foot noted by Cole in the last line of the first page of his letter to Dr. Mehl. As best I could determine from the description of the dig in his letter, Cole had poured plaster on everything (literally, directly on the bones) and then probably covered the right front foot with rock and dirt as he dug around to remove the larger bones of the right front limb. This pile of chalk and the plaster cap protected the bones for many years before it finally weathered away, and it enabled them to survive in place until they were found. It was simply a matter of a lot of good luck, good eyes, and being in the right place at the right time on the part of Tom Caggiano.

We got the rest of the group together and did a thorough search of the area. Although we found bits of plaster still clinging to some inoceramid shell fragments, we were unable to locate any additional remains of the *Niobrarasaurus*. If anything else had been left on the site by Cole, it had probably washed down the nearby gully long ago. Another heavy rain and the four little bones found that day by Tom would have suffered the same fate.

And there you have the rest of the story of *Niobrarasaurus coleii* . . . newly arrived back in Kansas where she or he had spent most of the last 86 million years and more recently was reunited with part of her or his right front foot.

At most, six sets of dinosaur remains have been found in the Smoky Hill Chalk over the past 130 years. If Sternberg's *Hierosaurus* remains are counted as one dinosaur, this figures out roughly to be a new discovery about every thirty years. While it is likely that additional remains of dinosaurs will be found buried in the middle of the Western Interior Sea, they will never be more than a rare occurrence. They are, however, valuable additions to our knowledge of dinosaurs during the Late Cretaceous. As noted by Carpenter et al. (1995), they represent "the best known assemblage presently available" from this relatively unknown time interval in the terrestrial Late Cretaceous of North America.

Thirteen

The Big Picture

The Western Interior Sea during the Late Cretaceous was a big place, extending from the present day Gulf of Mexico to the Arctic Ocean. Its width varied considerably through time, depending on changes in sea level and the movement of continents (plate tectonics), as well as other planet-wide events we don't yet fully understand. My rough estimate of the size of the sea during its maximum expansion near the end of the Turonian (90–89 mya) would be a length of about 5000 km (3000 mi.) from north to south, and a width of about 1600 km (1000 mi.) from east to west, across the center of North America. The sea narrowed at the north (Arctic) and south (Gulf) ends, but even so would have covered something in the neighborhood of 1 to 2 million square miles of what is mostly dry land today. As a comparison, this area would be equivalent to roughly half that of the United States. Although relatively small and much shallower in comparison to the Atlantic or Pacific Oceans, the Western Interior Sea was still a big place.

In previous chapters, I have described many of the animals whose remains have been discovered in the Smoky Hill Chalk, and to a lesser extent, elsewhere in the Cretaceous marine deposits in Kansas. Most of these animals lived in the Western Interior Sea during a relatively brief (in geologic terms) five-million-year span of time (87–82 mya) during the latter half of the Late Cretaceous. Five million years, however, is a long period of time by human standards, and is roughly the same amount of time that humans in some form have existed on this planet. Another way of looking at this time span is to think of it in terms of days—approximately 1.8 BILLION of them. That is a lot of sunrises and sunsets, and is enough time for many changes to occur in the environment that affected life in the oceans of Earth.

The deposition of the Smoky Hill Chalk occurred some 20 million years before the end the Cretaceous, when sea levels were high and the Western Interior Sea was slowly decreasing in size from what had been its greatest expansion over the middle of North America at the end of the Turonian. For life in the sea, times were relatively good. Some major marine groups were already extinct, including the ichthyosaurs and the giant pliosaurids (most recently, *Brachauchenius lucasi* in the Turonian of Kansas), while others, including the bony fish, turtles, mosasaurs, pterosaurs and birds, were evolving rapidly. We know relatively little about what was going on during the period when the Fort Hays Limestone (early Coniacian) was deposited prior to the Smoky Hill Chalk because only a few fossils have been discovered in this unit so far (Shimada, 1996; Shimada and Everhart, 2003). However, it is likely that the fauna was not too different from that of the chalk. The marine animals living near the middle of the Western Interior Sea were hundreds of miles from the nearest land and relatively isolated from its influences. The birds, pterosaurs and egg-laying marine reptiles (turtles) still were connected to the land, but what I consider to be the major players in this ecosystem lived their lives entirely at sea. On the following pages, I will describe the changes in the fauna that occurred as this epicontinental sea narrowed and became shallower.

As discussed briefly in Chapter 3, the productivity of the Western Interior Sea, as that of modern oceans and seas, was based on the abundance of microscopic, single-celled algae that used photosynthesis to convert sunlight into biomass for growth and reproduction. In the process, these cells (called coccolithophores) also released oxygen, and fixed carbon dioxide as calcium carbonate to construct tiny wheel-like structures (coccoliths) that they used to cover themselves. Besides being the major food source for microscopic and macroscopic consumers at the base of the food chain, vast quantities of their coccoliths settled to the sea floor over that same five-million-year period to form the calcareous ooze that eventually become the Smoky Hill Chalk. It was this chalk that faithfully preserved the wealth of vertebrate remains that we find today. By the early part of the Campanian (about 82 mya), the Western Interior Sea had decreased in size (width and depth) to the point that what is now western Kansas was much closer to land, and deposition of the Pierre Shale began as muds from the rock and soil that were being eroded from nearby land, especially from the Rocky Mountains rising to the west. However, life continued to flourish in the sea at the time of this transition and relatively little change has been noted in the fauna (Carpenter, 1990; 2003).

The five-million-year deposition of the Smoky Hill Chalk may be broken up into shorter periods of time which are defined here as the end of the Coniacian (a period of about 1.2 million years), all of the Santonian (a period of about 2.3 million years), and the beginning of the Campanian (a period of about 1.5 million years). The fauna living in the sea during that time changed slowly as new species evolved and others became extinct, or retreated to other places because of changing conditions.

Cope and Marsh, who described most of the vertebrate fauna from the chalk, were mainly interested in discovering new species, and bigger and better specimens, and apparently paid little attention to the stratigraphy. To some degree, this was due to a lack of knowledge regarding the geology of Kansas, but it is also apparent that knowing where a specimen came from within a formation wasn't considered to be important at the time. In his fictional book, *Buffalo Land*, Webb (1872) provided one of the first illustrations of a reconstructed fauna from the Smoky Hill Chalk. Although the print (Fig. 13.1) is not attributed to an artist by name, his work was likely influenced by Webb's contacts with E. D. Cope (Davidson, 2003). It is also apparent from the association of species illustrated in the fanciful scene that it is representative of the upper chalk (early Campanian). *Elasmosaurus, Liodon (Tylosaurus) proriger, Polycotylus, Protostega,* and the two species of *Pteranodon* shown were all first described and named by Cope. While *Elasmosaurus* is actually from the Pierre Shale, the formation was not recognized as being distinct from the Niobrara Chalk at the time the book was published.

As noted in Chapter 2, Logan (1897) was the first to refer to the chalk as the "Pteranodon Beds." Williston (1897; 1898b) informally divided the Pteranodon Beds into the lower Rudistes Beds and upper Hesperornis Beds and was among the first to report that some species occurred at different stratigraphic levels within the formation. Russell (1988) listed marine vertebrates from the Cretaceous of North America, including the fauna of the Kansas Niobrara, by their geologic and stratigraphic occurrence. Stewart (1990a) provided a comprehensive listing of species found in the Smoky Hill Chalk by their occurrence in six biostratigraphic zones based on Hattin's (1982) earlier work, and implored others to provide accurate stratigraphic information with newly discovered specimens. Bennett (2000) inferred the stratigraphic occurrence of *Pteranodon* specimens from the location where they had been collected, and Everhart (2001) further refined the ranges of three mosasaur species from new material. Others have documented recent discoveries of various species in the chalk (see previous chapters) and improved our knowledge of the life in the oceans of Kansas during the Late Cretaceous.

THE SEA WHICH ONCE COVERED THE PLAINS.

Elasmosaurus platyurus.   2. Liodon proriger.   3, 4, 5. Ornithochirus umbrosus.   6. Ornithochirus harpyia.
7. Protostega jigas.   8. Polycotylus latipinnis.

Table 13.1 provides a summary of most of the species that have become known from the Smoky Hill Chalk of Kansas since 1868. It is based on my collecting experience as well as the previous work by Hattin (1982), Russell (1988), Stewart (1990a), Carpenter (1990; 2003) and others. I have reduced the six biostratigraphic zones of Stewart (1990a) to the three time periods mentioned above: late Coniacian, Santonian, and early Campanian. This simplifies the data and is reasonably accurate for most species. I will go into more detail on some of the issues regarding the occurrence of the various species in the discussion that follows.

Figure 13.1. One of the first illustrations depicting several species of marine reptiles and pterosaurs from the Western Interior Sea (Buffalo Land, W. E. Webb, 1872). The caption reads: "The sea which once covered the plains. 1. Elasmosaurus platyurus 2. Liodon [Tylosaurus] proriger 3, 4, 5. Ornithochirus umbrosus 6. Ornithochirus harpyia 7. Protostega jigas [gigas] 8. Polycotylus latipinnis." Note that the pterosaur (4) is shown with teeth (see Chapter 10).

Late Coniacian

During the late Coniacian (87.0–85.8 mya), the middle portion of the Western Interior Sea was probably at its deepest, around 200 meters (Hattin, 1982). This is not very deep compared to what we know about modern oceans, but it was certainly a much different habitat than present at the time closer to the eastern and western shores, or along the Gulf of Mexico. It was deep enough, however, that very little, if any, light reached the bottom. Hattin (ibid.) also noted that the water just above the sea bottom was poorly circulated and may have been anoxic (with low or no oxygen) at times. In any case, the sea floor was not a very hospitable environment and would certainly not have supported the number of species present in shallower, better-circulated and well-oxygenated waters.

TABLE 13.1.
A partial listing of invertebrate and vertebrate species from the Smoky Hill Chalk by stage (Adapted from lists by Hattin (1982), Russell (1988), Stewart (1990 and pers. comm., 1992), Carpenter (1990; 2003), other individual papers, and my personal observations).

Late Coniacian 87–85.8 mya (1.2 million years)	Santonian 85.8–83.5 mya (2.3 million years)	Early Campanian 83.5–82 mya (1.5 million years)
Invertebrates	**Invertebrates**	**Invertebrates**
Volviceramus grandis		
	Cladoceramus undulatoplicatus	
Platyceramus platinus	*Platyceramus platinus*	*Platyceramus platinus*
Pseudoperna congesta	*Pseudoperna congesta*	*Pseudoperna congesta*
Durania maxima		*Durania maxima*
Tusoteuthis longa	*Tusoteuthis longa*	*Tusoteuthis longa*
	Unitacrinus socialis	
	Clioscaphites vermiformis	
	Clioscaphites choteauensis	
Spinaptychus (2 new species)	*Spinaptychus sternbergi*	*Rugaptychus* sp.
Unidentified ammonite (cast)	*Baculites* sp. (molds)	*Baculites* sp. (molds)
Sharks / Rays	**Sharks / Rays**	**Sharks / Rays**
Ptychodus anonymus	*Ptychodus mortoni* (?)	
Ptychodus mortoni		
Ptychodus martini		
Ptychodus occidentalis		
Ptychodus polygyrus		
Rhinobatos incertus	*Rhinobatos incertus*	*Rhinobatos incertus*
Scapanorhynchus raphiodon		
Pseudocorax laevis	*Pseudocorax laevis*	
Squalicorax falcatus	*Squalicorax falcatus*	
	Squalicorax kaupi	*Squalicorax kaupi*
		Squalicorax pristodontus
Johnlongia sp.		
Cretolamna appendiculata	*Cretolamna appendiculata*	*Cretolamna appendiculata*
Cretoxyrhina mantelli	*Cretoxyrhina mantelli*	*Cretoxyrhina mantelli*
Bony Fish	**Bony Fish**	**Bony Fish**
Belonostomus sp.		
Micropycnodon kansasensis	Unidentified pycnodont	*Hadrodus marshi*
Lepisosteus sp.		
Protosphyraena nitida		
Protosphyraena perniciosa		*Protosphyraena perniciosa* (?)
	Protosphyraena tenuis	*Protosphyraena tenuis*
Protosphyraena gladius	*Protosphyraena gladius*	*Protosphyraena gladius*

Paraliodesmus guadagnii	*Paraliodesmus guadagnii*	
Xiphactinus audax	*Xiphactinus audax*	*Xiphactinus audax*
Ichthyodectes ctenodon	*Ichthyodectes ctenodon*	*Ichthyodectes ctenodon*
Gillicus arcuatus	*Gillicus arcuatus*	*Gillicus arcuatus*
	Prosaurodon pygmaeus	*Saurocephalus lanciformis*
Saurodon leanus	*Saurodon leanus*	
Urenchelys abditus (eel)	*Leptecodon* sp.	
	Apateodus sp.	*Apateodus* sp.
Apsopelix anglicus	*Apsopelix anglicus*	*Apsopelix anglicus*
Pachyrhizodus minimus	*Pachyrhizodus minimus*	*Pachyrhizodus minimus*
Pachyrhizodus caninus	*Pachyrhizodus caninus*	*Pachyrhizodus caninus*
Pachyrhizodus leptopsis	*Pachyrhizodus leptopsis*	
Unidentified albulid	Unidentified albulid	
Cimolichthys nepaholica	*Cimolichthys nepaholica*	*Cimolichthys nepaholica*
Enchodus shumardi	*Enchodus shumardi*	*Enchodus shumardi*
Enchodus gladiolus	*Enchodus gladiolus*	*Enchodus gladiolus*
Enchodus petrosus	*Enchodus petrosus*	*Enchodus petrosus*
Enchodus dirus	*Enchodus dirus*	
Stratodus apicalis	*Stratodus apicalis*	*Stratodus apicalis*
Thryptodus zitteli		
Martinichthys ziphioides		
Martinichthys brevis		
"Other plethodids"	*Niobrara encarsia*	*Pentanogmius evolutus*
	Zanclites xenurus	
	Caproberyx sp.	
	Trachichthyoides sp.	
Holocentrids	Holocentrids	*Kansius sternbergi*
Unidentified coelacanth		*Aethocephalichthys hyainarhinos*

Turtles	**Turtles**	**Turtles**
Toxochelys latiremis	*Toxochelys latiremis*	*Toxochelys latiremis*
Porthochelys laticeps	*Ctenochelys stenopora*	*Bothremys barberi*
Chelosphargis advena		*Protostega gigas*

Plesiosaurs	**Plesiosaurs**	**Plesiosaurs**
Unidentified *polycotylid*	Unidentified *polycotylid*	*Polycotylus latipinnis*
		Dolichorhynchops osborni
	Styxosaurus snowii (?)	*Styxosaurus snowii*
		"*Elasmosaurus*" *sternbergi*
		Unidentified elasmosaur

Mosasaurs	**Mosasaurs**	**Mosasaurs**
Clidastes liodontus	*Clidastes liodontus*	*Clidastes liodontus*
		Clidastes propython

TABLE 13.1. (continued)

Mosasaurs	Mosasaurs	Mosasaurs
Platecarpus tympaniticus	*Platecarpus tympaniticus*	*Platecarpus tympaniticus*
	Platecarpus planifrons	
Tylosaurus n. sp.		
Tylosaurus nepaeolicus	*Tylosaurus proriger*	*Tylosaurus proriger*
	Ectenosaurus clidastoides (?)	*Ectenosaurus clidastoides*
		Halisaurus sternbergi

Pterosaurs	Pterosaurs	Pterosaurs
Pteranodon sternbergi	*Pteranodon longiceps*	*Pteranodon longiceps*
	Nyctosaurus gracilis	*Nyctosaurus gracilis*

Birds	Birds	Birds
Ichthyornis dispar	*Ichthyornis dispar*	*Ichthyornis dispar*
		Guildavis tener
		Apatornis celer
		Iaceornis marshi
		Baptornis advenus
		Parahesperornis alexi
		Hesperornis regalis

Dinosaurs	Dinosaurs	Dinosaurs
Niobrarasaurus coleii	*Niobrarasaurus coleii*	*Claosaurus agilis*
"*Hierosaurus sternbergi*"	Nodosauridae *incertae sedis*	

Without much competition for resources and with few predators, the inoceramids that could live there grew to be large and abundant, feeding on the continuous rain of organic debris from overhead. There were relatively few vertebrates living or feeding on the sea bottom, and we know little about most of them.

Volviceramus grandis was the largest of the clams in the lower chalk, reaching the size and relative shape of a bathroom basin. The shells of one species of rudist clam (*Durania maxima*) usually occurred as solitary individuals, but they are sometimes found in groups that have grown together in a small reef-like mass (Fig. 13.2). The outer surfaces of most shells (and any other solid surface) were usually covered with oysters (*Pseudoperna congesta*), with new generations of these oysters building on top of the older shells. Even the bones of dead animals were sometimes briefly colo-

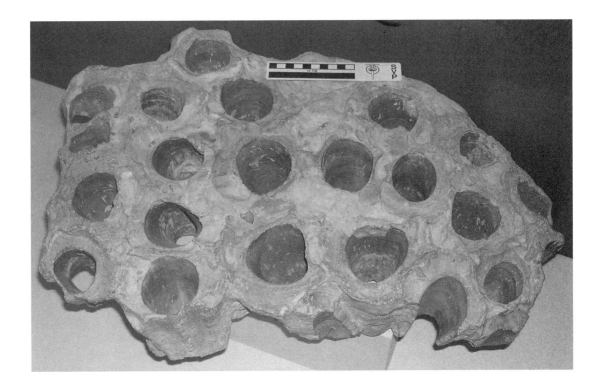

Figure 13.2. Exhibit in the Sternberg Museum showing an example of a rudist clam (Durania maxima) in the rare associated form. Most rudist remains from Smoky Hill Chalk are those of individuals.

nized between the time that they were defleshed and when they were covered by the chalky ooze. It was not a good place to attach, however, because the bones were covered by sediments within the space of a couple of years and any oysters that were living there were smothered while they were relatively young. Suitable sites for attachment were at a premium and only the inoceramids and the rudists figured out how to keep their shells above the soft and oxygen-poor mud bottom.

Ammonites were probably fairly numerous, but because their aragonite shells were usually dissolved before they could be preserved as fossils, we know little about them. The only evidence of their presence is a few partial external casts of their shells, and the delicate aptychi (a paired jaw structure; e.g., *Spinaptychus* n. sp.) which were not dissolved. A single species of giant squid (*Tusoteuthis longa*) probably completed with the ammonites for prey and may have grown as large as 10+ m (35+ ft.) in length. It is important to note that like modern squids, most of the length of a Late Cretaceous *Tusoteuthis* was in the arms, and the body was less than one-fifth of its total length. While it is interesting to think of these giant squid attacking mosasaurs and visa versa, no evidence exists in either case. More than likely, smaller squid were an abundant food source for many other predators (e.g., *Cimolichthys*, Chapter 5).

During the early deposition of the chalk (late Coniacian) we see the greatest variety of shark species, including the shell crushers

Figure 13.3. Close-up of the anterior teeth of a Cretoxyrhina mantelli "mummy" in the collection of the Sternberg Museum of Natural History. The specimen (FHSM VP-2187) is about 4.6 m (15 ft.) in length, and preserves the calcified cartilage of the cranium and jaws, including a complete dentition, as well as most of the calcified centra in the vertebral column.

(ptychodontids), and it appears that the giant ginsu shark, *Cretoxyrhina mantelli,* had reached its maximum size and abundance, based on the number and size of the shed teeth that we find. Almost all of the *Cretoxyrhina* "shark mummies" that I am aware of come from the lower chalk (Fig. 13.3). In these rare cases (Chapter 4), a major portion of the body of the shark, including the calcified cartilage of the braincase, jaws, fins and vertebrae, complete sets of teeth, and dermal denticles (shagreen) are preserved. *Cretolamna appendiculata* and *Scapanorhynchus raphiodon* teeth are found relatively rarely in the low chalk, while *Squalicorax falcatus* seems to be abundant, represented by shed teeth (an estimated 2/3 of all teeth in collections from the Smoky Hill Chalk), bite marks on the bones of many vertebrate specimens and occasional, well-preserved "mummies."

The ptychodontid sharks (*Ptychodus mortoni,* etc.) are relatively abundant during this period, at the same time that the bowl-shaped inoceramid, *Volviceramus grandis,* is the most common bivalve. Given their crushing dentitions, it seems apparent that the ptychodontids were feeding on hard-shelled prey on the bottom of the sea, although catching ammonites closer to the surface cannot be ruled out. Two genera of bony fish (*Martinichthys* and *Thryptodus*) that were possibly feeding on bivalves, or the oysters growing on them, were also fairly common at this time, and then disappeared as *V. grandis* became extinct. Little is known about these odd fish with their "battering-ram" noses (e.g., Figs. 5.7 and 5.8).

There are few complete specimens of small fish preserved from the late Coniacian chalk, although tiny fish bones, teeth and vertebrae are found occasionally in coprolites, as well as in acid washed samples of chalk. The tiny, distinctive jaws of *Enchodus shumardi* with their long front fangs are seen occasionally when layers of chalk are split apart. They represent individuals that would have been less than 15 cm (6 in.) in length. *Apsopelix anglicus* (Fig. 3.4) and *Pachyrhizodus minimus* are the smallest fish for which we have complete specimens from the chalk (excluding the tiny fish found preserved inside clam shells), ranging from about 0.3 m (12 in.) to 1 m (40 in.) in length. For unknown reasons, these two small fish seem to preserve far better than the larger individuals of other species.

Occasionally the fragmentary remains of other small fish have been found in the lower chalk. The distinctive, shell-crushing jaw plates of a small pycnodont (*Micropycnodon kansasensis*) occur occasionally, and are collected only in the lower chalk and underlying Fort Hays Limestone. While it is possible that they were feeding on the epifauna that was attached to the large inoceramids on the sea bottom, Earl Manning (pers. comm., 2004) believes it is more likely that these remains were from shallow-water fish that strayed into the middle of the deep sea and died there. Pycnodont remains are most often found in shallow water deposits, from nearer to shore. Another fish that may have made a "wrong turn" and ended up in deep water where it died was a small coelacanth specimen (LACM 131958) found by my wife, Pam, in 1990. To date, it is the only coelacanth that I am aware of from the Smoky Hill Chalk.

Among the medium-sized bony fish, *Enchodus petrosus*,

Figure 13.4. This relatively large (0.5 m [20 in.]) and complete specimen of Apsopelix anglicus *is in the fossil fish exhibit at the Sternberg Museum of Natural History. Many specimens of this species are less than 0. 3 m (12 in.) in length.*

Cimolichthys nepaholica, Saurodon leanus, Gillicus arcuatus, and several species of *Protosphyraena* appear to be the most common lower chalk varieties. *Enchodus,* with its over-sized palatine and anterior dentary fangs, occurs from the beginning of the Late Cretaceous and, unlike most Late Cretaceous fish, apparently survived for a time after the end of the Cretaceous. Its close relative, *Cimolichthys,* lived through the deposition of the chalk and into the Pierre Shale (middle Campanian) and has been found occasionally with the preserved remains of its last meal, including *Enchodus,* or in one case, a squid (*Tusoteuthis longa*). *Saurodon* was a long, eel-like fish and is unusual for the sword-like beak (predentary) attached to the front of its lower jaw. *Protosphyraena* also possessed a marlin-like snout, although in its case, it was attached as would be expected to the upper jaw. In addition, *Protosphyraena* had large, forward-pointing, blade-like teeth, and two species (*P. perniciosa* and *P. tenuis*) had long, saw-toothed fins. Whether this assortment of dangerous looking attachments were offensive or defensive weapons, or were weapons at all, we do not know. *Gillicus,* on the other hand, was a fairly large fish (up to 2 m [6 ft.]) with tiny teeth, and no other obvious armament. While its main claim to fame was ending up as preserved stomach contents in a giant fish called *Xiphactinus,* I have no doubt that it was a serious menace to smaller fish, and a successful predator in its own right.

The largest of the predatory bony fish during the late Coniacian were *Xiphactinus audax* and *Pachyrhizodus caninus.* Both of these species had large jaws and heavy teeth for preying on other animals. A *Pachyrhizodus* specimen (FHSM VP-2189) at the Sternberg Museum has a skull and teeth that are nearly as large as those of any *Xiphactinus* that I have ever seen. The body of *Pachyrhizodus,* however, is much shorter and more compact than that of *Xiphactinus.* Both would have been fearsome threats to squid, smaller fish and the occasional baby marine reptile. We know from the number of *Xiphactinus* specimens preserved with large prey inside (Chapter 5) that sometimes things went wrong, and the predator died before the meal could be digested. I am not aware of any specimens of *Pachyrhizodus caninus* that have been found with preserved stomach contents.

The late Coniacian represents a sort of "coming out" party for the mosasaurs (Chapter 9). Although the first mosasaur remains in Kansas occur in the middle Turonian Carlile Shale, they are few and far between (Martin and Stewart, 1977). Within a few million years, by the beginning of deposition of the chalk, mosasaurs are the second most common vertebrate fossil found (fish being the most common) in the chalk. Four species are found in the lower chalk: *Clidastes liodontus, Platecarpus tympaniticus, Tylosaurus nepaeolicus* and a second, undescribed species of *Tylosaurus.* The two species of *Tylosaurus* are the most abundant in terms of the number of specimens found (pers. obs.) and are also the largest of the three genera. Based on the identification of a badly scavenged

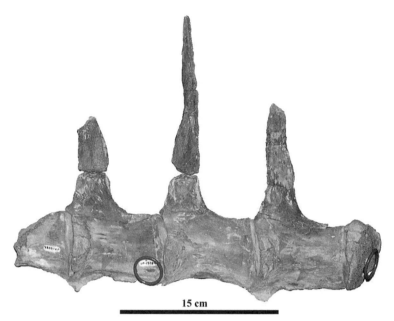

Figure 13.5. Cretoxyrhina mantelli *was the largest shark in the Western Interior Sea during deposition of the Smoky Hill Chalk and was capable of biting through the large bones of marine reptiles such as mosasaurs. This photograph shows a dorsal view of five vertebrae (FHSM VP-13283) severed from the middle of the back of a 6 m (20 ft.) long mosasaur. The anterior-most (left) and posterior-most (right) vertebrae have been severed by the shark's bite. Circles show the location of fragments of shark teeth that are still embedded in the bone. This piece of the mosasaur was swallowed by the shark and partially digested.*

15 cm

series of mosasaur vertebrae (Everhart, 2004a), *Ectenosaurus* may have been a rare inhabitant of the Western Interior Sea as early as the end of the Coniacian. Contrary to earlier reports by Williston (1898b) and others, mosasaurs of all ages are found in the lower chalk (Everhart, 2002), ranging from babies (2 m in length or less) to full-grown adults (7+ m).

The most common mosasaur specimens that are collected in the lower chalk are the pieces (skulls, vertebrae, and limbs) that have been severed by large sharks, swallowed and partially digested (Chapter 4). We are fairly certain of this scenario because we also find the broken tips of shark teeth embedded in these bones (Shimada, 1997), and know that the only predator capable of doing that kind of damage to a mosasaur carcass at the time was the giant ginsu shark, *Cretoxyrhina mantelli* (Fig. 13.5). While it may have been a bad time for the new guys (mosasaurs) as they spread across the Western Interior Sea and encountered the ginsu sharks, the mosasaurs were probably also feeding on baby sharks. Within a period of about five million years (by the early Coniacian), mosasaurs were much larger and more numerous, and the ginsu shark became extinct.

Sharks and mosasaurs may have made life tough not just on each other, but also on the few species of plesiosaurs living in the Western Interior Sea (Chapter 7). Until recently (Everhart, 2003), no remains of any plesiosaurs had been documented from the late Coniacian or Santonian of the Smoky Hill Chalk. The fragmentary, partially digested remains which have been found in the last fifteen years indicate that few plesiosaurs ventured into the middle of the sea, and those that did were likely to have been preyed upon by sharks. It is equally possible, however, that the very limited remains of plesiosaurs (six specimens—probably polycotylid) that have been found in this stratigraphic interval are pieces scavenged from carcasses that had floated in from other places.

Elasmosaurids were certainly still living around the edges of the Western Interior Sea, closer to shore and a possibly safer habitat, but have not yet been found in the late Coniacian chalk. A shark specimen (KUVP 68979; probably *Cretoxyrhina*) collected by H. T. Martin from the lower chalk near Hackberry Creek in Trego County provides indirect evidence of the presence of elasmosaurids during that time. Moodie (1912) briefly mentioned the incomplete remains of a large shark that contained "many hundreds of greatly abraded, very smooth and polished stones." Based on the size of the 41 calcified vertebral centra (10 cm [4 in.]) that were preserved, I estimate that the shark was at least 6 m (20 ft.) in length. It apparently died with more than 120 black chert gastroliths as gut contents (Everhart, 2000). The larger stones are about 7 cm (2.6 in.) long. Since sharks do not normally have gastroliths and elasmosaurids usually have large numbers of them (ibid.), it appears likely that the shark fed on an elasmosaurid before it died (Shimada, 1997).

Although the remains of an unidentified pterosaur reported from the Cenomanian Greenhorn Limestone represent the earliest occurrence of this group in Kansas (Liggett et al., 1997), the oldest *Pteranodon* remains occur in the chalk near the end of the Coniacian. The remains are few and fragmentary, and as discussed in Chapter 10, may represent individuals that died during migrations across the sea rather than those who were actively feeding in it. It is also likely that large hurricanes/typhoons occurred over the Western Interior Sea just as they do in our modern oceans. We do know that the bodies of dinosaurs living along the shores of the Western Interior Sea during this time could be transported hundreds of miles out to sea to where they sank and were buried in the Smoky Hill Chalk. The same process could have happened to the pteranodons and the plesiosaurs mentioned above. The remains of pterosaurs occur much more frequently in the upper chalk when the nearest land would have been much closer.

The only remains of birds are found in the lower chalk are those of the pigeon-sized *Ichthyornis dispar,* and they are limited to small fragments which appear to have fallen off or been removed from floating carcasses as they decomposed. So far as I am aware, such specimens have been very rare. Clarke (2004) recently reported that the majority of the *Ichthyornis* bones (mostly from the upper chalk) she examined in the Marsh collection at the Yale Peabody Museum were from the limbs (wings and legs), and inferred that many of those bones fell off of floating carcasses. Regardless of how the remains ended up in the chalk, it seems unlikely that there were many birds flying over the middle of the Western Interior Sea during the late Coniacian.

Santonian

The middle portion of the Smoky Hill Chalk was deposited during the Santonian, roughly 85.8 to 83.5 million years ago. From the fossil record, it appears that there was a major species turnover during the early part of Santonian. Unfortunately, the record for vertebrates is not as complete for the middle of the chalk as it is for the beginning and the end. In part, this may be due to a thick layer of resistant chalk near the middle that effectively tends to cap the underlying layers and results in relatively tall, vertical exposures that are more difficult to search and to collect. In any case, we know a bit less about the Santonian vertebrates in the chalk than about those of the late Coniacian or early Campanian.

By the beginning of the Santonian, the large bowl-shaped inoceramid, *Volviceramus grandis*, had been replaced by another large species (*Cladoceramus undulatoplicatus*). The new species was relatively thin-shelled and had a "rippled" appearance. It shared the sea floor with *Platyceramus platinus* clams that were somewhat larger than during the late Coniacian. The few remaining rudist clams were gone, but would reappear briefly during the early Campanian. *C. undulatoplicatus*, however, became extinct within the early Santonian, and *P. platinus* continued to get larger. While inoceramids were present almost to the top of the chalk, *P. platinus* became less abundant after about the middle of the Santonian.

Following the extinction of *C. undulatoplicatus* sometime before the middle of the Santonian, an unusual crinoid, *Unitacrinus socialis*, made a brief appearance in the Western Interior Sea and around the world. From the remains that are found of this crinoid in Kansas, it appears that they always lived together in a group. Most of the fossil crinoids (sea lilies) that we are familiar with from the Paleozoic were anchored to the sea bottom by a long stem and holdfasts. It appears that *Unitacrinus* was a free-swimming species, much like the few remaining modern crinoids. All that is visible in the preserved specimens is the calyx and a mass of tangled arms. It appears likely that they somehow floated at or near the surface with their long arms extending downward to feed. How they died is unknown, but when they died the colonies remained relatively intact on the way to the sea bottom, where dozens of individuals were covered and preserved *en masse*.

Two small, coiled scaphitid ammonites occurred near the middle of the Santonian: *Clioscaphites vermiformis* and *C. choteauensis* (Fig. 13.6). They may have occurred earlier but bottom conditions may have not been favorable for their preservation. Like their larger cousins, the ammonites that are found in the late Coniacian, the aragonite shells of these cephalopods were dissolved before they could be preserved. However, the process apparently occurred at a slow enough pace that the shape of the outer surface of the shell was retained as a finely detailed mold that is found occasionally when layers of chalk are carefully split. Although these scaphi-

Figure 13.6. An external cast of Clioscaphites choteauensis *from the Smoky Hill Chalk (early Santonian). This specimen in the Sternberg Museum of Natural History is about 10 cm (4 in) in diameter.*

tids were probably predators on smaller prey or scavengers, they were also a likely source of food for many of the larger predators, including mosasaurs.

Most of the ptychodontid sharks were extinct in the Western Interior Sea by the early Santonian. One species, *Ptychodus mortoni,* survived for a while longer in small numbers. We don't know whether this extinction is due to climate change, loss of their favored prey, or some other factor. *Cretoxyrhina* and *Cretolamna* were present, although in somewhat lower numbers as measured by the number of their teeth found in the Santonian chalk. *Squalicorax falcatus* appears to have been the most common shark at the time, again measured in terms of shed teeth, and appears to have become slightly larger. Its teeth and bite marks continue to be associated with the remains of other vertebrates, including other sharks, and it seems likely that these sharks were active scavengers. Another species, *S. kaupi,* first appeared in the chalk in small numbers during the early Santonian (pers. obs.) and replaced *S. falcatus* by the early Campanian. The other, smaller varieties of sharks (*Pseudocorax, Scapanorhynchus*) that have been found in the late Coniacian do not occur in the Santonian. Shimada (pers. comm., 2004) has recently recovered the teeth of the guitarfish, *Rhinobatos incertus,* and several other unreported species from an unusual bone bed collected from near Monument Rock in the middle of the chalk.

Most of the larger bony fish found in the late Coniacian continued into the Santonian. *Protosphyraena perniciosa* and *P. nitida*, however, appear to have become extinct at the end of the Coniacian, along with the plethodids, *Martinichthys* and *Thryptodus*. The remains of the three ichthyodectid species (*Xiphactinus*, *Ichthyodectes*, and *Gillicus*) are common, although most specimens of *Gillicus* are represented by only a detached caudal fin (pers. obs.—the remains of feeding by an unknown predator). The 4-m (13-ft.) fish-in-a-fish specimen at the Sternberg Museum comes from the early to middle Santonian, along with a larger (5.2 m [17 ft.]) *Xiphactinus* with a partially digested *Gillicus* inside that I found in 1996. Specimens of *Pachyrhizodus*, *Cimolichthys*, and *Enchodus* are also relatively common in the middle chalk.

Stewart (1990b; 1990c) reported on the discovery of numerous small (less than 12 cm [5 in.]) holocentrid (squirrelfish) bony fish that lived and died inside the giant inoceramid clams (*Platyceramus platinus*). Such preservation is apparently not unusual in a relatively narrow interval of chalk near Hattin's (1982) marker unit 8. The fish (Fig. 13.7) have not yet been formally described, but represent sizes and species not commonly found in the Smoky Hill Chalk. It seems likely that small schools (up to about a hundred individuals of a single species in at least one specimen—KUVP 49403) of these fish were living inside the clams and were trapped there when the clams died (Stewart, 1990b). The remains were effectively sealed inside the closed shell and protected from scavengers. Stewart (ibid.) also notes that other small species, including *Kansius sternbergi*, *Urenchelys abditus*, the small amioid *Par-*

Figure 13.7. Close-up shows the partial remains of several small (less than 12 cm [5 in.]) holocentrid fish preserved inside a Platyceramus platinus *shell from the Smoky Hill Chalk (early Santonian). The skeleton of a fairly complete specimen of this undescribed species is shown at the top of the photo (anterior to the left). (Scale = mm)*

aliodesmus guadagnii, and the eel-like dercetid, *Leptecodon rectus*, were found inside the shells of *Platyceramus platinus* from the late Coniacian and Santonian.

Toxochelys latiremis was the most common sea turtle found in the chalk through the Santonian and into the overlying early Campanian Pierre Shale. Several other species of small sea turtles lived in the Western Interior Sea (Chapter 6), but their remains are too rare and too fragmentary to say much about their occurrence. The scattered, partial remains of smaller sea turtles in the chalk provide evidence, however, that very few, if any, of these turtles died of old age. They appear to have been a favored prey of both *Cretoxyrhina* and *Squalicorax* sharks, and it seems likely that mosasaurs would have also eaten them.

Plesiosaurs appear to be largely absent during most of the Santonian in Kansas. Stewart (1990a) notes the occurrence of a possible polycotylid in the middle Santonian, but I am otherwise unaware of other specimens.

Relatively few remains of mosasaurs have been collected from the middle chalk. An early example of *Tylosaurus proriger* (FFHM 1997-10) was reported by Everhart (2001) from the middle Santonian of Gove County (Fig. 13.8). The most complete specimen known of *Platecarpus planifrons* (FHSM VP-13910) is also from the early Santonian (ibid.) of Gove County. *Platecarpus tympaniticus* appears to have been the most common mosasaur in the Santonian while *Tylosaurus proriger* became larger and approached 9 m (30 ft.) in length. The much smaller *Clidastes* was rare but still present in small numbers in the relatively deep water of the Western Interior Sea. They were much more common in the near shore environment of the Gulf Coast at that time. The nearly complete specimen of *Ectenosaurus clidastoides* (FHSM VP-401; Fig. 9.5) in

Figure 13.8. Dorso-left lateral view of the skull of the earliest known example of Tylosaurus proriger *(FFHM 1997–10), in the collection of the Fick Fossil and History Museum in Oakley, Kansas. The remains were found in the early to middle Santonian chalk of Gove County. The articulated skull is 1.2 m (4 ft.) in length and represents a mosasaur that would have been about 8.6 m (28 ft.) long. The scleral ring around the left eye and the tympanic membrane on the left quadrate were both preserved intact in this specimen.*

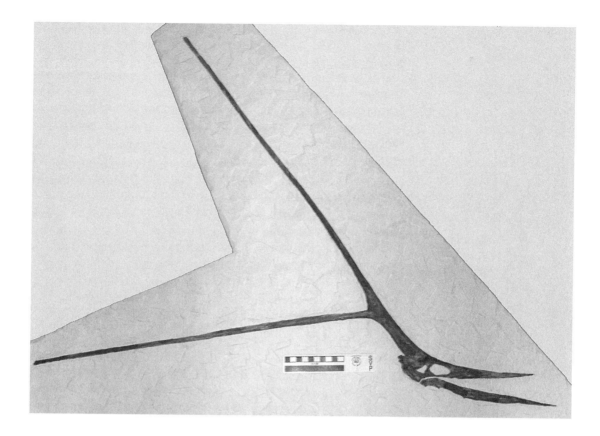

the Sternberg Museum, discovered in northwest Trego County in 1963, most likely came from chalk deposited near the end of Santonian.

The type specimen of the strangely crested *Pteranodon sternbergi* is also from the Santonian chalk. The size of the skull indicates that it came from a large individual, with a wingspan of at least 7 m (24 ft.). The smaller pterosaur, *Nyctosaurus gracilis*, also appears for the first time in the late Santonian. Bennett (2003) reported on the discovery of two unusual *Nyctosaurus* specimens with very large crests in the Santonian chalk of Trego County (Fig. 13.9). The occurrence of pterosaur specimens becomes much more frequent as you go higher in the chalk as the Western Interior Sea becomes shallower and land becomes closer. While this may mean that there were larger populations of pterosaurs flying over the Western Interior Sea at that time, I think the most likely reason is that land was much closer to where these remains occurred. The question of whether these specimens represent animals that died during routine feeding at sea, or instead constitute losses during migration or storms, is likely to remain unanswered for now. As the sea became narrower, it is also possible that more floating carcasses of pterosaurs could be carried into the middle, although one would also expect to see the remains of more terrestrial animals (dinosaurs) if that was the case.

Figure 13.9. Cast of the skull in right lateral view of the smaller of two crested Nyctosaurus gracilis *specimens found recently by Kenneth Jenkins in the Santonian chalk of Trego County. The skull measures 24.5 cm (9.6 in), while the crest is almost three times longer (71.7 cm [28 in.]). Cast by Kenneth Jenkins (Sternberg Museum of Natural History).*

Ichthyornis is the only bird represented in collections from the Santonian chalk. Several of these are fairly complete specimens, including the type specimen of *Ichthyornis dispar* (YPM 1450; See Clarke, 2004) and a more recently collected set of remains in the Sternberg collection (FHSM VP-2503; Fig. 11.6). Both of these specimens are from the middle chalk of Rooks County and appear to represent remains that had dropped to the sea floor relatively soon after death, before their wings and legs had a chance to fall off or be detached by scavengers.

Early Campanian

By the early Campanian (83.5–82 mya), there were very few large invertebrates (notably inoceramid clams) living on the sea bottom where the Smoky Hill Chalk was being deposited. Although uncommon, *Platyceramus platinus* still served as a substrate for attachment of oysters. *Pseudoperna congesta* appears to be on the decline during the early Campanian and was replaced other species. Environmental conditions on the sea floor may have changed to the point that it was no longer suitable for the giant bivalves. The continued presence of ammonites is indicated by the preservation of their aptychi (*Rugaptychus* sp.), along with external casts of *Baculites* sp. Squid (*Tusoteuthis longa*) remains are found throughout the chalk and into the overlying Pierre Shale. Vertebrate remains (Fig. 13.10) are unusually well preserved (uncrushed and relatively undisturbed) in the early Campanian chalk as compared to preservation in the lower and middle chalk. The incidence of scavenging by sharks on vertebrate remains appears to be much less based on fewer observed bite marks.

Cretoxyrhina mantelli makes its last appearance in the Western Interior sometime before the end of deposition of the chalk. My

Figure 13.10. Two dorsal vertebrae from a medium-sized mosasaur (probably Platecarpus tympaniticus) *that I discovered in the upper chalk (early Campanian) of Logan County. The uncrushed condition is typical of remains found in the upper chalk and was the reason that many of the early collectors favored working there rather than lower in the formation. Compare these vertebrae with examples of flattened vertebrae from the late Coniacian (Fig. 13.5). This photograph also shows the characteristic rounded condyle and cupped cotyle of mosasaur vertebrae. (Scale = cm)*

wife and I have collected their teeth as high in the chalk as Hattin's (1982) marker unit 21. It has not been found in the overlying Pierre Shale. *Squalicorax kaupi* replaced *S. falcatus* as the most common shark, and in turn was replaced by *S. pristodontus* later in the Campanian. The only two *S. pristodontus* teeth (FHSM VP-15010 and VP-15011) that I know of from the Smoky Hill Chalk were collected by a friend of mine (Pete Bussen) from Logan County right below the contact with the Pierre Shale. This species becomes more common and larger in the middle of the Campanian. While *S. pristodontus* replaced *Cretoxyrhina mantelli* as the dominant large shark in the Western Interior Sea, it never approached the size or achieved the bone-severing bite of the ginsu shark.

Although many of the more common fish species found in the lower chalk continued through to the overlying Pierre Shale (Carpenter 1990; 2003), their remains are not preserved as frequently in the upper levels of the chalk.

Protostega gigas, a giant turtle with a body about the size of a small car, seems to appear out of nowhere during the early Campanian. It is likely that this species migrated into the Western Interior Sea from the Gulf coast and decided to stay. My wife found the complete left side of the plastron of a sub-adult *Protostega* (FHSM VP-13448) in the upper chalk near the Santonian-Campanian boundary in 1994. This specimen may represent an "earliest" occurrence of the species in Kansas, but we do not have enough accurate stratigraphic data on other specimens to be certain. Although there were no bite marks visible on the delicate bones of the plastron, the fact that there were no other remains of the turtle in the immediate area suggests that the turtle's body had been dismembered by scavengers. Both *Cretoxyrhina* and *Squalicorax* are known to have scavenged the remains of *Protostega* (Chapter 6). A related, but larger species of sea turtle (*Archelon ischyros*) appears later in the Campanian and is found in the Pierre Shale of South Dakota. Although *Archelon* probably swam over what is now Kansas during that same period, the upper levels of the Pierre Shale that would have preserved their remains have been eroded away.

Conditions for plesiosaurs appear to have improved in the Campanian. Two species of polycotylids were present in the early Campanian: *Polycotylus latipinnis* Cope and *Dolichorhynchops osborni* Williston. Although they are never common, their remains do occur for the first time in the chalk as reasonably complete skeletons (Chapter 8). At least one is also known to have been found as stomach contents of a large *Tylosaurus proriger* (Everhart, 2004b). Elasmosaurids appear for the first time in the chalk and are represented by ten fragmentary specimens (Table 7.1). All except one (USNM 11910) of the specimens were found before 1900 and none have been found since 1927. Two of the more interesting ones are *Styxosaurus snowii* (KUVP 1301) and the giant "*Elasmosaurus*" *sternbergi* (KUVP 1312). Neither specimen is anywhere near complete. The type specimen of *Styxosaurus snowii* (Williston, 1890a) includes the only skull of an elasmosaurid

Figure 13.11. The centrum of this large dorsal vertebra of Elasmosaurus sternbergi *(KUVP 1312) from the early Campanian chalk of Logan County measures about 16 cm (6.5 in.) across the widest part and is at least (33%) wider than the largest dorsal vertebra of E. platyurus. Note that the picture shows the anterior (front) of the vertebra with the dorsal process to the right. The vertebra has been crushed from front to back and was originally somewhat thicker. (Scale = cm)*

known from the chalk, or from Kansas for that matter. The skull is 48 cm (19 in.) in length and is attached to the anterior portion of the neck containing 28 cervical vertebrae (Williston, 1890b). According to Charles H. Sternberg, the remains of "*E.*" *sternbergi* (Fig. 13.11) were originally much more complete before being nearly destroyed during the construction of a building.

Mosasaurs were much larger and more diverse in the upper chalk, although most of those changes are better exhibited in specimens from just above the contact of the Smoky Hill Chalk and the Pierre Shale. The largest known *Tylosaurus proriger* specimen from the early Campanian chalk is about 10 m (33 ft.) in length. The slightly younger "Bunker" *Tylosaurus proriger* specimen (KUVP 5033) from near the base of the Sharon Springs Member of the Pierre Shale is about 12 m (40 ft.) in length. *Platecarpus tympaniticus* remains the most common mosasaur in the upper chalk, and *Clidastes liodontus* is present in larger numbers. Neither continued very far into the Campanian. While *Ectenosaurus clidastoides, Clidastes propython* and *Halisaurus sternbergi* occur in the upper chalk (early Campanian), *Globidens* (FHSM VP-13828) occurs in Kansas for the first time in the overlying early Campanian Pierre Shale along with *Plioplatecarpus* (pers. obs.). *Mosasaurus* and *Prognathodon* occur at a slightly higher level in the Pierre Shale than is preserved in Kansas. The number and size of mosasaurs found in the early Campanian indicate that even as it narrowed, the Western Interior Sea was still teeming with abundant life and was capable of supporting a large number of very large predators.

Pteranodon longiceps appears to have replaced *P. sternbergi* by the latter part of the Santonian in the Smoky Hill Chalk and continues upward into the Pierre Shale. Full grown, presumed males of the species have a wingspread of at least 7 m (24 ft.), although the average wingspread of the male specimens (Bennett, 1992) that have been collected is about 5.6 m (18 ft.). A few specimens preserve stomach contents (Chapter 11) and indicate that small fish were the usual prey of pteranodons. The remains of a smaller pterosaur (*Nyctosaurus*) are also found more frequently in the upper chalk and this may indicate of a significant reduction in the flying distance from the nearest land.

More bird species occur in the early Campanian chalk than during the Santonian or late Coniacian. Besides *Ichthyornis dispar,* the list of flying birds includes *Guildavis tener, Apatornis celer,* and a recently named species, *Iaceornis marshi* (Clarke, 2004). The first examples of the large and flightless *Hesperornis regalis* also occur in the upper chalk. Martin (1984) described a new species of more primitive hesperornithid from the upper chalk (*Parahesperornis alexi*—KUVP 2287) that was about 30 percent smaller than the type specimen of *Hesperornis regalis,* but larger than *Baptornis advenus.* The new species had originally been referred to *Hesperornis gracilis* by Williston (1898a). The fact that more numerous remains of *Hesperornis* are found further north in Canadian deposits clearly indicates a connection of some sort with cooler waters. The arrival of these hesperornithids in more southern waters may indicate a cooler climate, closer shores, adaptations to changes in prey, or other factors. The presence of these diving birds certainly seems to distinguish the upper chalk fauna from that of the Coniacian and Santonian. While hesperornithids fed primarily on small fish and cephalopods, it is likely that they provided yet another food source for hungry mosasaurs (Chapter 8).

The preceding "big picture" demonstrates that the fauna of the Western Interior Sea (and the sea itself) was undergoing fairly constant changes during the five million years when the Smoky Hill Chalk was deposited. While we know a lot about the marine creatures of this "ocean of Kansas," there is clearly much more work to be done in the chalk to discover and describe what species were living there during the Late Cretaceous.

Epilogue:
Where Did It Go?

What happened to the oceans of Kansas? Where did all that water go? The answer is simple and complex at the same time. The simple answer is that near the end of the Cretaceous, the land beneath the Western Interior Sea rose, and as a result the ocean receded to its present shorelines.

Long before the end of the Cretaceous, however, the collision of two tectonic plates along the western edge of North America began the final process of closing the shallow sea covering the Midwest for the final time. Beginning during the Jurassic, the Farallon Plate subducted beneath (slid under) the western edge of the North American Plate as it was being pushed westward by the spreading of the Atlantic Ocean. Over a period of millions of years, the collision created the Rocky Mountains and slowly uplifted much of the Midwest. This process would eventually create the Great Plains by raising what had been sea bottom thousands of feet above sea level. During the Cretaceous, a nearly continuous series of volcanic eruptions caused by this collision of tectonic plates dumped many cubic miles of volcanic ash over the expanse of the shallow sea covering the midsection of North America. Some of these volcanic eruptions were much larger than any ever observed by man. Marine life (and dinosaurs) persevered through the hundreds of major eruptions and resulting ash falls that periodically blanketed much of the western half of North America and the Western Interior Sea. So far as I am aware, there are few, if any, records of a mass mortality of marine life associated with these catastrophic events.

About the middle of the Campanian (about 75 mya), a stony asteroid about 2 km (1.5 mi.) in diameter impacted along the eastern edge of the Western Interior Sea in what is now north-western Iowa near the small town of Manson. There is no evidence of the crater now on the surface. It was reduced by millions of years of erosion following the impact, and then ground down further by the glaciers that covered Iowa during the Ice Ages. The crater was so well hidden beneath relatively flat farm fields, in fact, that it remained unrecognized until the 1990s. Since then, studies have determined that it is a crater left by the impact of a large meteorite during the latter stages of the Cretaceous. It is still uncertain if the Manson impact point was in shallow water or on the nearby land, but the resulting crater was nearly 35 km (22 mi.) across. The environmental effects of this impact are still largely unstudied, but it is likely they were catastrophic for all forms of terrestrial and marine life within hundreds of miles of "ground zero." Based on studies involving the shape of the buried crater, it appears that the meteor came from the southeast at a low angle. Ejecta from the impact has been noted to the north-west of the crater in South Dakota from the shallow marine Crow Creek Member of the Pierre Shale, and there are indications of a major tsunami that would have inundated most of the coastal areas along the Western Interior Sea. While the effects were certainly devastating, they were probably much less so than those of the impact in the Yucatan that would follow in about ten million years at the end of the Cretaceous. The rich fossil record in marine deposits to the northwest in the Pierre Shale, however, does indicate that fish, birds, and marine reptiles returned to the seaway. In 2003, I was able to participate in the dig of a large elasmosaur in the Mobridge Chalk Member (middle Maastrichtian— dated at 68 mya) of the Pierre Shale in northeast Nebraska, not far from the Manson Crater. From the size and condition of the elasmosaur and the larger number of other remains that were found (mosasaurs, fish, invertebrates), it certainly appeared to me that the life in the sea was back to normal by that time.

By the late Maastrichtian (about 68 mya), however, in response to the uplift of the Rocky Mountains, the Western Interior Sea had been roughly divided through the middle into "Gulf" and "Arctic" remnants, with the land across northern Oklahoma and Kansas probably among the first areas to dry out. As the land continued to rise, the new shorelines continued to retreat further to the south and north. At the same time, formations of shale, sandstone, chalk, and limestone that had been underwater for millions of years were exposed to weathering and began to erode away. Over time, literally thousands of feet of marine sediments would disappear completely, leaving little or no evidence of the Cretaceous sea that had once covered most of the state. Enough remained, however, to give us a glimpse of the richness of life that existed in the oceans of Kansas.

So what happened at the very end of the Cretaceous? Did the asteroid/comet impact in the Yucatan Peninsula of Mexico cause the extinction of the dinosaurs, pteranodons, mosasaurs, and plesiosaurs? That question has yet to be completely resolved. My personal opinion is that it was a worldwide, long-term (millions of years) combination of events, including continental drift, that resulted in changes in oceanic currents (El Niño on a planetary scale), climate (especially temperature decreases), and increased volcanic activity around the world that significantly affected the environment on the land and in the ocean. At some point, large portions of the earth's ecology collapsed and species that were unable to adapt became extinct. While the Yucatan impact may have been the final straw, my gut feeling is that the extinction that we presume occurred at or near what we call the Cretaceous-Tertiary boundary would have happened anyway.

As a final note in that regard, I would leave you with a bit of mathematical juggling that puts the Yucatan impact into better perspective (at least for me). There have been reams of data and many papers and books produced about this event, popularizing the phrase "nuclear winter" and generally comparing it to the effect of thousands of the largest nuclear weapons going off at the same moment and place. While it was certainly the "end of the world" for anything living within hundreds or thousands of miles of the impact, or possibly all of North America, in my opinion it was much less so compared to the overwhelming size and mass of the earth.

Imagine a nearly smooth globe of rock, roughly 66 inches (1.7 m) in diameter. This is the earth reduced to a scale of one inch for every 120 miles. At this scale, the earth's 24,900-mile circumference at the equator would be reduced to slightly more than 17 feet (5.2 m). The atmosphere around our micro-planet is contained almost entirely within an inch (2.5 cm) of the surface. The average depth of the ocean at this scale is only 1/50 of an inch. Within the upper layer of the ocean and the base of the atmosphere, the whole life-zone containing the planet's biosphere is but a thin film.

Coming out of space at very high speed, a tiny rock, about the size of a BB (1/10 inch or 2.5 mm) impacts the surface of our micro-planet at the edge of the ocean. The impact generates enough heat to vaporize the rock and some of the surface of the little planet. The resulting crater is a circle about an inch (25 mm) wide and 1/120 of an inch deep. The diameter of the crater is 1/207 of the circumference of the globe and the area (less than one square inch) is about 0.00005 percent of the micro-planet's surface. Seen from even a short distance away (several feet), the impact hardly leaves a blemish on the surface of our little model.

In real life, was it devastating to at least some portion of the life on Earth? Certainly. Was it the sole cause of the major extinction at the end of the Cretaceous? I don't think so. Whatever killed off the marine reptiles and the dinosaurs of the Late Cretaceous was far more complex than just a big rock falling out of the sky.

1. Introduction

Buchanan, R. C., ed. 1984. *Kansas Geology: An Introduction to Landscapes, Rocks, Minerals and Fossils.* Lawrence: University Press of Kansas. 208 pp.

Buchanan, R. C., and J. R. McCauley. 1987. *Roadside Kansas: A Traveler's Guide to Its Geology and Landmarks.* Lawrence: University Press of Kansas. 365 pp.

Carpenter, K., D. Dilkes, and D. B. Weishampel. 1995. "The Dinosaurs of the Niobrara Chalk Formation (Upper Cretaceous, Kansas)." *Journal of Vertebrate Paleontology* 15 (2): 275–297.

Cope, E. D. 1872. [Sketch of an expedition in the valley of the Smoky Hill River in Kansas]. *Proceedings of the American Philosophical Society* 12 (87): 174–176.

Everhart, M. J., P. A. Everhart, and K. Ewell. 2004. "A Marine Ichthyofauna from the Upper Dakota Sandstone (Late Cretaceous)." Abstracts of oral presentations and posters, Joint Annual Meeting of the Kansas and Missouri Academies of Science, p. 48.

Hattin, D. E. 1982. *Stratigraphy and Depositional Environment of the Smoky Hill Chalk Member, Niobrara Chalk (Upper Cretaceous) of the Type Area, Western Kansas.* Kansas Geological Survey Bulletin, no. 225. Lawrence: Kansas Geological Survey, University of Kansas. 108 pp.

Hattin, D. E., and C. T. Siemers. 1978. *Upper Cretaceous Stratigraphy and Depositional Environments of Western Kansas.* Guidebook Series, no. 3. Lawrence: Kansas Geological Survey. 55 pp. (Reprinted 1987.)

Leidy, J. 1870. [Remarks on *Elasmosaurus platyurus*]. *Proceedings of the Academy of Natural Sciences of Philadelphia* 22: 9–10.

Lesquereux, L. 1868. "On Some Cretaceous Fossil Plants From Nebraska." *American Journal of Science,* Series 2, 46 (136): 91–105.

Mehl, M. G. 1941. "*Dakotasuchus kingi,* a Crocodile from the Dakota of Kansas." *Denison University Bulletin, Journal of the Scientific Laboratory* 36: 47–66. 3 figs., 2 pls.

Merriam, D. F. 1963. *The Geologic History of Kansas.* State Geological Survey of Kansas, Bulletin no. 162. Lawrence: University of Kansas. 317 pp. (Reprinted 1988.)

Rogers, K. 1991. *The Sternberg Fossil Hunters: A Dinosaur Dynasty*. Missoula, Mont.: Mountain Press Publishing Company, 288 pp.

Scott, R. W. 1970. *Paleoecology and Paleontology of the Lower Cretaceous Kiowa Formation, Kansas*. University of Kansas Paleontological Contributions, Article 52. Lawrence: University of Kansas Paleontological Institute. 94 pp.

Vaughn, P. P. 1956. "A Second Specimen of the Cretaceous Crocodile *Dakotasuchus* from Kansas." *Kansas Academy of Science, Transactions* 59 (3): 379–381.

2. Our Discovery of the Western Interior Sea

Almy, K. J. 1987. "Thof's Dragon and the Letters of Capt. Theophilus Turner, M.D., U.S. Army." *Kansas History Magazine* 10 (3): 170–200.

Anonymous. 1868. "Excursion to Monument." *Topeka Weekly Leader* 3 (36): 2 (August 13).

Baur, G. 1892. "On the Morphology of the Skull of the Mosasauridae." *Journal of Morphology* 7 (1): 1–22. 2 pls.

Cope, E. D. 1868a. "Remarks on a New Enaliosaurian, *Elasmosaurus platyurus*." *Proceedings of the Academy of Natural Sciences of Philadelphia* 20: 92–93.

———. 1868b. "Note on the fossil Reptiles, Near Fort Wallace." In J. L. LeConte, *Notes on the Geology of the Survey for the Extension of the Union Pacific Railway, E. D., from the Smoky Hill River, Kansas, to the Rio Grande*, p. 68. Philadelphia: Review Printing House. 76 pp. with folded map.

———. 1869a. [Remarks on *Holops brevispinus, Ornithotarsus immanis*, and *Macrosaurus proriger*]. *Proceedings of the Academy of Natural Sciences of Philadelphia* 11 (81): 123.

———. 1869b. "Synopsis of the Extinct Batrachia and Reptilia of North America: Part I." *Transactions of the American Philosophical Society*, New Series, 14: 1–235. 51 figs., 11 pls.

———. 1870. "Synopsis of the Extinct Batrachia and Reptilia of North America." *Transactions of the American Philosophical Society*, New Series, 14: 1–252 + i–viii. 55 figs., 14 pls.

———. 1871. "On the Fossil Reptiles and Fishes of the Cretaceous Rocks of Kansas." In *Special Reports, 4th Annual Report of the U.S. Geological Survey of the Territories (Hayden)*, Article 6, pp. 385–424 of Part 4. 511 pp.

———. 1872a. "Note on Some Cretaceous Vertebrata in the State Agricultural College of Kansas." *Proceedings of the American Philosophical Society* 12 (87): 168–170.

———. 1872b. [Sketch of an expedition in the valley of the Smoky Hill River in Kansas]. *Proceedings of the American Philosophical Society* 12 (87): 174–176.

———. 1872c. "On the Geology and Paleontology of the Cretaceous Strata of Kansas." *Preliminary Report of the United States Geological Survey of Montana and Portions of the Adjacent Territories*, Part III: *Paleontology*, pp. 318–349. Washington, D.C.: Government Printing Office.

———. 1872d. "Catalogue of the Pythonomorpha Found in the Cretaceous Strata of Kansas." *Proceedings of the Academy of Natural Sciences of Philadelphia* 12 (87): 264 and 12 (88): 265–287.

———. 1875. *The Vertebrata of the Cretaceous Formations of the West*.

Report of the U.S. Geological Survey Territories (Hayden) 2. Washington, D.C.: Government Printing Office. 302 pp., 57 pls.

Everhart, M. J. 2001. "Revisions to the Biostratigraphy of the Mosasauridae (Squamata) in the Smoky Hill Chalk Member of the Niobrara Chalk (Late Cretaceous) of Kansas." *Kansas Academy of Science, Transactions* 104 (1–2): 56–75.

Goldfuss, A. 1845. "Der Schädelbau des *Mosasaurus*, durch Beschreibung einer neuen Art dieser Gattung erläutert." *Nova Acta Academa Caesar Leopoldino-Carolinae Germanicae Natura Curiosorum* 21: 1–28, pls. VI–IX.

Harlan, R. 1824. "On a New Fossil Genus of the Order Enalio sauri, (of Conybeare)." *Proceedings of the Academy of Natural Sciences of Philadelphia*, Series 1, 3 (pt. 2): 331–337; pl. 12, figs. 1–5.

———. 1834. "Notice of the Discovery of the Remains of *Ichthyosaurus* in Missouri N. A." *Transactions of the American Philosophical Society* 4: 405–409, pl. 20.

Hattin, D. E. 1982. *Stratigraphy and Depositional Environment of the Smoky Hill Chalk Member, Niobrara Chalk (Upper Cretaceous) of the Type Area, Western Kansas*. Kansas Geological Survey Bulletin, no. 225. Lawrence: Kansas Geological Survey, University of Kansas. 108 pp.

Hayden, F. V. 1872. *Final Report of the United States Geological Survey of Nebraska and Portions of the Adjacent Territories*. Washington, D.C.: Government Printing Office. 264 pp.

Hays, I. 1830. "Description of a Fragment of the Head of a New Fossil Animal, Discovered in a Marl Pit, near Moorestown, New Jersey." *Transactions of the American Philosophical Society*, Series 2, 3 (18): 471–477, pl. 16.

LeConte, J. L. 1868. *Notes on the Geology of the Survey for the Extension of the Union Pacific Railway, E. D., from the Smoky Hill River, Kansas, to the Rio Grande*. Philadelphia: Review Printing House. 76 pp. with folded map.

Leidy, J. 1856. "Remarks on Certain Extinct Species of Fishes." *Proceedings of the Academy of Natural Sciences of Philadelphia* 8: 301–302.

———. 1868. [Photographs of fossil bones]. *Proceedings of the Academy of Natural Sciences of Philadelphia* 20: 316.

Lesquereux, L. 1868. "On Some Cretaceous Fossil Plants from Nebraska." *American Journal of Science*, Series 2, 46 (136): 91–105.

Logan, W. N. 1897. "The Upper Cretaceous of Kansas: With an Introduction by Erasmus Haworth." *University Geological Survey of Kansas* 2: 194–234.

Meek, F. B., and F. V. Hayden. 1861. "Descriptions of New Lower Silurian (Primordial), Jurassic, Cretaceous, and Tertiary Fossils, Collected in Nebraska." *Proceedings of the Academy of Natural Sciences of Philadelphia* 13: 415–447.

Mitchell, S. L. 1818. "Observations on the Geology of North America, Illustrated by the Description of Various Organic Remains Found in That Part of the World." In G. Cuvier, *Essay on the Theory of the Earth*, pp. 319–431, pls. 6–8. New York: Kirk and Mercein. xxiii + 431 pp., 8 pls.

Moore, R. C., and W. P. Hays. 1917. "Oil and Gas Resources of Kansas." *Kansas Geological Survey Bulletin* 3. 391 pp.

Moulton, G. E., ed. 1983–1997. *The Journals of the Lewis and Clark Expedition*. Vols. 1–11. Lincoln and London: University of Nebraska Press.

Mudge, B. F. 1866a. "Discovery of Fossil Footmarks In the Liassic (?) Formation in Kansas." *American Journal of Science,* Series 2, 41 (122): 174–176.

———. 1866b. *First Annual Report on the Geology of Kansas.* Lawrence: State Printer. 57 pp.

Peterson, J. M. 1987. "Science in Kansas: The Early Years, 1804–1875." *Kansas History Magazine* 10 (3): 201–240.

Rogers, K. 1991. *The Sternberg Fossil Hunters: A Dinosaur Dynasty.* Missoula, Mont.: Mountain Press Publishing Company, 288 pp.

Russell, D. A. 1967. *Systematics and Morphology of American Mosasaurs.* New Haven, Conn.: Peabody Museum of Natural History, Yale University, Bulletin, no. 23. New Haven, Conn.: Peabody Museum of Natural History, Yale University. 241 pp.

———. 1970. *The Vertebrate Fauna of the Selma Formation of Alabama.* Part VII: *The Mosasaurs.* Fieldiana, Geology Memoirs 3 (7): 369–380. Chicago: Chicago Natural History Museum.

Schumacher, B. A. 1993. "Biostratigraphy of Mosasauridae (Squamata, Varanoidea) from the Smoky Hill Chalk Member, Niobrara Chalk (Upper Cretaceous) of Western Kansas." M.S. thesis, Fort Hays State University. 68 pp.

Sheldon, M. A. 1996. "Stratigraphic Distribution of Mosasaurs in the Niobrara Formation of Kansas." *Paludicola* 1: 21–31.

Spamer, E. E., R. M. McCourt, R. Middleton, E. Gilmore, and S. B. Duran. 2000. "A National Treasure: Accounting for the Natural History Specimens from the Lewis and Clark Expedition (Western North America, 1803–1806) in the Academy of Natural Sciences of Philadelphia." *Proceedings of the Academy of Natural Sciences of Philadelphia* 150: 47–58.

Stewart, J. D. 1988. "The Stratigraphic Distribution of Late Cretaceous *Protosphyraena* in Kansas and Alabama: Geology, Paleontology and Biostratigraphy of Western Kansas." In M. E. Nelson, ed., *Geology, Paleontology and Biostratigraphy of Western Kansas: Articles in Honor of Myrl V. Walker,* pp. 80–94. Fort Hays Studies, Third Series, no. 10, Science. Hays, Kans.: Fort Hays State University.

———. 1990. "Niobrara Formation Vertebrate Stratigraphy." In S. C. Bennett, ed., *Niobrara Chalk Excursion Guidebook,* pp. 19–30. Lawrence: University of Kansas Museum of Natural History and Kansas Geological Survey.

Webb, W. E. 1872. *Buffalo Land: An Authentic Account of the Discoveries, Adventures, and Mishaps of a Scientific and Sporting Party in the Wild West.* Philadelphia: Hubbard Bros. 503 pp.

Williston, S. W. 1895. "New or Little-Known Extinct Vertebrates." *Kansas University Quarterly* 3 (3): 165–176.

———. 1897. "The Kansas Niobrara Cretaceous." *University Geological Survey of Kansas* 2: 235–246.

———. 1898a. "Addenda to Part I." *University Geological Survey of Kansas* 4: 28–32.

———. 1898b. "Mosasaurs." *University Geological Survey of Kansas* 4 (5): 81–221.

Zakrzewski, R. J. 1996. "Geologic Studies in Western Kansas in the 19th Century." *Kansas Academy of Science, Transactions* 99 (3–4): 124–133.

3. Invertebrates, Plants, and Trace Fossils

Brown, R. W. 1940. "Fossil Pearls from the Colorado Group of Western Kansas." *Washington Academy of Science* 30 (9): 365–374.

Buckland, W. 1829. "On the Discovery of a New Species of Pterodactyle; and Also of the Faeces of the *Ichthyosaurus;* and of a Black Substance Resembling Sepia, or Indian Ink, in the Lias at Lyme Regis." *Proceedings of the Geological Society of London* 1: 96–98.

Carpenter, K. 1996. "Sharon Springs Member, Pierre Shale (Lower Campanian) Depositional Environment and Origin of its Vertebrate Fauna, with a Review of North American Plesiosaurs." Ph.D. dissertation, University of Colorado. 251 pp.

Elias, M. K. 1933. "Cephalopods of the Pierre Formation of Wallace County, Kansas, and Adjacent Area." *University of Kansas Science Bulletin* 21 (9): 289–263.

Everhart, M. J. 1999. "Evidence of Feeding on Mosasaurs by the Late Cretaceous Lamniform Shark, *Cretoxyrhina mantelli.*" *Journal of Vertebrate Paleontology* 17 (Supplement to Number 3): 43A–44A.

———. 2001. "Revisions to the Biostratigraphy of the Mosasauridae (Squamata) in the Smoky Hill Chalk Member of the Niobrara Chalk (Late Cretaceous) of Kansas." *Kansas Academy of Science, Transactions* 104 (1–2): 56–75.

———. 2003. "First Records of Plesiosaur Remains in the Lower Smoky Hill Chalk Member (Upper Coniacian) of the Niobrara Formation in Western Kansas." *Kansas Academy of Science, Transactions* 106 (3–4): 139–148.

———. 2004a. "Plesiosaurs as the Food of Mosasaurs: New Data on the Stomach Contents of a *Tylosaurus proriger* (Squamata; Mosasauridae) from the Niobrara Formation of Western Kansas." *The Mosasaur* 7: 41–46.

———. 2004b. "Late Cretaceous Interaction between Predators and Prey: Evidence of Feeding by Two Species of Shark on a Mosasaur." *PalArch Vertebrate Paleontology Series* 1 (1): 1–7.

Everhart, M. J., and P. A. Everhart. 1992. "Oyster-Shell Concentrations: A Stratigraphic Marker in the Smoky Hill Chalk (Upper Cretaceous) of Western Kansas." *Kansas Academy of Science, Transactions* 11 (Abstracts): 12.

Gill, J. R., W. A. Cobban, and L. G. Schultz. 1972. *Stratigraphy and Composition of the Sharon Springs Member of the Pierre Shale in Western Kansas.* Geological Survey Professional Paper 728. Washington, D.C.: U.S. Government Printing Office. 50 pp., 2 pls.

Green, R. G. 1977. "*Niobrarateuthis walkeri,* a New Species of Teuthid from the Upper Cretaceous Niobrara Formation of Kansas." *Journal of Paleontology* 51 (5): 992–995.

Grinnell, G. B. 1876. "On a New Crinoid from the Cretaceous Formation of the West." *American Journal of Science,* Series 3, 12 (3): 81–83.

Hattin, D. E. 1982. *Stratigraphy and Depositional Environment of the Smoky Hill Chalk Member, Niobrara Chalk (Upper Cretaceous) of the Type Area, Western Kansas.* Kansas Geological Survey Bulletin, no. 225. Lawrence: Kansas Geological Survey, University of Kansas. 108 pp.

———. 1988. "Rudists as Historians: Smoky Hill Member of Niobrara Chalk (Upper Cretaceous) of Kansas." *Fort Hays Studies,* Third Series, 10: 4–22.

———. 1996. "Fossilized Regurgitate from Smoky Hill Chalk Member of Niobrara Chalk (Upper Cretaceous) of Kansas, USA." *Cretaceous Research* 17: 443–450.

Hawkins, T. 1834. *Memoirs of Ichthyosauri and Plesiosauri, Extinct Monsters of the Ancient Earth.* London: Relfe and Fletcher. 58 pp., 28 pls.

Hawn, F. 1858. "The Trias of Kansas." *Transactions of the Academy of Science of St. Louis* 1 (2): 171–172.

Hayden, F. V. 1873. *Report of the United States Geological Survey of the Territories.*

Kauffman, E. G. 1990. "Giant Fossil Inoceramid Bivalve Pearls." In A. J. Boucot, *Evolutionary Paleobiology of Behavior and Coevolution,* pp. 66–68. Amsterdam: Elsevier.

LeConte, J. L. 1868. *Notes on the Geology of the Survey for the Extension of the Union Pacific Railway, E. D., from the Smoky Hill River, Kansas, to the Rio Grande.* Philadelphia: Review Printing House. 76 pp. with folded map.

Lehman, U. 1979. "The Jaws and Radula of the Jurassic Ammonite *Dactylioceras.*" *Paleontology* 22 (1): 265–271, pl. 27.

Lesquereux, L. 1868. "On Some Cretaceous Fossil Plants From Nebraska." *American Journal of Science,* Series 2, 46 (136): 91–105.

Logan, W. N. 1897. "The Upper Cretaceous of Kansas: With an Introduction by Erasmus Haworth." *University Geological Survey of Kansas* 2: 195–234.

———. 1898. "The Invertebrates of the Benton, Niobrara and Fort Pierre Groups." *University Geological Survey of Kansas* 4: 432–518, pls. LXXXVI–CXX.

———. 1899. "Some Additions to the Cretaceous Invertebrates of Kansas." *Kansas University Quarterly* 8 (2): 87–98, pls. XX–XXIII.

Marsh, O. C. 1871a. "Scientific Expedition to the Rocky Mountains." *American Journal of Science,* Series 3, 1 (2): 142–143.

———. 1871b. "On the Geology of the Eastern Uintah Mountains." *American Journal of Science,* Series 3, 1 (3): 191–198.

Martin, J. E., and P. R. Bjork. 1987. "Gastric Residues Associated with a Mosasaur from the Late Cretaceous (Campanian) Pierre Shale in South Dakota." *Dakoterra* 3: 68–72.

McAllister, J. A. 1985. "Reevaluation of the Formation of Spiral Coprolites." *University of Kansas Paleontology Contributions,* Paper 114. 12 pp.

Meek, F. B., and F. V. Hayden. 1859. "On the So-Called Triassic Rocks of Kansas and Nebraska." *American Journal of Science,* Series 2, 27 (79): 31–35.

———. 1861. "Descriptions of New Lower Silurian (Primordial), Jurassic, Cretaceous, and Tertiary Fossils, Collected in Nebraska." *Proceedings of the Philadelphia Academy of Natural Science* 13: 415–447.

Miller, H. W., Jr. 1957a. "*Niobrarateuthis bonneri,* a New Genus and Species of Squid from the Niobrara Formation of Kansas." *Journal of Paleontology* 31 (5): 809–811.

———. 1957b. "Intestinal Casts in *Pachyrhizodus,* an Elopid Fish, from the Niobrara Formation of Kansas." *Kansas Academy of Science, Transactions* 60 (4): 399–401.

———. 1968a. *Invertebrate Fauna and Environment of Deposition of the Niobrara (Cretaceous) of Kansas.* Fort Hays Studies, Science Series, no. 8. Fort Hays, Kans.: Fort Hays Kansas State College. 90 pp.

———. 1968b. "*Enchoteuthis melanae* and *Kansasteuthis lindneri,* New

Genera and Species of Teuthids, and a Septid from the Niobrara Formation of Kansas." *Kansas Academy of Science, Transactions* 71 (2): 176–183.

———. 1969. "Additions to the Fauna of the Niobrara Formation of Kansas." *Kansas Academy of Science, Transactions* 72 (4): 533–546.

Miller, H. W., Jr., G. F. Sternberg, and M. V. Walker. 1957. "Uintacrinus Localities in the Niobrara Formation of Kansas." *Kansas Academy of Science, Transactions* 60 (2): 163–166. 2 figs.

Morton, N. 1981. "Aptychi: The Myth of the Ammonite Operculum." *Lethaia* 14 (1): 57–61.

Mudge, B. F. 1866. *First Annual Report on the Geology of Kansas.* Lawrence: State Printer. 57 pp.

———. 1876. "Notes on the Tertiary and Cretaceous Periods of Kansas." *Bulletin of the U.S. Geological Survey of the Territories (Hayden)* 2 (3): 211–221. Washington, D.C.: Government Printing Office.

———. 1877. "Notes on the Tertiary and Cretaceous Periods of Kansas." In *Ninth Annual Report of the U.S. Geological and Geographical Survey of the Territories (Hayden) for 1875,* Part I: *Geology,* pp. 277–294. Washington, D.C.: Government Printing Office. 827 pp.

Nicholls, E. L., and H. Isaak. 1987. "Stratigraphic and Taxonomic Significance of *Tusoteuthis longa* Logan (Coleoidea, Teuthida) from the Pembina Member, Pierre Shale (Campanian), of Manitoba." *Journal of Paleontology* 61 (4): 727–737.

Nicollet, J. N. 1843. *Report Intended to Illustrate a Map of the Hydrographical Basin of the Upper Mississippi River.* 26th Congress, 2nd Session, Senate Document 237, serial 380. 170 pp. with folded map.

Shimada, K. 1997. "Shark-Tooth-Bearing Coprolite from the Carlile Shale (Upper Cretaceous), Ellis County, Kansas." *Kansas Academy of Science, Transactions* 100 (3–4): 133–138.

Stenzel, H. B. 1971. "Oysters." In R. C. Moore, *Treatise on Invertebrate Paleontology,* Part N, vol. 3: *Mollusca 6: Bivalvia.* Lawrence: Geological Society of America and University of Kansas Press.

Sternberg, C. H. 1917. *Hunting Dinosaurs in the Badlands of the Red Deer River, Alberta, Canada.* San Diego, Calif.: published by the author.

———. 1922. "Explorations of the Permian of Texas and the Chalk of Kansas, 1918." *Kansas Academy of Science, Transactions* 30 (1): 119–120.

Sternberg, M. 1920. *George Miller Sternberg: A Biography.* Chicago: American Medical Association. 331 pp., 10 pls.

Stewart, J. D. 1976. "Teuthids of the North American Late Cretaceous." *Kansas Academy of Science, Transactions* 79 (3–4): 74.

———. 1978. "Enterospirae (Fossil Intestines) from the Upper Cretaceous Niobrara Formation of Western Kansas." *University of Kansas Paleontological Contributions* 89: 9–16.

———. 1990. "Niobrara Formation Vertebrate Stratigraphy." In S. C. Bennett, ed., *Niobrara Chalk Excursion Guidebook,* pp. 19–30. Lawrence: University of Kansas Museum of Natural History and the Kansas Geological Survey.

Stewart, J. D., and K. Carpenter. 1990. "Examples of Vertebrate Predation on Cephalopods in the Late Cretaceous of the Western Interior." In A. J. Boucout, ed., *Evolutionary Paleobiology of Behavior and Coevolution,* pp. 203–208. New York: Elsevier.

Williston, S. W. 1897. "The Kansas Niobrara Cretaceous." *University Geological Survey of Kansas* 2: 235–246.

———. 1898. "Mosasaurs." *University Geological Survey of Kansas* 4 (5): 81–347, pls. 10–72.

———. 1902. "On Certain Homoplastic Characters in Aquatic Air-Breathing Vertebrates." *Kansas University Science Bulletin* 1 (9): 259–266.

Zakrzewski, R. J. 1996. "Geologic Studies in Western Kansas in the 19th Century." *Kansas Academy of Science, Transactions* 99 (3–4): 124–133.

4. Sharks

Agassiz, J. L. R. 1833–1844. *Recherches sur les Poissons Fossiles.* Neuchâtel and Soleure. 3: vii + 390 + 32 pp.

Beamon, J. C. 1999. "Depositional Environment and Fossil Biota of a Thin Clastic Unit of the Kiowa Formation, Lower Cretaceous (Albian), McPherson County, Kansas." M.S. thesis, Fort Hays State University. 97 pp.

Cappetta, H. 1973. "Selachians from the Carlile Shale (Turonian) of South Dakota." *Journal of Paleontology* 47: 504–514.

Carpenter, K., D. Dilkes, and D. B. Weishampel. 1995. "The Dinosaurs of the Niobrara Chalk Formation (Upper Cretaceous, Kansas)." *Journal of Vertebrate Paleontology* 15 (2): 275–297.

Cicimurri, D. J. 2004. "Late Cretaceous Chondrichthyans from the Carlile Shale (Middle Turonian to Early Coniacian) of the Black Hills Region, South Dakota and Wyoming." *The Mountain Geologist* 41 (1): 1–16.

Cope, E. D. 1872a. "On the Geology and Paleontology of the Cretaceous Strata of Kansas." In *Preliminary Report of the United States Geological Survey of Montana and Portions of the Adjacent Territories,* Part III: *Paleontology,* pp. 318–349. Washington, D.C.: Government Printing Office.

———. 1872b. "On the Families of Fishes of the Cretaceous Formation in Kansas." *Proceedings of the American Philosophical Society* 12 (88): 327–357.

———. 1874. "Review of the Vertebrata of the Cretaceous Period Found West of the Mississippi River." *U.S. Geological Survey of the Territories, Bulletin* 1 (2): 3–48.

———. 1875. *The Vertebrata of the Cretaceous Formations of the West.* Report of the U.S. Geological Survey Territories (Hayden) 2. Washington, D.C.: Government Printing Office. 302 pp., 57 pls.

Corrado, C. A., D. A. Wilhelm, K. Shimada, and M. J. Everhart. 2003. "A New Skeleton of the Late Cretaceous Lamniform Shark, *Cretoxyrhina mantelli,* from Western Kansas." *Journal of Vertebrate Paleontology* 23 (Supplement to Number 3): 43A.

Druckenmiller, P. S., A. J. Daun, J. L. Skulan, and J. C. Pladziewicz. 1993. "Stomach Contents in the Upper Cretaceous Shark *Squalicorax falcatus.*" *Journal of Vertebrate Paleontology* 13 (Supplement to Number 3): 33A.

Eastman, C. R. 1895. "Bietrage Kenntniss Gattung *Oxyrhina* mit besonderer Berucksichtigung von *Oxyrhina mantelli* Agassiz." *Palaeontographica* 41: 149–192.

Everhart, M. J. 1999. "Evidence of Feeding on Mosasaurs by the Late Cretaceous Lamniform Shark, *Cretoxyrhina mantelli.*" *Journal of Vertebrate Paleontology* 17 (Supplement to Number 3): 43A–44A.

———. 2002. "New Data on Cranial Measurements and Body Length of the Mosasaur, *Tylosaurus nepaeolicus* (Squamata; Mosasauridae), from the Niobrara Formation of Western Kansas." *Kansas Academy of Science, Transactions* 105 (1–2): 33–43.

———. 2004a. "First Record of the Hybodont Shark Genus,'*Polyacrodus*' sp. (Chondrichthyes; Polyacrodontidae) from the Kiowa Formation (Lower Cretaceous) of McPherson County, Kansas." *Kansas Academy of Science, Transactions* 107 (1–2): 39–43.

———. 2004b. "Plesiosaurs as the Food of Mosasaurs: New Data on the Stomach Contents of a *Tylosaurus proriger* (Squamata; Mosasauridae) from the Niobrara Formation of Western Kansas." *The Mosasaur* 7: 41–46.

———. 2004c. "Late Cretaceous Interaction between Predators and Prey: Evidence of Feeding by Two Species of Shark on a Mosasaur." *PalArch Vertebrate Paleontology Series* 1 (1): 1–7.

Everhart, M. J., and T. Caggiano. 2004. "An Associated Dentition of the Late Cretaceous Elasmobranch, *Ptychodus anonymus* Williston 1900." *Paludicola* 4 (4): 125–136.

Everhart, M. J. and M. K. Darnell. 2004. Note on the Occurrence of *Ptychodus mammillaris* (Elasmobranchii) in the Fairport Chalk Member of the Carlile Shale (Upper Cretaceous) of Ellis County, Kansas. Kansas Academy of Science, Transactions 107 (3–4): 126–130.

Everhart, M. J. and P. A. Everhart. 2003. "First Report of the Paleozoic Shark *Ctenacanthus amblyxiphias* Cope 1891 from the Lower Permian of Morris County, Kansas." *Kansas Academy of Science, Transactions* (Abstracts) 22: 13.

Everhart, M. J., P. A. Everhart, and K. Ewell. 2004. "A Marine Ichthyofauna from the Upper Dakota Sandstone (Late Cretaceous)." Abstracts of oral presentations and posters, Joint Annual Meeting of the Kansas and Missouri Academies of Science, p. 48.

Everhart, M. J., P. Everhart, E. M. Manning, and D. E. Hattin. 2003. "A Middle Turonian Marine Fish Fauna from the Upper Blue Hill Shale Member, Carlile Shale, of North Central Kansas." *Journal of Vertebrate Paleontology* 23 (Supplement to Number 3): 49A.

Ewell, K., and M. J. Everhart. 2004. "A Paleozoic Shark Fauna from the Council Grove Group (Lower Permian)." Abstracts of oral presentations and posters, Joint Annual Meeting of the Kansas and Missouri Academies of Science, pp. 48–49.

Hamm, S. A. and M. J. Everhart. 1999. The Occurrence of a Rare Ptychodontid shark from the Smoky Hill Chalk (Upper Cretaceous) of Western Kansas. *Kansas Academy of Science, Transactions* (Abstracts) 18: 34.

———. 2001. "Notes on the Occurrence of Nodosaurs (Ankylosauridae) in the Smoky Hill Chalk (Upper Cretaceous) of Western Kansas." *Journal of Vertebrate Paleontology* 21 (Supplement to Number 3): 58A.

Hamm, S. A., and K. Shimada. 2002. "Associated Tooth Set of the Late Cretaceous Lamniform Shark, *Scapanorhynchus raphiodon* (Mitsukurinidae), from the Niobrara Chalk of Western Kansas." *Kansas Academy of Science, Transactions* 105 (1–2): 18–26.

Hattin, D. E. 1962. *Stratigraphy of the Carlile Shale (Upper Cretaceous) in*

Kansas. State Geological Survey of Kansas, Bulletin no. 156. Lawrence: University of Kansas. 155 pp., 2 pls.

———. 1982. *Stratigraphy and depositional environment of the Smoky Hill Chalk Member, Niobrara Chalk (Upper Cretaceous) of the type area, western Kansas*. Kansas Geological Survey Bulletin no. 225. Lawrence, Kan.: Kansas Geological Survey, University of Kansas, 108 pp.

———. 1996. "Fossilized Regurgitate from Smoky Hill Chalk Member of Niobrara Chalk (Upper Cretaceous) of Kansas, USA." *Cretaceous Research* 17: 443–450.

Herman, J. 1977. *Les sélaciens des terrains néocrétacés et paléocenes de Belgique et des contrées limitrophes. Eléments d'une biostratigraphique inter-continentale*. Mémoires pour sérvir a l'explication des Cartes géologiques et miniéres de la Belgique. Service Géologique de Belgique, Mémoire 15, 401 pp.

Kauffman, E. G. 1972. "*Ptychodus* Predation upon a Cretaceous *Inoceramus*." *Palaeontology* 15 (3): 439–444.

Lane, H. H. 1944. "A Survey of the Fossil Vertebrates of Kansas. Part I: The Fishes." *Kansas Academy of Science, Transactions* 47 (2): 129–176. 7 pls.

LeConte, J. L. 1868. *Notes on the Geology of the Survey for the Extension of the Union Pacific Railway, E. D., from the Smoky Hill River, Kansas, to the Rio Grande*. Philadelphia: Review Printing House. 76 pp. with folded map.

Leidy, J. 1859. [*Xystracanthus, Cladodus,* and *Petalodus* from the Carboniferous of Kansas]. *Proceedings of the Academy of Natural Sciences of Philadelphia* 11: 3.

———. 1868. "Notice of American Species of *Ptychodus*." *Proceedings of the Academy of Natural Sciences of Philadelphia* 20: 205–208.

———. 1873. *Contributions to the Extinct Vertebrate Fauna of the Western Interior Territories*. Report of the U.S. Geological Survey of the Territories (Hayden) 1. Washington, D.C.: Government Printing Office. 358 pp., 37 pls.

Marsh, O. C. 1871. "Note on a New and Gigantic Species of Pterodactyle." *American Journal of Science,* Series 3, 1 (6): 472.

Martin, L. D. and B. M. Rothschild. 1989. "Paleopathology and Diving Mosasaurs." *American Scientist* 77: 460–467.

Meyer, R. L. 1974. "Late Cretaceous Elasmobranchs from the Mississippi and East Texas Embayments of the Gulf Coastal Plain." Ph.D. dissertation, Southern Methodist University, Dallas. 400 pp.

Mudge, B. F. 1876. "Notes on the Tertiary and Cretaceous Periods of Kansas." *Bulletin of the U.S. Geological Survey Territories (Hayden)* 2 (3): 211–221. Washington, D.C.: Government Printing Office.

———. 1877a. "Annual Report of the Committee on Geology, for the Year Ending November 1, 1876." *Kansas Academy of Science, Transactions,* Ninth Annual Meeting, pp. 4–5.

———. 1877b. "Notes on the Tertiary and Cretaceous Periods of Kansas." In *Ninth Report of the U.S. Geological and Geographical Survey of the Territories (Hayden),* Part I: Geology, pp. 277–294. 827 pp.

Schultze, H.-P., J. D. Stewart, A. M. Neuner, and R. W. Coldiron. 1982. *Type and Figured Specimens of Fossil Vertebrates in the Collection of the University of Kansas Museum of Natural History,* Part I: *Fishes.* University of Kansas Museum of Natural History, Miscellaneous Publications, No. 73. Lawrence: University of Kansas Museum of Natural History. 53 pp.

Schwimmer, D. R., J. D. Stewart, and G. D. Williams. 1997. "Scavenging by Sharks of the Genus *Squalicorax* in the Late Cretaceous of North America." *Palaios* 12: 71–83.

Scott, R. W. 1970. "Paleoecology and Paleontology of the Lower Cretaceous Kiowa Formation, Kansas." *The University of Kansas Paleontological Contributions,* Article 52. 94 pp.

Shimada, K. 1996. "Selachians from the Fort Hays Limestone Member of the Niobrara Chalk (Upper Cretaceous), Ellis County, Kansas." *Kansas Academy of Science, Transactions* 99 (1–2): 1–15.

———. 1997a. "Stratigraphic Record of the Late Cretaceous Lamniform Shark, *Cretoxyrhina mantelli* (Agassiz), in Kansas." *Kansas Academy of Science, Transactions* 100 (3–4): 139–149.

———. 1997b. "Dentition of the Late Cretaceous Lamniform Shark, *Cretoxyrhina mantelli,* from the Niobrara Chalk of Kansas." *Journal of Vertebrate Paleontology* 17 (2): 269–279.

———. 1997c. "Paleoecological Relationships of the Late Cretaceous Lamniform Shark, *Cretoxyrhina mantelli* (Agassiz)." *Journal of Paleontology* 71 (5): 926–933.

———. 1997d. "Gigantic Lamnoid Shark Vertebra from the Lower Cretaceous Kiowa Shale of Kansas." *Journal of Paleontology* 71 (3): 522–524.

Shimada, K., K. Ewell, and M. J. Everhart. 2004. "The First Record of the Lamniform Shark Genus, *Johnlongia,* from the Niobrara Chalk (Upper Cretaceous), Western Kansas." *Kansas Academy of Science, Transactions* 107 (3–4): 131–135.

Shimada, K., and G. E. Hooks, III. 2004. "Shark-Bitten Protostegid Turtles from the Upper Cretaceous Mooreville Chalk, Alabama." *Journal of Paleontology* 78 (1): 205–210.

Siverson, M. 1996. "Lamniform Sharks of the Mid Cretaceous Alinga Formation and Beedagong Claystone, Western Australia." *Palaeontology* 39: 813–849.

Spamer, E. E., E. Daeschler, and L. G. Vostreys-Shapiro. 1995. "A Study of Fossil Vertebrate Types in the Academy of Natural Sciences of Philadelphia; Taxonomic, Systematic, and Historical Perspectives." *Academy of Natural Sciences of Philadelphia, Special Publication* 16. 434 pp.

Sternberg, C. H. 1907. "Some Animals Discovered in the Fossil Beds of Kansas." *Kansas Academy of Science, Transactions* 20: 122–124.

———. 1909. *The Life of a Fossil Hunter.* New York: Henry Holt and Company. 286 pp. (Reprinted 1990 by Indiana University Press.)

———. 1911. "In the Niobrara and Laramie Cretaceous." *Kansas Academy of Science, Transactions* 23: 70–74.

———. 1917. *Hunting Dinosaurs in the Badlands of the Red Deer River, Alberta, Canada.* San Diego, Calif.: published by the author.

Sternberg, M. 1920. *George Miller Sternberg: A Biography.* Chicago: American Medical Association. 331 pp., 10 pls.

Stewart, J. D. 1980. "Reevaluation of the Phylogenetic Position of the Ptychodontidae." *Kansas Academy of Science, Transactions* (Abstracts) 83 (3): 154.

———. 1990. "Niobrara Formation Vertebrate Stratigraphy." In S. C. Bennett, ed., *Niobrara Chalk Excursion Guidebook,* pp. 19–30.

Lawrence: University of Kansas Museum of Natural History and the Kansas Geological Survey.

Varricchio, D. J. 2001. "Gut Contents from a Cretaceous Tyrannosaurid; Implications for Theropod Dinosaur Digestive Tracts." *Journal of Paleontology* 75 (2): 401–406.

Welton, B. J., and R. F. Farish. 1993. *The Collector's Guide to Fossil Sharks and Rays from the Cretaceous of Texas.* Dallas, Tex.: Horton Printing Company. 204 pp.

Williston, S. W. 1898. "Mosasaurs." *University Geological Survey of Kansas* 4: 81–347, pls. 10–72.

———. 1900. "Cretaceous Fishes: Selachians and Pycnodonts." *University Geological Survey of Kansas* 6 (2): 237–256, pls. 24–32.

Woodward, A. S. 1887. "On the Definition and Affinities of the Selachian Genus *Ptychodus* Agassiz." *Quarterly Journal of the Geological Society* 13: 121–131, with pl. X.

5. FISHES

Bardack, D. 1965. *Anatomy and Evolution of Chirocentrid Fishes.* University of Kansas Paleontological Contributions, Article 10. Lawrence: University of Kansas Paleontological Institute. 88 pp. 2 pls.

———. 1976. "Paracanthopterygian and Acanthopterygian Fishes from the Upper Cretaceous of Kansas." *Fieldiana Geology* 33 (20): 355–374.

Beamon, J. C. 1999. "Depositional Environment and Fossil Biota of a Thin Clastic Unit of the Kiowa Formation, Lower Cretaceous (Albian), McPherson County, Kansas." M.A. thesis, Fort Hays State University, Hays, Kans., 97 pp.

Carpenter, K. 1996. "Sharon Springs Member, Pierre Shale (Lower Campanian) Depositional Environment and Origin of Its Vertebrate Fauna, with a Review of North American Plesiosaurs." Ph.D. dissertation, University of Colorado. 251 pp.

Cicimurri, D. J., and M. J. Everhart. 2001. "An Elasmosaur with Stomach Contents and Gastroliths from the Pierre Shale (Late Cretaceous) of Kansas." *Kansas Academy of Science, Transactions* 104 (3–4): 129–143.

Cope, E. D. 1868. "On a New Large Enaliosaur." *American Journal of Science,* Series 2, 46 (137): 263–264.

———. 1871. "On the Fossil Reptiles and Fishes of the Cretaceous Rocks of Kansas." In *Special Reports, 4th Annual Report of the U.S. Geological Survey of the Territories (Hayden),* Article 6, pp. 385–424 of Part 4. 511 pp.

———. 1872a. "Note on Some Cretaceous Vertebrata in the State Agricultural College of Kansas." *Proceedings of the American Philosophical Society* 12 (87): 168–170.

———. 1872b. "On the Families of Fishes of the Cretaceous Formation in Kansas." *Proceedings of the American Philosophical Society,* 12 (88): 327–357.

———. 1872c. "On the Geology and Paleontology of the Cretaceous Strata of Kansas." *Preliminary Report of the United States Geological Survey of Montana and Portions of the Adjacent Territories,* Part III: *Paleontology,* pp. 318–349. Washington, D.C.: Government Printing Office.

———. 1873a. [On an extinct genus of saurodont fishes]. *Proceedings of the Academy of Natural Sciences of Philadelphia* 24: 280–281.

———. 1873b. "On Two New Species of Saurodontidae." *Proceedings of the Academy of Natural Sciences of Philadelphia* 25: 337–339.

———. 1874. "Review of the Vertebrata of the Cretaceous Period Found West of the Mississippi River." *U.S. Geological Survey of the Territories, Bulletin* 1 (2): 3–48.

———. 1875. *The Vertebrata of the Cretaceous Formations of the West.* Report of the U.S. Geological Survey of the Territories (Hayden) 2. Washington, D.C.: Government Printing Office. 302 pp., 57 pls.

———. 1877a. "On Some New or Little Known Reptiles and Fishes of the Cretaceous No. 3 of Kansas." *Proceedings of the American Philosophical Society* 17 (100): 176–181.

———. 1877b. "On the Genus *Erisichthe*." *Bulletin of the U.S. Geological and Geographical Survey*, Article XXXV, 3 (4): 821–823.

———. 1886. "Note on *Erisichthe*." *Geology Magazine* 3: 239.

Druckenmiller, P. S., A. J. Daun, J. L. Skulan, and J. C. Pladziewicz. 1993. "Stomach Contents in the Upper Cretaceous Shark *Squalicorax falcatus*." *Journal of Vertebrate Paleontology* 13 (Supplement to Number 3): 33A.

Dunkle, D. H. 1969. "A New Amioid Fish from the Upper Cretaceous of Kansas." *Kirtlandia*, Cleveland Museum of Natural History, 7: 1–6.

Dunkle, D. H., and C. W. Hibbard. 1946. "Some Comments upon the Structure of a Pycnodontid Fish from the Upper Cretaceous of Kansas." *University of Kansas Science Bulletin* 31 (8): 161–181, pt. 1.

Everhart, M. J. 2002. "New Data on Cranial Measurements and Body Length of the Mosasaur *Tylosaurus nepaeolicus* (Squamata; Mosasauridae), from the Niobrara of Western Kansas." *Kansas Academy of Science, Transactions* 105 (1–2): 33–43.

Everhart, M. J., and P. A. Everhart. 1992. "Oyster-Shell Concentrations: A Stratigraphic Marker in the Smoky Hill Chalk (Upper Cretaceous) of Western Kansas." *Kansas Academy of Science, Transactions* (Abstracts) 11: 12.

———. 1993. "Notes on the Biostratigraphy of the Plethodid *Martinichthys* in the Smoky Hill Chalk (Upper Cretaceous) of Western Kansas." *Kansas Academy of Science, Transactions* (Abstracts) 12: 36.

Everhart, M. J., P. A. Everhart, and K. Ewell. 2004. "A Marine Ichthyofauna from the Upper Dakota Sandstone (Late Cretaceous)." Abstracts of oral presentations and posters, Joint Annual Meeting of the Kansas and Missouri Academies of Science, p. 48.

Everhart, M. J., P. Everhart, E. M. Manning, and D. E. Hattin. 2003. "A Middle Turonian Marine Fish Fauna from the Upper Blue Hill Shale Member, Carlile Shale, of North Central Kansas." *Journal of Vertebrate Paleontology* 23 (Supplement to Number 3): 49A.

Ewell, K., and M. J. Everhart. 2004. "A Paleozoic Shark Fauna from the Council Grove Group (Lower Permian)." Abstracts of oral presentations and posters, Joint Annual Meeting of the Kansas and Missouri Academies of Science, pp. 48–49.

Fielitz, C. 1999. "Phylogenetic Analysis of the Family Enchodontidae and Its Relationship to Recent Members of the Order Aulopiformes." Ph.D. dissertation, University of Kansas. 86 pp.

Fielitz, C., and K. Shimada. 1999. "A New Species of *Bananogmius* (Teleostei; Tselfatiformes) from the Upper Cretaceous Carlile Shale of Western Kansas." *Journal of Paleontology* 73 (3): 504–511.

Fielitz, C., J. D. Stewart, and J. Wiffen. 1999. "*Aethocephalichthys hyainarhinos* gen. et sp. nov., a New and Enigmatic Late Cretaceous

Actinopterygian from North America and New Zealand." In G. Arrantia and H.-P. Shultze, eds., *Mesozoic Fishes 2: Systematics and Fossil Record,* pp. 95–106, 7 figs. Proceedings of the International Meeting. Munich: F. Pfeil.

Goody, P. C. 1970. "The Cretaceous Teleostean Fish *Cimolichthys* from the Niobrara Formation of Kansas and the Pierre Shale of Wyoming." *American Museum Noviates* 2434: 1–29.

———. 1976. "*Enchodus* (Teleostei: Enchodontidae) from the Upper Cretaceous Pierre Shale of Wyoming and South Dakota with an Evaluation of the North American Enchodontid Species." *Palaeontographica Abteilung* A 152: 91–112.

Harlan, R. 1824. "On a New Fossil Genus of the Order Enalio sauri, (of Conybeare)." *Journal of the Academy of Natural Sciences of Philadelphia,* Series 1, 3 (pt. 2): 331–337, pl. 12, figs. 1–5.

Hattin, D. E. 1962. *Stratigraphy of the Carlile Shale (Upper Cretaceous) in Kansas.* State Geological Survey of Kansas, Bulletin no. 156. Lawrence: University of Kansas. 155 pp., 2 pls.

———. 1982. *Stratigraphy and Depositional Environment of the Smoky Hill Chalk Member, Niobrara Chalk (Upper Cretaceous) of the Type Area, Western Kansas.* Kansas Geological Survey Bulletin, no. 225. Lawrence: Kansas Geological Survey, University of Kansas. 108 pp.

Hay, O. P. 1898. "Observations on the Genus of Cretaceous Fishes Called by Professor Cope *Portheus.*" *Science* 7 (175): 646.

———. 1903. "On Certain Genera and Species of North American Cretaceous Actinopterous Fishes." *Bulletin of the American Museum of Natural History* 19: 1–95, pls. I–V. 72 text figs.

Hays, I. 1830. "Description of a Fragment of the Head of a New Fossil Animal, Discovered in a Marl Pit, near Moorestown, New Jersey." *Transactions of the American Philosophical Society,* Series 2, 3 (18): 471–477, pl. 16.

Hibbard, C. W. 1939. "A New Pycnodont Fish from the Upper Cretaceous of Russell County, Kansas." *University of Kansas Science Bulletin* 26 (9): 373–375. 1 pl.

Hibbard, C. W., and A. Graffham. 1941. "A New Pycnodont Fish from the Upper Cretaceous of Rooks County, Kansas." *University of Kansas Science Bulletin* 27 (5): 71–77. 1 fig.

Hussakof, L. 1929. "A New Teleostean Fish from the Niobrara of Kansas." *American Museum Novitates* 357: 1–4. 2 figs.

Jordan, D. S. 1924. "A Collection of Fossil Fishes in the University of Kansas from the Niobrara Formation of the Cretaceous." *University of Kansas Science Bulletin* 15 (2): 219–245.

Kauffman, E. G. 1990. "Cretaceous Fish Predation on a Large Squid." In A. J. Boucot, ed., *Evolutionary Paleobiology of Behavior and Coevolution,* pp. 195–196. Amsterdam: Elsevier.

Leidy, J. 1856. "Notes on the Fishes in the Collection of the Academy of Natural Sciences of Philadelphia." *Proceedings of the Academy of Natural Sciences of Philadelphia* 8: 221.

———. 1857. "Remarks on Saurocephalus and Its Allies." *Transactions of the American Philosophical Society* 11: 91–95, with pl. VI.

———. 1859. [*Xystracanthus, Cladodus,* and *Petalodus* from the Carboniferous of Kansas]. *Proceedings of the Academy of Natural Sciences of Philadelphia* 11: 3.

———. 1865. "Cretaceous Reptiles of the United States." *Smithsonian Contributions to Knowledge* 14 (6): 1–135, pls. I–XX.

————. 1868. "Notice of American Species of *Ptychodus*." *Proceedings of the Academy of Natural Sciences of Philadelphia* 20: 205–208.

————. 1870. [Remarks on ichthyodorulites and on certain fossil Mammalia]. *Proceedings of the Academy of Natural Sciences of Philadelphia* 22: 12–13.

Liggett, G. A. 2001. *Dinosaurs to Dung Beetles: Expeditions Through Time. Guide to the Sternberg Museum of Natural History.* Hays, Kans.: Sternberg Museum of Natural History. 127 pp.

Loomis, F. B. 1900. "Die anatomie und die verwandtschaft der Ganoid- und Knochen-fische aus der Kreide-Formation von Kansas, U.S.A." *Palaeontographica* 46: 213–283.

Mantell, G. 1822. *The Fossils of the South Downs; or, Illustrations of the Geology of Sussex.* London: Lupton Relfe. xiv + 327 pp., 42 pls.

Marsh, O. C. 1871. "Scientific Expedition to the Rocky Mountains." *American Journal of Science,* Series 3, 1 (2): 142–143.

McClung, C. E. 1908. "Ichthyological Notes on the Kansas Cretaceous, I." *University of Kansas Science Bulletin,* 4: 235–246, pls. X–XIII. 10 text-figs.

————. 1926. "*Martinichthys,* a New Genus of Cretaceous Fish from Kansas, with Descriptions of Six New Species." *Proceedings of American Philosophical Society,* 65 (5—Supplement): 20–26. 2 pls.

Miller, H. W. 1957. "Intestinal Casts in *Pachyrhizodus,* an Elopid Fish, from the Niobrara Formation of Kansas." *Kansas Academy of Science, Transactions* 60 (4): 399–401.

Mudge, B. F. 1866. "Discovery of Fossil Footmarks in the Liassic(?) Formation in Kansas." *American Journal of Science,* Series 2, 41 (122): 174–176.

————. 1874. "Rare Forms of Fish in Kansas." *Kansas Academy of Science, Transactions* 3: 121–122.

————. 1876. "Notes on the Tertiary and Cretaceous Periods of Kansas." *Bulletin of the U.S. Geological and Geographical Survey of the Territories (Hayden)* 2 (3): 211–221. Washington, D.C.: Government Printing Office.

Newton, E. T. 1878. "Remarks on *Saurocephalus,* and on the Species Which Have Been Referred to This Genus." *Quarterly Journal of the Geological Society of London* 34: 786–796.

Osborn, H. F. 1904. "The Great Cretaceous Fish *Portheus molossus* Cope." *Bulletin of the American Museum of Natural of Natural History* 20 (31): 377–381, pl. 10.

Rogers, K. 1991. *The Sternberg Fossil Hunters: A Dinosaur Dynasty.* Missoula, Mont.: Mountain Press Publishing Company. 288 pp.

Romer, A. S. 1966. *Vertebrate Paleontology.* 3rd ed. Chicago: University of Chicago Press. 468 pp.

Russell, D. A. 1988. *A Check List of North American Marine Cretaceous Vertebrates Including Fresh Water Fishes.* Occasional Paper of the Tyrrell Museum of Palaeontology, no. 4. Drumheller, Alta.: Tyrrell Museum of Palaeontology. 57 pp.

Schwimmer, D. R., J. D. Stewart, and G. D. Williams. 1997. "*Xiphactinus vetus* and the Distribution of *Xiphactinus* Species in the Eastern United States." *Journal of Vertebrate Paleontology* 17 (3): 610–615.

Shimada, K. 1997. "Paleoecological Relationships of the Late Cretaceous Lamniform Shark *Cretoxyrhina mantelli* (Agassiz)." *Journal of Paleontology* 71 (5): 926–933.

Shimada, K., and M. J. Everhart. 2004. "Shark Bitten *Xiphactinus audax* (Teleostei: Ichthyodectiformes) from the Niobrara Chalk (Upper Cretaceous) of Kansas." *The Mosasaur* 7: 35–39.

Shimada, K. and B. A. Schumacher. 2003. "The Earliest Record of the Late Cretaceous Fish *Thryptodus* (Teleostei: Tselfatiiformes) from Central Kansas." *Kansas Academy of Science, Transactions* 106 (1–2): 54–58.

Shor, E. N. 1971. *Fossils and Flies: The Life of a Compleat Scientist— Samuel Wendell Williston, 1851–1918.* Norman: University of Oklahoma Press. 285 pp.

Spamer, E. E., R. M. McCourt, R. Middleton, E. Gilmore, and S. B. Duran. 2000. "A National Treasure: Accounting for the Natural History Specimens from the Lewis and Clark Expedition (Western North America, 1803–1806) in the Academy of Natural Sciences of Philadelphia." *Proceedings of the Academy of Natural Sciences of Philadelphia* 150: 47–58.

Sternberg, M. L. 1920. *George Miller Sternberg: A Biography.* Chicago: American Medical Association. i–ix. 331 pp., 10 pls.

Stewart, A. 1898. "A Contribution to the Knowledge of the Ichthyic Fauna of the Kansas Cretaceous." *Kansas University Quarterly* 7 (1): 22–29, pl. 1.

———. 1899. "*Pachyrhizodus minimus,* a New Species of Fish from the Cretaceous of Kansas." *Kansas University Quarterly* 8 (1): 37–38.

———. 1899. "Notes on the Osteology of *Anogmius polymicrodus* Stewart." *Kansas University Quarterly* 8 (3): 118–121, pl. XXXI.

———. 1899. "*Leptichthys,* a New Genus of Fishes from the Cretaceous of Kansas." *American Geologist* 24 (2): 78–79.

———. 1900. "Teleosts of the Upper Cretaceous." *The University Geological Survey of Kansas* 6: 257–403, pls. 33–78. 6 figs.

Stewart, J. D. 1979. "Biostratigraphic Distribution of Species of *Protosphyraena* (Osteichthyes: Actinopterygii) in the Niobrara and Pierre Formations of Kansas." In *Proceedings of the Nebraska Academy of Sciences and Affiliated Societies,* 89th Annual Meeting, pp. 51–52.

———. 1988. "The Stratigraphic Distribution of Late Cretaceous *Protosphyraena* in Kansas and Alabama, Geology." In M. E. Nelson, ed., *Geology, Paleontology and Biostratigraphy of Western Kansas: Articles in Honor of Myrl V. Walker,* pp. 80–94. Fort Hays Studies, Third Series, no. 10, Science. Hays, Kans.: Fort Hays State University.

———. 1990a. "Niobrara Formation Vertebrate Stratigraphy." In S. C. Bennett, ed., *Niobrara Chalk Excursion Guidebook,* pp. 19–30. Lawrence: University of Kansas Museum of Natural History and Kansas Geological Survey.

———. 1990b. "Preliminary Account of Holecostome-Inoceramid Commensalism in the Upper Cretaceous of Kansas." In A. J. Boucot, ed., *Evolutionary Paleobiology of Behavior and Coevolution,* pp. 51–59. New York: Elsevier.

———. 1996. "Cretaceous Acanthomorphs of North America." In G. Arratia and G. Viohl, eds., *Mesozoic Fishes: Systematics and Paleoecology,* pp. 283–294. Munich: F. Pfeil. 2 figs.

———. 1999. "A New Genus of Saurodontidae (Teleostei: Ichthyodectiformes) from Upper Cretaceous Rocks of the Western Interior of North America." In G. Arrantia, and H.-P. Schultze, eds., *Mesozoic Fishes 2: Systematics and Fossil Record,* pp. 335–360. Munich: F. Pfeil.

Stewart, J. D., and G. L. Bell, Jr. 1994. "North America's Oldest Mosasaurs

Are Teleosts." *Contributions to Science (Natural History Museum of Los Angeles County)* 441: 1–9.

Stewart, J. D., and K. Carpenter. 1990. "Examples of Vertebrate Predation on Cephalopods in the Late Cretaceous of the Western Interior." In A. J. Boucout, ed., *Evolutionary Paleobiology of Behavior and Co-evolution*, pp. 203–208. New York: Elsevier.

Stewart, J. D., P. A. Everhart, and M. J. Everhart. 1991. "Small Coelacanths from Upper Cretaceous Rocks of Kansas." *Journal of Vertebrate Paleontology* 11 (Supplement to Number 3): 56A.

Stewart, J. D., and V. Friedman. 2001. "Oldest North American Record of Saurodontidae (Teleostei: Ichthyodectiformes)." *Journal of Vertebrate Paleontology* 21 (Supplement to Number 3): 104A.

Taverne, L. 1999. "Revision de *Zanclites xenurus*, Teleosteen (Pisces, Tselfatiiformes) marin du Santonian (Crétacé supérieur) du Kansas (États-Unis)." *Belgian Journal of Zoology* 129 (2): 421–438.

———. 2000a. "Révision du genre *Martinichthys*, poisson marin (Teleostei, Tselfatiiformes) du Crétacé supérieur du Kansas (États-Unis)." *Geobios* 33 (2): 211–222. Translated by Jean-Michel Benoit, 2003.

———. 2000b. "Osteology et position systematique du genre *Plethodus*, et des nouveaux genres *Dixonangomius* et *Pentanogmius*, poissons du Crétacé (Telostei, Tselfatiiformes)." *Biologisch Jaarboek Dodonaea* 67 (1): 94–123.

———. 2001a. "Révision du genre *Bananogmius* (Teleostei, Tselfatiiformes), poisson marin du Crétacé supérieur d'Amérique du Nord et d'Europe." *Geodiversitas* 23 (1): 17–40.

———. 2001b. "Révision de *Niobrara encarsia* téléostéen (Osteichthyes, Tselfatiiformes) du Crétacé supérieur marin du Kansas (États-Unis)." *Belgian Journal of Zoology* 131 (1): 3–16.

———. 2001c. "Révision de *Syntegmodus altus* (Teleostei, Tselfathformes), poisson marin du Crétacé supérieur du Kansas (États-Unis)." *Cybium* 25 (3): 251–260.

———. 2002a. "Révision de *Luxilites striolatus* poisson marin (Teleostei, Tselfatiiformes) du Crétacé supérieur du Kansas (États-Unis)." *Belgian Journal of Zoology* 132 (1): 25–34.

———. 2002b. "Étude de *Pseudanogmius maiseyi* gen. et. sp. nov., poisson marin (Teleostei, Tselfatiiformes) du Crétacé supérieur du Kansas (États-Unis)." *Geobios* 35: 605–614.

———. 2003. "Redescription critique des genres *Thryptodus*, *Pseudothryptodus* et *Paranogmius*, poissons marin (Teleostei, Tselfatiiformes) du Crétacé supérieur des États-Unis d'Egypte et de Libye." *Belgian Journal of Zoology* 133 (2): 163–173.

———. 2004. "Ostéologie de *Pentanogmius evolutus* (Cope, 1877) n. comb. (Teleostei, Tselfatiiformes) du Crétacé supérieur marin des États-Unis. Remarques sur la systématique du genre *Pentanogmius* Taverne 2000." *Geodiversitas* 26 (1): 89–113.

Thurmond, J. T. 1969. "Notes on Mosasaurs from Texas." *Texas Journal of Science* 21 (1): 69–79.

Wiley, E. O., and J. D. Stewart. 1977. "A Gar (*Lepisosteus* sp.) from the Marine Cretaceous Niobrara Formation of Western Kansas." *Copeia* 4: 761–762.

———. 1981. "*Urenchelys abditus*, New Species, the First Undoubted Eel (Telostei: Anguilliformes) from the Cretaceous of North America." *Journal of Vertebrate Paleontology* 1 (1): 43–47.

Williston, S. W. 1898. "Addenda to Part I." *University Geological Survey of Kansas* 4: 28–32.

———. 1900. "Cretaceous Fishes: Selachians and Pycnodonts." *University Geological Survey of Kansas* 6: 237–256, with pls.

Woodward, A. S. 1895. *Catalogue of the Fossil Fishes in the British Museum: Part 3.* British Museum of Natural History, London. i–xliii, 1–544 pp., pls.

Zielinski, S. L. 1994. "First Report of Pycnodontidae (Osteichthyes) from the Blue Hill Shale Member of the Carlile Shale (Upper Cretaceous; Middle Turonian), Ellis County, Kansas." *Kansas Academy of Science, Transactions* (Abstracts) 13: 44.

6. TURTLES

Agassiz, Louis. 1849. [Remarks on crocodiles of the green sand of New Jersey and on *Atlantochelys*]. *Proceedings of the Academy of Natural Sciences of Philadelphia* 4: 169.

Almy, K. J. 1987. "Thof's Dragon and the Letters of Capt. Theophilus Turner, M.D., U.S. Army." *Kansas History Magazine* 10 (3): 170–200.

Beamon, J. C. 1999. "Depositional Environment and Fossil Biota of a Thin Clastic Unit of the Kiowa Formation, Lower Cretaceous (Albian), McPherson County, Kansas." M.S. thesis, Fort Hays State University, Hays, Kansas. 97 pp.

Carpenter, K. 1996. "Sharon Springs Member, Pierre Shale (Lower Campanian) Depositional Environment and Origin of Its Vertebrate Fauna, with a Review of North American Plesiosaurs." Ph.D. dissertation, University of Colorado. 251 pp.

Case, E. C. 1898. "*Toxochelys.*" *University Geological Survey of Kansas* 4: 370–385, pls. 79–84.

Cope, E. D. 1870. "Observations on the Reptilia of the Triassic Formations of the Atlantic Regions of the United States." *Proceedings of the American Philosophical Society* 11 (84): 444–446.

———. 1872a. [Sketch of an expedition in the valley of the Smoky Hill River in Kansas]. *Proceedings of the American Philosophical Society* 12 (87): 174–176.

———. 1872b. "On a New Testudinate from the Chalk of Kansas." *Proceedings of the American Philosophical Society* 12 (88): 308–310.

———. 1872c. "A Description of the Genus *Protostega,* a Form of Extinct Testudinata." *Proceedings of the American Philosophical Society* 12 (88): 422–433.

———. 1872d. "On the Geology and Paleontology of the Cretaceous Strata of Kansas." *Annual Report of the U.S. Geological Survey of the Territories* 5: 318–349.

———. 1872e. [On a species of *Clidastes* and *Plesiosaurus gulo* Cope]. *Proceedings of the Academy of Natural Sciences of Philadelphia* 24: 127–129.

———. 1873. [On *Toxochelys latiremis*]. *Proceedings of the Academy of Natural Sciences of Philadelphia* 25: 10.

———. 1875. *The Vertebrata of the Cretaceous Formations of the West.* Report of the U.S. Geological Survey of the Territories (Hayden) 2. Washington, D.C.: Government Printing Office. 302 pp., 57 pls.

———. 1877. "On Some New or Little Known Reptiles and Fishes of the Cretaceous of Kansas." *Proceedings of the American Philosophical Society* 17 (100): 176–181.

Dollo, L. 1887. "Le hainosaure et les nouveaux vertébrés fossiles du Musée de Bruxelles." *Revue des Questions Scientifiques* 21: 504–539 and 22: 70–112.

Druckenmiller, P. S., A. J. Daun, J. L. Skulan, and J. C. Pladziewicz. 1993. "Stomach Contents in the Upper Cretaceous Shark *Squalicorax falcatus.*" *Journal of Vertebrate Paleontology* 13 (Supplement to Number 3): 33A.

Elliot, D. K., G. V. Irby, and J. H. Hutchinson. 1997. "*Desmatochelys lowi,* a Marine Turtle from the Upper Cretaceous." In J. M. Callaway and E. L. Nicholls, eds., *Ancient Marine Reptiles,* pp. 243–258. San Diego, Calif.: Academic Press.

Gaffney, E. S., and R. Zangerl. 1968. "A Revision of the Chelonian Genus *Bothremys* (Pleurodira: Pelomedusidae)." *Fieldiana, Geology* 16 (7): 193–239.

Hattin, D. E. 1982. *Stratigraphy and Depositional Environment of the Smoky Hill Chalk Member, Niobrara Chalk (Upper Cretaceous) of the Type Area, Western Kansas.* Kansas Geological Survey Bulletin, no. 225. Lawrence: Kansas Geological Survey, University of Kansas. 108 pp.

Hay, O. P. 1895. On Certain Portions of the Skeleton of *Protostega gigas.* Field Columbian Museum Publication no. 7, Zoological Series 1 (2): 57–62, pls. 4–5. Chicago: Field Columbian Museum.

———. 1896. On the Skeleton of *Toxochelys latiremis.* Field Columbian Museum Publication no. 13, Zoological Series 1 (5): 101–106, pls. 14–15. Chicago: Field Columbian Museum.

———. 1908. *The Fossil Turtles of North America.* Carnegie Institution of Washington, Publication no. 75. Washington, D.C.: Published by the Carnegie Institution of Washington. 568 pp., 113 pls.

Hirayama, R. 1997. "Distribution and Diversity of Cretaceous chelonioids." In J. M. Callaway and E. L. Nicholls, eds., *Ancient Marine Reptiles,* pp. 225–241. San Diego, Calif.: Academic Press.

Hooks, G. E., III. 1998. "Systematic Revision of the Protostegidae, with a Redescription of *Carcarichelys gemma* Zangerl, 1957." *Journal of Vertebrate Paleontology* 18 (1): 85–98.

Lane, H. H. 1946. "A Survey of the Fossil Vertebrates of Kansas. Part III: The Reptiles." *Kansas Academy of Science, Transactions* 49 (3): 289–332, figs. 1–7.

Leidy, J. 1865. "Cretaceous Reptiles of the United States." *Smithsonian Contributions to Knowledge* 14 (192): 1–135, pls. 1–20.

———. 1873. *Contributions to the Extinct Vertebrate Fauna of the Western Interior Territories.* Report of the U.S. Geological Survey of the Territories (Hayden) 1. Washington, D.C.: Government Printing Office. 358 pp., 37 pls.

Manning, E. M. 1994. "Dr. William Spillman (1806–1886), Pioneer Paleontologist of Mississippi." *Mississippi Geology* 15 (4): 64–69.

Nicholls, E. L. 1988. "New Material of *Toxochelys latiremis* Cope, and a Revision of the Genus *Toxochelys* (Testudines, Chelonioidea)." *Journal of Vertebrate Paleontology,* 8 (2): 181–187.

———. 1992. "Note on the Occurrence of the Marine Turtle *Desmatochelys* (Reptilia: Chelonioidea) from the Upper Cretaceous of Vancouver Island." *Canadian Journal of Earth Sciences* 29: 377–380.

———. 1997. "Introduction to Part III: Testudines." In J. M. Callaway and E. L. Nicholls, eds., *Ancient Marine Reptiles,* pp. 225–241. San Diego, Calif.: Academic Press.

Russell, D. A. 1993. "Vertebrates in the Western Interior Sea." In W. G. E. Caldwell and E. G. Kauffman, eds., *Evolution of the Western Interior Basin*, pp. 665–680. Geological Association of Canada, Special Paper 39. St. John's, Nfld.: Geological Association of Canada.

Schultze, H.-P., L. Hunt, J. Chorn, and A. M. Neuner. 1985. *Type and Figured Specimens of Fossil Vertebrates in the Collection of the University of Kansas Museum of Natural History.* Part II: *Fossil Amphibians and Reptiles.* Miscellaneous Publications of the University Kansas Museum of Natural History 77. Lawrence: University of Kansas Museum of Natural History. 66 pp.

Schwimmer, D. R., J. D. Stewart, and G. D. Williams. 1997. "Scavenging by Sharks of the Genus *Squalicorax* in the Late Cretaceous of North America." *Palaios* 12: 71–83.

Shimada, K., and G. E. Hooks, III. 2004. "Shark-Bitten Protostegid Turtles from the Upper Cretaceous Mooreville Chalk, Alabama." *Journal of Paleontology* 78 (1): 205–210.

Shor, E. N. 1971. *Fossils and Flies: The Life of a Compleat Scientist—Samuel Wendell Williston, 1851–1918.* Norman: University of Oklahoma Press. 285 pp.

Sternberg, C. H. 1884. "Directions for Collecting Vertebrate Fossils." *Kansas City Review of Science and Industry* 8 (4): 219–221.

———. 1905. "*Protostega gigas* and Other Cretaceous Reptiles and Fishes from the Kansas Chalk." *Kansas Academy of Science, Transactions* 19: 123–128.

———. 1909. *The Life of a Fossil Hunter.* New York: Henry Holt and Company. 286 pp.(Reprinted 1990 by Indiana University Press.)

Stewart, J. D. 1978. "Earliest Record of the Toxochelyidae." *Kansas Academy of Science, Transactions* 81 (2): 178.

———. 1990. "Niobrara Formation Vertebrate Stratigraphy." In S. C. Bennett, ed., *Niobrara Chalk Excursion Guidebook*, pp. 19–30. Lawrence: University of Kansas Museum of Natural History and Kansas Geological Survey.

Wieland, G. R. 1896. "*Archelon ischyros*: A New Gigantic Cryptodire Testudinate from the Fort Pierre Cretaceous of South Dakota." *American Journal of Science*, Series 4, 2 (12): 399–412, pl. V.

———. 1898. "The Protostegan Plastron." *American Journal of Science*, Series 4, 5: 15–20, with pl. II.

———. 1900. "Some Observations on Certain Well-Marked Stages in the Evolution of the Testudinate Humerus." *American Journal of Science*, Series 4, 9 (54): 413–424. 23 text figs.

———. 1902. "Notes on the Cretaceous Turtles, *Toxochelys* and *Archelon*, with a Classification of the Marine Testudinata." *American Journal of Science*, Series 4, 14: 95–108. 2 text figs.

———. 1905. "A New Niobrara *Toxochelys*." *American Journal of Science*, Series 4, 20 (119): 325–343, pl. X. 8 text figs.

———. 1906. "The Osteology of *Protostega*." *Memoirs of the Carnegie Museum* 2 (7): 279–305.

———. 1909. "Revision of the Protostegidae." *American Journal of Science*, Series 4, 27 (158): 101–130, pls. II–IV. 12 text figs.

Williston, S. W. 1894a. "A New Turtle from the Benton Cretaceous." *Kansas University Quarterly* 3 (1): 5–18, pls. 2–5.

———. 1894b. "On Various Vertebrate Remains from the Lowermost Cretaceous of Kansas." *Kansas University Quarterly* 3 (1): 1–4, pl. I.

———. 1898. "Turtles." *University Geological Survey of Kansas* 4: 349–369, pls. 73–78.

———. 1901. "A New Turtle from the Kansas Cretaceous." *Kansas Academy of Science, Transactions* 17: 195–199, pls. 18–22.

———. 1902. "On the Hind Limb of *Protostega*." *American Journal of Science*, Series 4, 13 (76): 276–278. 1 fig.

———. 1914. *Water Reptiles of the Past and Present*. Chicago: University of Chicago Press. 251 pp.

Zangerl, R. 1953. *The Vertebrate Fauna of the Selma Formation of Alabama*. Part IV: *The Turtles of the Family Toxochelyidae*. Fieldiana, Geology Memoirs 3 (4): 137–277. Chicago: Chicago Natural History Museum.

———. 1980. "Patterns of Phylogenic Differentiation in the Toxochelyid and Chelonid Sea Turtles." *American Zoologist* 20: 585–596.

Zangerl, R., and R. E. Sloan. 1960. "A New Description of *Desmatochelys lowi* (Williston): A Primitive Cheloniid Sea Turtle from the Cretaceous of South Dakota." *Fieldiana, Geology Memoirs* 14: 7–40.

7. Where the Elasmosaurs Roamed

Adams, D. A. 1997. "*Trinacromerum bonneri*, New Species, Last and Fastest Pliosaur of the Western Interior Seaway." *Texas Journal of Science* 49 (3): 179–198.

Alexander, R. M. 1989. *Dynamics of Dinosaurs and Other Extinct Giants*. Columbia University Press, New York, 167 pp.

Almy, K. J. 1987. "Thof's Dragon and the Letters of Capt. Theophilus Turner, M.D., U.S. Army." *Kansas History Magazine* 10 (3): 170–200.

Bakker, R. T. 1993. "Plesiosaur Extinction Cycles: Events That Mark the Beginning, Middle and End of the Cretaceous." In W. G. E. Caldwell and E. G. Kauffman, eds., *Evolution of the Western Interior Basin*, pp. 641–664. Geological Association of Canada, Special Paper 39. St. John's, Nfld.: Geological Association of Canada.

Ballou, W. H. 1890. [Reply by O. C. Marsh regarding charges made by E. D. Cope]. *New York Herald* 19 (508) (January 19, 1890): 11.

Bell, G. L., Jr., M. A. Sheldon, J. P. Lamb, and J. E. Martin. 1996. "The First Direct Evidence of Live Birth in Mosasauridae (Squamata): Exceptional Preservation in Cretaceous Pierre Shale of South Dakota." *Journal of Vertebrate Paleontology* 16 (Supplement to Number 3): 21A.

Bennett, S. C. 2000. "Inferring Stratigraphic Position of Fossil Vertebrates from the Niobrara Chalk of Western Kansas." *Current Research in Earth Sciences*, Kansas Geological Survey Bulletin 244 (1): 1–26.

Caldwell, M. W., and M. S. Y. Lee. 2001. "Live Birth in Cretaceous Marine Lizards (Mosasauroids)." *Proceedings of the Royal Society of London: Biological Sciences* 268 (1484): 2397–2401.

Carpenter, K. 1996. "A Review of Short-Necked Plesiosaurs from the Cretaceous of the Western Interior, North America." *Neues Jahrbuch für Geologie und Palaeontologie*, Abhandlungen (Stuttgart) 201 (2): 259–287.

———. 1999. "Revision of North American Elasmosaurs from the Cretaceous of the Western Interior." *Paludicola* 2 (2): 148–173.

Chatterjee, S., and W. J. Zinsmeister. 1982. "Late Cretaceous Marine Vertebrates from Seymour Island, Antarctic Peninsula." *Antarctic Journal* 17 (5): 66.

Cicimurri, D. J., and M. J. Everhart. 2001. "An Elasmosaur with Stomach Contents and Gastroliths from the Pierre Shale (Late Cretaceous) of Kansas." *Kansas Academy of Science, Transactions* 104 (3–4): 129–143.

Cope, E. D. 1868. "Remarks on a New Enaliosaurian, *Elasmosaurus platyurus.*" *Proceedings of the Academy of Natural Sciences of Philadelphia* 20: 92–93.

———. 1869. "Synopsis of the Extinct Batrachia and Reptilia of North America: Part I." *Transactions of the American Philosophical Society,* New Series, 14: 1–235. 51 figs., 11 pls.

———. 1870. "Synopsis of the Extinct Batrachia and Reptilia of North America: Part I." *Transactions of the American Philosophical Society,* New Series, 14: 1–235. 51 figs., 11 pls.

———. 1872. "On the Geology and Paleontology of the Cretaceous Strata of Kansas." In *Preliminary Report of the United States Geological Survey of Montana and Portions of the Adjacent Territories,* Part III: *Paleontology,* pp. 318–349. Washington, D.C.: Government Printing Office.

———. 1875. *The Vertebrata of the Cretaceous Formations of the West.* Report of the U.S. Geological Survey of the Territories (Hayden) 2. Washington, D.C.: Government Printing Office. 302 pp., 57 pls.

Cruikshank, A. R. I., P. G. Small, and M. A. Taylor. 1991. "Dorsal Nostrils and Hydrodynamically Driven Underwater Olfaction in Plesiosaurs." *Nature* 352: 62–64.

De la Beche, H. T., and W. D. Conybeare. 1821. "Notice of the Discovery of a New Animal, Forming a Link between the *Ichthyosaurus* and Crocodile, Together with General Remarks on the Osteology of *Ichthyosaurus.*" *Transactions of the Geological Society of London* 5: 559–594.

Ellis, R. 2003. *Sea Dragons: Predators of the Prehistoric Oceans.* Lawrence: University of Kansas Press. 313 pp.

Everhart, M. J. 2000. "Gastroliths Associated with Plesiosaur Remains in the Sharon Springs Member of the Pierre Shale (Late Cretaceous), Western Kansas." *Kansas Academy of Science, Transactions* 103 (1–2): 58–69.

———. 2002. "Remains of Immature Mosasaurs (Squamata; Mosasauridae) from the Niobrara Chalk (Late Cretaceous) Argue against Nearshore Nurseries." *Journal of Vertebrate Paleontology* 22 (Supplement to Number 3): 52A.

———. 2003. "First Records of Plesiosaur Remains in the Lower Smoky Hill Chalk Member (Upper Coniacian) of the Niobrara Formation in Western Kansas." *Kansas Academy of Science, Transactions* 106 (3–4): 139–148.

———. 2004a. "Plesiosaurs as the Food of Mosasaurs: New Data on the Stomach Contents of a *Tylosaurus proriger* (Squamata; Mosasauridae) from the Niobrara Formation of Western Kansas." *The Mosasaur* 7: 41–46.

———. 2004b. "Conchoidal Fractures Preserved on Elasmosaur Gastroliths are Evidence of Use in Processing Food." *Journal of Vertebrate Paleontology* 24 (Supplement to 3): 56A.

Hector, J. 1874. "On the Fossil Reptilia of New Zealand." *Transactions and Proceedings of the New Zealand Institute* 6: 333–364.

LeConte, J. L. 1868. *Notes on the Geology of the Survey for the Extension of the Union Pacific Railway, E. D., from the Smoky Hill River,*

Kansas, to the Rio Grande. Philadelphia: Review Printing House. 76 pp. with folded map.

Leidy, J. 1870. [Remarks on *Elasmosaurus platyurus*]. *Proceedings of the Academy of Natural Sciences of Philadelphia* 22: 9–10.

Lingham-Soliar, T. 2000. "Plesiosaur Locomotion: Is the Four-Wing Problem Real or Merely an Atheoretical Exercise?" *Neues Jahrbuch für Geologie und Paläontologie,* Abhandlungen (Stuttgart) 217: 45–87.

———. 2003. "Extinction of Ichthyosaurs: A Catastrophic or Evolutionary Paradigm?" *Neues Jahrbuch für Geologie und Palaeontologie,* Abhandlungen (Stuttgart) 228 (3): 421–452.

Long, J. A. 1998. *Dinosaurs of Australia and New Zealand and Other Animals of the Mesozoic Era.* Cambridge, Mass.: Harvard University Press. 188 pp.

Martin, L. D., and J. D. Stewart. 1977. "The Oldest (Turonian) Mosasaurs from Kansas." *Journal of Paleontology* 51 (5): 973–975.

Massare, J. A. 1987. "Tooth Morphology and Prey Preference of Mesozoic Marine Reptiles." *Journal of Vertebrate Paleontology* 7 (2): 121–137.

McGowan, C. 2001. *The Dragon Seekers.* Cambridge, Mass.: Perseus Publishing. 253 pp.

Mudge, B. F. 1876. "Notes on the Tertiary and Cretaceous Periods of Kansas." *Bulletin of the U.S. Geological Survey of the Territories (Hayden)* 2 (3): 211–221. Washington, D.C.: Government Printing Office.

———. 1877. "Notes on the Tertiary and Cretaceous Periods of Kansas." In *Ninth Annual Report of the U.S. Geological and Geographical Survey of the Territories,* Part I: *Geology,* pp. 277–294. Washington, D.C.: Government Printing Office. 827 pp.

Mulder, E. W. A., N. Bardet, P. Godefroit, and J. W. M. Jagt. 2003. "Elasmosaur Remains from the Maastrichtian Type Area, and a Review of the Latest Cretaceous Elasmosaurs (Reptilia, Plesiosauroidea)." In E. W. A. Mulder, ed., *On Latest Cretaceous Tetrapods from the Maastrichtian Type Area,* pp. 93–105, Publicaties van het Natuurhistorisch Genootshap in Limburg, Reeks XLIV, aflevering 1. Maastricht: Stichting Natuurpublicaties Limburg.

Nakaya, H. 1989. "Upper Cretaceous Elasmosaurid (Reptilia, Plesiosauria) from Hobetsu, Hokkaido, Northern Japan." *Translated Proceedings of the Paleontology Society of Japan,* New Series, 154: 96–116. 14 figs.

Nicholls, E. L. 1988. "Marine Vertebrates of the Pembina Member of the Pierre Shale (Campanian, Upper Cretaceous) of Manitoba and Their Significance to the Biogeography." Ph.D. dissertation, University of Calgary. 317 pp.

O'Neill, J. P. 1999. *The Great New England Sea Serpent.* East Peoria, Ill.: Versa Press. 256 pp.

Rothschild, B. M., and L. Martin. 1993. *Paleopathology: Disease in the Fossil Record.* Boca Raton, Fla.: CRC Press. 386 pp.

Russell, D. A. 1967. *Systematics and Morphology of American Mosasaurs.* New Haven, Conn.: Peabody Museum of Natural History, Yale University, Bulletin, no. 23. New Haven, Conn.: Peabody Museum of Natural History, Yale University. 241 pp.

———. 1993. "Vertebrates in the Western Interior Sea." In W. G. E. Caldwell and E. G. Kauffman, eds., *Evolution of the Western Interior Basin,* pp. 665–680. Geological Association of Canada, Special Paper 39. St. John's, Nfld.: Geological Association of Canada.

Sato, T. 2003. "*Terminonatator ponteixensis,* a New Elasmosaur (Reptilia:

Sauropterygia) from the Upper Cretaceous of Saskatchewan." *Journal of Vertebrate Paleontology* 23 (1): 89–103.

Shuler, E. W. 1950. "A New Elasmosaur from the Eagle Ford Shale of Texas: Part II." *Fondren Science Series* 4 (2): 1–33. 26 figs.

Sternberg, C. H. 1922. "Explorations of the Permian of Texas and the Chalk of Kansas, 1918." *Kansas Academy of Science, Transactions* 30 (1): 119–120.

———. 1917 (1932). *Hunting Dinosaurs in the Badlands of the Red Deer River, Alberta, Canada.* San Diego, Calif.: Published by the author. 261 pp.

Storrs, G. W. 1984. "*Elasmosaurus platyurus* and a Page from the Cope-Marsh War." *Discovery* 17 (2): 25–27.

———. 1993. "Function and Phylogeny in Sauropterygian (Diapsida) Evolution." *American Journal of Science* 293A: 63–90.

———. 1999. "An Examination of Plesiosauria (Diapsida: Sauropterygia) from the Niobrara Chalk (Upper Cretaceous) of Central North America." *University of Kansas Paleontological Contributions,* New Series, Number 11. 15 pp.

Taylor, M. A. 1981. "Plesiosaurs—Rigging and Ballasting." *Nature* 290: 628–629.

Webb, W. E. 1872. *Buffalo Land: An Authentic Account of the Discoveries, Adventures, and Mishaps of a Scientific and Sporting Party in the Wild West.* Philadelphia: Hubbard Brothers. 504 pp.

Welles, S. P. 1943. "Elasmosaurid Plesiosaurs with a Description of the New Material from California and Colorado." *University of California Memoirs* 13: 125–254, figs. 1–37, pls. 12–29.

———. 1952. *A Review of the North American Cretaceous Elasmosaurs.* University of California Publications in Geological Sciences, no. 29. 25 text figs. Berkeley: University of California Press.

———. 1962. "A New Species of Elasmosaur from the Aptian of Columbia and a Review of the Cretaceous Plesiosaurs." *University of California Publications in Geological Sciences* 44 (1). 96 pp., 4 pls., 23 text figs.

Whittle, C. H. and M. J. Everhart. 2000. "Apparent and Implied Evolutionary Trends in Lithophagic Vertebrates from New Mexico and Elsewhere." In S. G. Lucas and A. B. Heckert, eds., *Dinosaurs of New Mexico,* pp. 75–82. Albuquerque, N.M.: New Mexico Museum of Natural History and Science, Bulletin 17.

Williston, S. W. 1890. "A New Plesiosaur From the Niobrara Cretaceous of Kansas." *Kansas Academy of Science, Transactions* 7: 174–178. 2 figs.

———. 1893. "An Interesting Food Habit of the Plesiosaurs." *Kansas Academy of Science, Transactions* 13: 121–122. 1 pl.

———. 1906. "North American Plesiosaurs: *Elasmosaurus, Cimoliasaurus,* and *Polycotylus.*" *American Journal of Science,* Series 4, 21 (123): 221–234. 4 pls.

———. 1914. *Water Reptiles of the Past and Present.* Chicago: University of Chicago Press. 251 pp.

8. PLIOSAURS AND POLYCOTYLIDS

Adams, D. A. 1997. "*Trinacromerum bonneri,* New Species, Last and Fastest Pliosaur of the Western Interior Seaway." *Texas Journal of Science* 49 (3): 179–198.

Ballou, W. 1890. [O. C. Marsh's published letter about Cope.] *New York Herald* 19 (508): 11, cols. 5–6.

Bonner, O. W. 1964. "An Osteological Study of *Nyctosaurus* and *Trinacromerum* with a Description of a New Species of *Nyctosaurus.*" M.S. thesis, Fort Hays State University, Hays, Kansas. 63 pp.

Carpenter, K. 1996. "A Review of Short-Necked Plesiosaurs from the Cretaceous of the Western Interior, North America." *Neues Jahrbuch für Geologie und Palaeontologie*, Abhandlungen (Stuttgart) 201 (2): 259–287.

———. 1997. "Comparative Cranial Anatomy of two North American Cretaceous Plesiosaurs." In J. M. Calloway and E. L. Nicholls, eds., *Ancient Marine Reptiles,* pp. 191–216. San Diego, Calif.: Academic Press.

Cicimurri, D. J., and M. J. Everhart. 2001. "An Elasmosaur with Stomach Contents and Gastroliths from the Pierre Shale (Late Cretaceous) of Kansas." *Kansas Academy of Science, Transactions* 104 (3–4): 129–143.

Cope, E. D. 1868. "Remarks on a New Enaliosaurian, *Elasmosaurus platyurus.*" *Proceedings of the Academy of Natural Sciences of Philadelphia* 20: 92–93.

———. 1869. [Remarks on fossil reptiles, *Clidastes propython, Polycotylus latipinnis, Ornithotarsus immanis*]. *Proceedings of the American Philosophical Society* 11: 117.

———. 1871. "Synopsis of the Extinct Batrachia and Reptilia of North America." *Transactions of the American Philosophical Society,* New Series, 14: 1–252 + i–viii. 55 figs., 14 pls.

Cragin, F. W. 1888. "Preliminary Description of a New or Little Known Saurian from the Benton of Kansas." *American Geologist* 2: 404–407.

———. 1894. "Vertebrata from the Neocomian of Kansas." *Bulletin of the Washburn College Laboratory* 2 (9): 69–73. 2 pls.

Davidson, J. P. 2003. "Edward Drinker Cope, Professor Paleozoic and Buffalo Land." *Kansas Academy of Science, Transactions* 106 (3–4): 177–191.

Ellis, R. 2003. *Sea Dragons: Predators of the Prehistoric Oceans.* Lawrence: University of Kansas Press. 313 pp.

Everhart, M. J. 2003. "First Records of Plesiosaur Remains in the Lower Smoky Hill Chalk Member (Upper Coniacian) of the Niobrara Formation in Western Kansas." *Kansas Academy of Science, Transactions* 106 (3–4): 139–148.

———. 2004a. "New Data Regarding the Skull of *Dolichorhynchops osborni* (Plesiosauroidea: Polycotylidae) from Rediscovered Photos of the Harvard Museum of Comparative Zoology Specimen." *Paludicola* 4 (3): 74–80.

———. 2004b. "Plesiosaurs as the Food of Mosasaurs: New Data on the Stomach Contents of a *Tylosaurus proriger* (Squamata; Mosasauridae) from the Niobrara Formation of Western Kansas." *The Mosasaur* 7: 41–46.

Gilmore, C. W. 1921. "An Extinct Sea-Lizard from Western Kansas." *Scientific American* 124: 273, 280. 3 figs.

Hattin, D. E. 1982. *Stratigraphy and Depositional Environment of the Smoky Hill Chalk Member, Niobrara Chalk (Upper Cretaceous) of the Type Area, Western Kansas.* Kansas Geological Survey Bulletin, no. 225. Lawrence: Kansas Geological Survey, University of Kansas. 108 pp.

Liggett, G. A., S. C. Bennett, K. Shimada, and J. Huenergarde. 1997. "A Late Cretaceous (Cenomanian) Fauna in Russell County, KS." *Kansas Academy of Science, Transactions* (Abstracts) 16: 26.

Lingham-Soliar, T. 2003. "Extinction of Ichthyosaurs: A Catastrophic or Evolutionary Paradigm?" *Neues Jahrbuch für Geologie und Palaeontologie,* Abhandlungen (Stuttgart) 228 (3): 421–452.

Marsh, O. C. 1871. "Scientific Expedition to the Rocky Mountains." *American Journal of Science,* Series 3, 1 (2): 142–143.

Owen, R. 1842. "Report on British Fossil Reptiles." *Report of the Eleventh Meeting (at Plymouth, 1841), British Association for the Advancement of Science,* Part 2, pp. 60–204.

Peterson, J. M. 1987. "Science in Kansas: The Early Years, 1804–1875." *Kansas History Magazine* 10 (3): 201–240.

Riggs, E. S. 1944. "A New Polycotylid Plesiosaur." *University of Kansas Science Bulletin* 30: 77–87.

Rothschild, B. M., and L. D. Martin. 1993. *Paleopathology: Disease in the Fossil Record.* Boca Raton, Fla.: CRC Press. 386 pp.

Schultze, H.-P., L. Hunt, J. Chorn, and A. M. Neuner. 1985. *Type and Figured Specimens of Fossil Vertebrates in the Collection of the University of Kansas Museum of Natural History.* Part II: *Fossil Amphibians and Reptiles.* Miscellaneous Publications of the University Kansas Museum of Natural History 77. Lawrence: University of Kansas Museum of Natural History. 66 pp.

Schumacher, B. A., and M. J. Everhart. 2004. "A New Assessment of Plesiosaurs from the Old Fort Benton Group, Central Kansas." Abstracts of oral presentations and posters, Joint Annual Meeting of the Kansas and Missouri Academies of Science, p. 50.

Scott, R. W. 1970. *Paleoecology and Paleontology of the Lower Cretaceous Kiowa Formation, Kansas.* University of Kansas Paleontological Contributions, Article 52. Lawrence: University of Kansas Paleontological Institute. 94 pp.

Sternberg, C. H. 1922. "Explorations of the Permian of Texas and the Chalk of Kansas, 1918." *Kansas Academy of Science, Transactions* 30 (1): 119–120.

Sternberg, G. F., and M. V. Walker. 1957. "Report on a Plesiosaur Skeleton from Western Kansas." *Kansas Academy of Science, Transactions* 60 (1): 86–87.

Stewart, J. D. 1990. "Niobrara Formation Vertebrate Stratigraphy." In S. C. Bennett, ed., *Niobrara Chalk Excursion Guidebook,* pp. 19–30. Lawrence: University of Kansas Museum of Natural History and Kansas Geological Survey.

Storrs, G. W. 1999. "An Examination of Plesiosauria (Diapsida: Sauropterygia) from the Niobrara Chalk (Upper Cretaceous) of Central North America." *University of Kansas Paleontological Contributions,* New Series, no. 11. 15 pp.

Webb, W. E. 1872. *Buffalo Land: An Authentic Account of the Discoveries, Adventures, and Mishaps of a Scientific and Sporting Party In the Wild West.* Philadelphia: Hubbard Brothers. 503 pp.

Williston, S. W. 1902. "Restoration of *Dolichorhynchops osborni,* a new Cretaceous Plesiosaur." *Kansas University Science Bulletin* 1 (9): 241–244. 1 pl.

———. 1903. *North American Plesiosaurs.* Field Columbian Museum, Publication 73, Geological Series, 2 (1). Chicago: Field Columbian Museum. 79 pp., 29 pls.

—————. 1906. "North American Plesiosaurs: *Elasmosaurus, Cimoliasaurus,* and *Polycotylus.*" *American Journal of Science,* Series 4, 21 (123): 221–234. 4 pls.

—————. 1907. "The Skull of *Brachauchenius,* with Special Observations on the Relationships of the Plesiosaurs." *United States National Museum Proceedings* 32: 477–489, pls. 34–37.

9. ENTER THE MOSASAURS

Adams, D. A. 1997. "*Trinacromerum bonneri,* New Species, Last and Fastest Pliosaur of the Western Interior Seaway." *Texas Journal of Science* 49 (3): 179–198.

Almy, K. J. 1987. "Thof's Dragon and the Letters of Capt. Theophilus Turner, M.D., U.S. Army." *Kansas History Magazine* 10 (3): 170–200.

Bakker, R. T. 1993. "Plesiosaur Extinction Cycles: Events That Mark the Beginning, Middle and End of the Cretaceous." In W. G. E. Caldwell and E. G. Kauffman, eds., *Evolution of the Western Interior Basin,* pp. 641–664. Geological Association of Canada, Special Paper 39. St. John's, Nfld.: Geological Association of Canada.

Bardet, N., and J. W. M. Jagt. 1996. "*Mosasaurus hoffmani,* le 'Grand Animal fossile des Carrieres de Maestricht': deux siècles d'histoire." *Bulletin de Museum d'Histoire Naturelle,* Paris, 4th Series, 18: 569–593.

Bardet, N., J. W. M. Jagt, M. M. M. Kuypers, and R. W. Dortangs. 1998. "Shark Tooth Marks on a Vertebra of the Mosasaur *Plioplatecarpus marshi* from the Late Maastrichtian of Belgium." *Publicaties van het Natuurhistorisch Genootschap in Limburg* 41 (1): 52–55.

Bardet, N., and X. P. Suberbiola. 2001. "The Basal Mosasaurid *Halisaurus sternbergii* from the Late Cretaceous of Kansas (North America): A Review of the Uppsala Type Specimen." *Comptes Rendus de l'Académie des Sciences,* Series IIA, Earth and Planetary Science 332: 395–402.

Bardet, N., and C. Tunoglu. 2002. "The First Mosasaur (Squamata) from the Late Cretaceous of Turkey." *Journal of Vertebrate Paleontology* 22 (3): 712–715.

Baur, G. 1892. "On the Morphology of the Skull of the Mosasauridae." *Journal of Morphology* 7 (1): 1–22. 2 pls.

Bell, G. L., Jr. 1997a. "Part IV: Mosasauridae—Introduction." In J. M. Callaway and E. L. Nicholls, eds., *Ancient Marine Reptiles,* pp. 281–292. San Diego, Calif.: Academic Press. 501 pp.

—————. 1997b. "A Phylogenetic Revision of North American and Adriatic Mosasauroidea." In J. M. Callaway and E. L Nicholls, eds., *Ancient Marine Reptiles,* pp. 293–332. San Diego, Calif.: Academic Press. 501 pp.

Bell, G. L., Jr., and J. E. Martin. 1995. "Direct Evidence of Aggressive Intraspecific Competition in *Mosasaurus conodon* (Mosasauridae: Squamata)." *Journal of Vertebrate Paleontology* 15 (Supplement to Number 3): 18A.

Bell, G. L., Jr., M. A. Sheldon, J. P. Lamb, and J. E. Martin. 1996. "The First Direct Evidence of Live Birth in Mosasauridae (Squamata): Exceptional Preservation in Cretaceous Pierre Shale of South Dakota." *Journal of Vertebrate Paleontology* 16 (Supplement to Number 3): 21A.

Bell, G. L., Jr., and J. P. VonLoh. 1998. "New Records of Turonian Mosasauroids from the Western United States: Fossil Vertebrates of the Niobrara Formation in South Dakota." *Dakoterra* 5: 15–28.

Betts, C. W. 1871. "The Yale College Expedition of 1870." *Harper's New Monthly Magazine* 43 (257): 663–671.

Burnham, D. A. 1991. "A New Mosasaur from the Upper Demopolis Formation of Sumter County, Alabama." M.S. thesis, University of New Orleans. 63 pp.

Caldwell, M. W. 1999. "Squamate Phylogeny and the Relationships of Snakes and Mosasaurids." *Zoological Journal of the Linnean Society* 125: 115–147. 7 figs.

Caldwell, M. W., L. A. Budney, and D. O. Lamoureux. 2003. "Histology of Tooth Attachment Tissues in the Late Cretaceous Mosasaurid *Platecarpus.*" *Journal of Vertebrate Paleontology* 23 (3): 622–630.

Camp, C. L. 1942. *California Mosasaurs.* Berkeley: University of California Press. 67 pp.

Carpenter, K. 1996. "A Review of Short-Necked Plesiosaurs from the Cretaceous of the Western Interior, North America." *Neues Jahrbuch für Geologie und Palaeontologie,* Abhandlungen (Stuttgart) 201 (2): 259–287.

Carroll, R. L., and M. Debraga. 1992. "Aigialosaurs: Mid-Cretaceous Varanid Lizards." *Journal of Vertebrate Paleontology* 12 (1): 66–86.

Case, J. A., J. E. Martin, D. S. Chaney, M. Reguero, S. A. Marenssi, S. M. Santillana, and M. O. Woodburne. 2000. "The First Duck-Billed Dinosaur (Family Hadrosauridae) from Antarctica." *Journal of Vertebrate Paleontology* 20 (3): 612–614.

Chatterjee, S., and W. J. Zinsmeister. 1982. "Late Cretaceous Marine Vertebrates from Seymour Island, Antarctic Peninsula." *Antarctic Journal* 17 (5): 66.

Christiansen, P., and N. Bonde. 2002. "A New Species of Gigantic Mosasaur from the Late Cretaceous of Israel." *Journal of Vertebrate Paleontology* 22 (3): 629–644.

Cope, E. D. 1868a. "Remarks on a New Enaliosaurian, *Elasmosaurus platyurus.*" *Proceedings of the Academy of Natural Sciences of Philadelphia* 20: 92–93.

———. 1868b. "On Some Cretaceous Reptilia." *Proceedings of the Academy of Natural Sciences of Philadelphia* 20: 233–242.

———. 1869a. [Remarks on *Thoracosaurus brevispinus, Ornithotarsus immanis,* and *Macrosaurus proriger*]. *Proceedings of the Academy of Natural Sciences of Philadelphia* 11 (81): 123.

———. 1869b. "On the Reptilian Orders Pythonomorpha and Streptosauria." *Proceedings of the Boston Society of Natural History* 12: 250–266.

———. 1872. "On the Geology and Paleontology of the Cretaceous Strata of Kansas." In *Preliminary Report of the United States Geological Survey of Montana and Portions of the Adjacent Territories,* Part III: *Paleontology,* pp. 318–349. Washington, D.C.: Government Printing Office.

———. 1874. "Review of the Vertebrata of the Cretaceous Period Found West of the Mississippi River." *U.S. Geological Survey of the Territories,* Bulletin 1 (2): 3–48.

Dollo, L. 1887. "Le hainosaure et les nouveaux vertébrés fossiles du Musée de Bruxelles." *Revue des Questions Scientifiques* 21: 504–539 and 22: 70–112.

Dortangs, R. W., A. S. Schulp, E. W. A. Mulder, J. W. M. Jagt, H. H. G. Peeters, and D. Th. De Graaf. 2002. "A Large New Mosasaur from the Upper Cretaceous of the Netherlands." *Netherlands Journal of Geosciences/Geologie en Mijnbouw* 81 (1): 1–8.

Ellis, R. 2003. *Sea Dragons: Predators of the Prehistoric Oceans.* Lawrence: University of Kansas Press. 313 pp.

Everhart, M. J. 1999. "Evidence of Feeding on Mosasaurs by the Late Cretaceous Lamniform Shark, *Cretoxyrhina mantelli.*" *Journal of Vertebrate Paleontology* 17 (Supplement to Number 3): 43A–44A.

———. 2001. "Revisions to the Biostratigraphy of the Mosasauridae (Squamata) in the Smoky Hill Chalk Member of the Niobrara Chalk (Late Cretaceous) of Kansas." *Kansas Academy of Science, Transactions* 104 (1–2): 56–75.

———. 2002a. "New Data on Cranial Measurements and Body Length of the Mosasaur, *Tylosaurus nepaeolicus* (Squamata; Mosasauridae), from the Niobrara Formation of Western Kansas." *Kansas Academy of Science, Transactions* 105 (1–2): 33–43.

———. 2002b. "Remains of Immature Mosasaurs (Squamata; Mosasauridae) from the Niobrara Chalk (Late Cretaceous) Argue against Nearshore Nurseries." *Journal of Vertebrate Paleontology* 22 (Supplement to Number 3): 52A.

———. 2004a. "Plesiosaurs as the Food of Mosasaurs: New Data on the Gut Contents of a *Tylosaurus proriger* (Squamata; Mosasauridae) from the Niobrara Formation of Western Kansas." *The Mosasaur* 7: 41–46.

———. 2004b. "*Tylosaurus novum* sp.: An Update on an Unnamed Species of Basal Mosasaur." In *First Mosasaur Meeting Abstract Book and Field Guide,* pp. 35–39. Maastricht: Maastricht Museum of Natural History. 107 pp.

Everhart, M. J., and P. A. Everhart. 1996. "First Report of the Shell Crushing Mosasaur, *Globidens* sp., from the Sharon Springs Member of the Pierre Shale (Upper Cretaceous) of Western Kansas." *Kansas Academy of Science, Transactions* (Abstracts) 15: 17.

———. 1997. "Earliest Occurrence of the Mosasaur *Tylosaurus proriger* (Mosasauridae: Squamata) in the Smoky Hill Chalk (Niobrara Formation, Upper Cretaceous) of Western Kansas." *Journal of Vertebrate Paleontology* 17 (Supplement to Number 3): 44a.

Everhart, M. J., P. A. Everhart, and J. Bourdon. 1997. "Earliest Documented Occurrence of the Mosasaur, *Clidastes liodontus,* in the Smoky Hill Chalk (Upper Cretaceous) of Western Kansas." *Kansas Academy of Science, Transactions* (Abstracts) 16: 14.

Everhart, M. J., and S. E. Johnson. 2001. "The Occurrence of the Mosasaur, *Platecarpus planifrons,* in the Smoky Hill Chalk (Upper Cretaceous) of Western Kansas." *Journal of Vertebrate Paleontology* 21 (Supplement to Number 3): 48A.

Gilmore, C. W. 1912. "A New Mosasauroid Reptile from the Cretaceous of Alabama." *Proceedings of the U.S. National Museum* 40 (1870): 489–484.

Goldfuss, A. 1845. "Der Schädelbau des *Mosasaurus,* durch beschreibung einer neuen art dieser gattung erläutert." *Nova Acta Academa Caesar Leopoldino-Carolinae Germanicae Natura Curiosorum* 21: 1–28.

Harlan, R. 1834. "Notice of the Discovery of the Remains of the *Ichthyosaurus* in Missouri, N. A." *Transactions of the American Philosophical Society* 4: 405–409.

Hattin, D. E. 1982. *Stratigraphy and Depositional Environment of the Smoky Hill Chalk Member, Niobrara Chalk (Upper Cretaceous) of the Type Area, Western Kansas.* Kansas Geological Survey Bulletin, no. 225. Lawrence: Kansas Geological Survey, University of Kansas. 108 pp.

Holmes, R., M. W. Caldwell, and S. L. Cumbaa. 1999. "A New Specimen of *Plioplatecarpus* (Mosasauridae) from the Lower Maastrichtian of Alberta: Comments on Allometry, Functional Morphology, and Paleoecology." *Canadian Journal of Earth Science* 36: 363–369.

Kase, T., P. A. Johnston, A. Seilacher, and J. B. Boyce. 1998. "Alleged Mosasaur Bite Marks on Late Cretaceous Ammonites Are Limpet (Patellogastropod) Home Scars." *Geology* 26 (10): 947–950.

Kass, M. S. 1999. "*Prognathodon stadtmani* (Mosasauridae): A New Species from the Mancos Shale (Lower Campanian) of Western Colorado." In D. D. Gillette, ed., *Vertebrate Paleontology in Utah*, 275–294. Utah Geological Survey, Miscellaneous Publication, no. 99. Salt Lake City: Utah Geological Survey.

Kauffman, E. G., and R. V. Kesling. 1960. *An Upper Cretaceous Ammonite Bitten by a Mosasaur.* University of Michigan Contributions from the Museum of Paleontology, no. 15 (9): 193–248. Ann Arbor: Museum of Paleontology, University of Michigan. 9 pls., 7 figs.

Lee, M. S., G. L. Bell, Jr., and M. W. Caldwell. 1999. "The Origin of Snake Feeding." *Nature* 400: 655–659.

Lindgren, J., and M. Siverson. 2004. "*Halisaurus sternbergi,* a Small Mosasaur with an Intercontinental Distribution." In J. Lindgren, *Early Campanian Mosasaurs (Reptilia: Mosasauridae) from the Kristianstad Basin, Southern Sweden,* Paper IV. Litholund theses no. 4, Department of Geology, Lund University, Sweden. 11 pp.

Lingham-Soliar, T. 1992. "A New Mode of Locomotion in Mosasaurs: Subaqueous Flying in *Plioplatecarpus marshii.*" *Journal of Vertebrate Paleontology* 12 (4): 405–421.

———. 1994. "The Mosasaur '*Angolasaurus*' *bocagei* (Reptilia: Mosasauridae) from the Turonian of Angola Re-interpreted as the Earliest Member of the Genus *Platecarpus.*" *Paläontology Z* 68 (1–2): 267–282.

———. 1998. "A New Mosasaur *Pluridens walkeri* from the Upper Cretaceous Maastrichtian of the Lullemmeden Basin, Southwest Niger." *Journal of Vertebrate Paleontology* 18 (4): 709–717.

———. 1999a. "A Functional Analysis of the Skull of *Goronyosaurus nigeriensis* (Squamata: Mosasauridae) and Its Bearing on the Predatory Behavior and Evolution of This Enigmatic Taxon." *Neues Jahrbuch für Geologie und Palaeontologie,* Abhandlungen (Stuttgart) 213 (3): 355–374.

———. 1999b. "What Happened 65 Million Years Ago: The Study of Giant Marine Reptiles Throws New Light on the Last Major Mass Extinction." *Science Spectra* 17: 20–29.

———. 2002. "First Occurrence of Premaxillary Caniniform Teeth in the Varanoidea: Presence in the Extinct Mosasaur *Goronyosaurus* (Squamata: Mosasauridae) and Its Functional and Paleoecological Considerations." *Lethaia* 35: 187–190.

———. 2003. "Extinction of Ichthyosaurs: A Catastrophic or Evolutionary Paradigm?" *Neues Jahrbuch für Geologie und Palaeontologie,* Abhandlungen (Stuttgart) 228 (3): 421–452.

Manning, E. M. 1994. "Dr. William Spillman (1806–1886), Pioneer Paleontologist of Mississippi." *Mississippi Geology* 15 (4): 64–69.

Marsh, O. C. 1869. "Notice of Some New Mosasauroid Reptiles from the Green-Sand of New Jersey." *American Journal of Science* 48 (144): 392–397.

———. 1871a. "Scientific Expedition to the Rocky Mountains." *American Journal of Science,* Series 3, 1 (6): 142–143.

———. 1871b. "Notice of Some New Fossil Reptiles from the Cretaceous and Tertiary Formations." *American Journal of Science,* Series 3, 1 (6): 447–459.

———. 1872a. "Discovery of the Dermal Scutes of Mosasaurid Reptiles." *American Journal of Science,* Series 3, 3 (16): 290–292.

———. 1872b. "On the Structure of the Skull and Limbs in Mosasaurid Reptiles, with Descriptions of New Genera and Species." *American Journal of Science,* Series 3, 3 (18): 448–464, pls. 10–13.

Martin, J. E., and P. R. Bjork. 1987. "Gastric Residues Associated with a Mosasaur from the Late Cretaceous (Campanian) Pierre Shale in South Dakota." *Dakoterra* 3: 68–72.

Martin, J. E. and J. E. Fox. 2004. "Molluscs in the Stomach Contents of Globidens, a Shell-Crushing Mosasaur, from the Late Cretaceous Pierre Shale, Big Bend Area of the Missouri River, Central South Dakota." Abstracts with programs, Geological Society of America, 2004 Rocky Mountain and Cordilleran Regions Joint Meeting, 36 (4): 80.

Martin, L. D., and J. D. Stewart. 1977. "The Oldest (Turonian) Mosasaurs from Kansas." *Journal of Paleontology* 51 (5): 973–975.

Massare, J. A. 1987. "Tooth Morphology and Prey Preference of Mesozoic Marine Reptiles." *Journal of Vertebrate Paleontology* 7 (2): 121–137.

Merriam, J. C. 1894. "Über die Pythonomorphen der Kansas-Kreide." *Palaeontographica* 41. 39 pp., 4 pls.

Mitchell, S. L. 1818. "Observations on the Geology of North America, Illustrated by the Description of Various Organic Remains Found in That Part of the World." In G. Cuvier, *Essay on the Theory of the Earth,* pp. 319–431, pls. 6–8. New York: Kirk and Mercein.

Moulton, G. E., ed. 1983–1997. *The Journals of the Lewis and Clark Expedition.* Vols. 1–11. Lincoln and London: University of Nebraska Press.

Mudge, B. F. 1876. "Notes on the Tertiary and Cretaceous Periods of Kansas." *Bulletin of the U.S. Geological Survey of the Territories (Hayden)* 2 (3): 211–221. Washington, D.C.: Government Printing Office.

Mulder, E. W. A. 2003. "On the Latest Cretaceous Tetrapods from the Maastrichtian Type Area." *Publicaties van het Natuurhistorisch Genootschap in Limburg, Reeks* XLIV, aflevering 1. Maastricht: Stichting Natuurpublicaties Limburg.

Nicholls, E. L., and S. J. Godfrey. 1994. "Subaqueous Flight in Mosasaurs: A Discussion." *Journal of Vertebrate Paleontology* 14 (3): 450–452.

Novas, F. E., M. Fernández, Z. B. Gasparini, J. M. Lirio, H. J. Nuñez, and P. Puerta. 2002. "*Lakumasaurus antarcticus,* n. gen. et sp., a New Mosasaur (Reptilia, Squamata) from the Upper Cretaceous of Antarctica." *Ameghiniana* 39 (2): 245–249.

Osborn, H. F. 1899. "A Complete Mosasaur Skeleton, Osseous and Cartilaginous." *Bulletin of the Peabody Museum of Natural History* 1 (4): 167–188.

Ott, C. J., A. D. B. Behlke, and D. K. Kelly. 2002. "An Unusually Large Specimen of *Clidastes* (Mosasauroidae) from the Niobrara Chalk of Western Kansas." *Kansas Academy of Science, Transactions* (Abstracts) 21: 32.

Rothschild, B. M., and L. Martin. 1987. "Avascular Necrosis: Occurrence in Diving Cretaceous Mosasaurs." *Science* 236: 75–77.

Russell, D. A. 1967. *Systematics and Morphology of American Mosasaurs.* Peabody Museum of Natural History, Yale University, Bulletin no. 23. New Haven, Conn.: Peabody Museum of Natural History, Yale University. 241 pp.

———. 1970. *The Vertebrate Fauna of the Selma Formation of Alabama.* Part VII: *The Mosasaurs. Fieldiana Geology* Memoirs, 3 (7). Chicago: Chicago Natural History Museum.

———. 1975. "A New Species of *Globidens* from South Dakota." *Fieldiana Geology* 33 (13): 235–256.

———. 1988. "A Checklist of North American Marine Cretaceous Vertebrates Including Fresh Water Fishes." Occasional Paper of the Tyrrell Museum of Paleontology, no. 4. 57 pp.

———. 1993. "Vertebrates in the Western Interior Sea." In W. G. E. Caldwell and E. G. Kauffman, eds., *Evolution of the Western Interior Basin*, pp. 665–680. Geological Association of Canada, Special Paper 39. St. John's, Nfld.: Geological Association of Canada.

Schumacher, B. A. 1993. "Biostratigraphy of Mosasauridae (Squamata, Varanoidea) from the Smoky Hill Chalk Member, Niobrara Chalk (Upper Cretaceous) of Western Kansas." M.A. thesis, Fort Hays State University, Hays, Kansas. 68 pp.

Schumacher, B. A. and D. W. Varner. 1996. "Mosasaur Caudal Anatomy." *Journal of Vertebrate Paleontology* 16 (Supplement to Number 3): 63A.

Schwimmer, D. R. 2002. *King of the Crocodilians: The Paleobiology of Deinosuchus.* Bloomington: Indiana University Press. 220 pp.

Schwimmer, D. R., J. D. Stewart, and G. D. Williams. 1997. "Scavenging by Sharks of the Genus *Squalicorax* in the Late Cretaceous of North America." *Palaios* 12: 71–83.

Sheldon, M. A. 1996. "Stratigraphic Distribution of Mosasaurs in the Niobrara Formation of Kansas." *Paludicola* 1: 21–31.

Shimada, K. 1996. "Ichthyosaur (Reptilia: Ichthyosauria) Vertebra from the Kiowa Shale (Lower Cretaceous: Upper Albian), Clark County, Kansas." *Kansas Academy of Science, Transactions* 99 (1–2): 39–44.

———. 1997. "Paleoecological Relationships of the Late Cretaceous Lamniform Shark, *Cretoxyrhina mantelli* (Agassiz)." *Journal of Paleontology* 71 (5): 926–933.

Shor, E. N. 1971. *Fossils and Flies: The Life of a Compleat Scientist—Samuel Wendell Williston, 1851–1918.* Norman: University of Oklahoma Press. 285 pp.

Simpson, G. G. 1942. "The Beginnings of Vertebrate Paleontology in North America." *Proceedings of the American Philosophical Society* 86 (11): 130–188.

Snow, F. H. 1878. "On the Dermal Covering of a Mosasauroid Reptile." *Kansas Academy of Science, Transactions* 6: 54–58, figs. 1–2.

Sternberg, C. H. 1922. "Field Work in Kansas and Texas." *Kansas Academy of Science, Transactions* 30 (2): 339–348.

Stewart, J. D. 1988. "The Stratigraphic Distribution of Late Cretaceous *Protosphyraena* in Kansas and Alabama, Geology." In M. E. Nelson, ed., *Geology, Paleontology and Biostratigraphy of Western Kansas: Articles in Honor of Myrl V. Walker*, pp. 80–94. Fort Hays Studies, Third Series, no. 10, Science. Hays, Kans.: Fort Hays State University.

Stewart, J. D. 1990. "Niobrara Formation Vertebrate Stratigraphy." In S. C. Bennett, ed., *Niobrara Chalk Excursion Guidebook*, pp. 19–30. Lawrence: University of Kansas Museum of Natural History and Kansas Geological Survey.

Stewart, J. D., and G. L. Bell, Jr. 1994. "North America's Oldest Mosasaurs Are Teleosts." *Contributions to Science (Natural History Museum of Los Angeles County)* 441: 1–9.

Varricchio, D. J. 2001. "Gut Contents from a Cretaceous Tyrannosaurid; Implications for Theropod Dinosaur Digestive Tracts." *Journal of Paleontology* 75 (2): 401–406.

Williston, S. W. 1891. "The Skull and Hind Extremity of *Pteranodon*." *American Naturalist* 25 (300): 1124–1126.

———. 1897. "The Kansas Niobrara Cretaceous." *University Geological Survey of Kansas* 2: 235–246.

———. 1898a. "Mosasaurs." *University Geological Survey of Kansas* 4 (5): 81–347, pls. 10–72.

———. 1898b. "Editorial Notes." *Kansas University Quarterly* 7 (4): 235.

———. 1899. "Some Additional Characters of the Mosasaurs." *Kansas University Quarterly* 8 (1): 39–41. 1 pl.

———. "Notes on some New or Little-Known Extinct Reptiles." *Kansas University Science Bulletin* 1 (9): 247–254. 2 pls.

———. 1904. "The Relationships and Habits of the Mosasaurs." *Journal of Geology* 12: 43–51.

———. 1914. *Water Reptiles of the Past and Present.* Chicago: University of Chicago Press. 251 pp.

Wiman, C. J. 1920. "Some Reptiles from the Niobrara Group in Kansas." *Bulletin of the Geological Institute of Uppsala* 18: 9–18, pls. II–IV. 9 figs.

10. Pteranodons

Anonymous. 1872. "On Two New Ornithosaurians from Kansas." *American Journal of Science,* Series 3, 3 (17): 374–375.

Bennett, S. C. 1987. "New Evidence on the Tail of the Pterosaur *Pteranodon* (Archosauria: Pterosauria)." In P. J. Currie and E. H. Koster, eds., *Fourth Symposium on Mesozoic Terrestrial Ecosystems,* pp. 18–23. Occasional Papers of the Tyrrell Museum of Palaeontology, no. 3. Drumheller, Alta.: Tyrrell Museum of Palaeontology.

———. 1990. "Inferring Stratigraphic Position of Fossil Vertebrates from the Niobrara Chalk of Western Kansas." In S. C. Bennett, ed., *Niobrara Chalk Excursion Guidebook*, pp. 43–72. Lawrence: University of Kansas Museum of Natural History and Kansas Geological Survey.

———. 1992. "Sexual Dimorphism of *Pteranodon* and Other Pterosaurs, with Comments on Cranial Crests." *Journal of Vertebrate Paleontology* 12 (4): 422–434.

———. 1994. *Taxonomy and Systematics of the Late Cretaceous Pterosaur Pteranodon (Pterosauria, Pterodactyloida).* Occasional Papers of the Natural History Museum, University of Kansas, no. 169. Lawrence: Natural History Museum, University of Kansas. 70 pp.

———. 2000. "New Information on the Skeletons of *Nyctosaurus.*" *Journal of Vertebrate Paleontology* 20 (Supplement to Number 3): 29A.

———. 2001. "The Osteology and Functional Morphology of the Late Cretaceous Pterosaur *Pteranodon*: Parts I and II." *Palaeontographica,* Abteilung A, 260: 1–112, 113–153.

———. 2003. "New Crested Specimens of the Late Cretaceous Pterosaur *Nyctosaurus*." *Paläontologische Zeitschrift* 77: 61–75.

Betts, C. W. 1871. "The Yale College Expedition of 1870." *Harper's New Monthly Magazine* 43 (257): 663–671.

Bonner, O. W. 1964. "An Osteological Study of *Nyctosaurus* and *Trinacromerum* with a Description of a New Species of *Nyctosaurus*." M.S. thesis, Fort Hays State University. 63 pp.

Brower, J. C. 1983. "The Aerodynamics of *Pteranodon* and *Nyctosaurus*, Two Large Pterosaurs from the Upper Cretaceous of Kansas." *Journal of Vertebrate Paleontology* 3 (2): 84–124.

Brown, B. 1943. "Flying Reptiles." *Natural History* 52: 104–111.

Carpenter, K. 1996. "Sharon Springs Member, Pierre Shale (Lower Campanian) Depositional Environment and Origin of Its Vertebrate Fauna, with a Review of North American Plesiosaurs." Ph.D. dissertation, University of Colorado. 251 pp.

Cope, E. D. 1872a. "On the Geology and Paleontology of the Cretaceous Strata of Kansas." *Annual Report of the U.S. Geological Survey of the Territories* 5: 318–349.

———. 1872b. "On Two New Ornithosaurians from Kansas." *Proceedings of the American Philosophical Society* 12 (88): 420–422.

———. 1874. "Review of the Vertebrata of the Cretaceous Period Found West of the Mississippi River." *Bulletin of the U.S. Geological Survey of the Territories* 1 (2): 3–48.

———. 1875. *The Vertebrata of the Cretaceous Formations of the West.* Report of the U.S. Geological Survey Territories (Hayden) 2. Washington, D.C.: Government Printing Office. 302 pp., 57 pls.

Eaton, G. F. 1903. "The Characters of *Pteranodon*." *American Journal of Science,* Series 4, 16 (91): 82–86, pls. 6–7.

———. 1904. "The Characters of *Pteranodon* (Second Paper)." *American Journal of Science,* Series 4, 17 (100): 318–320, pls. 19–20.

———. 1910. "Osteology of *Pteranodon*." *Memoirs of the Connecticut Academy of Arts and Sciences* 2: 1–38, pls. 1–31.

Hankin, E. H., and D. M. S. Watson. 1914. "On the Flight of Pterodactyls." *The Aeronautical Journal* 18: 324–335.

Harksen, J. C. 1966. "*Pteranodon sternbergi*, a New Fossil Pterodactyl from the Niobrara Cretaceous of Kansas." *Proceedings South Dakota Academy of Science* 45: 74–77.

Lane, H. H. 1946. "A Survey of the Fossil Vertebrates of Kansas. Part III: The Reptiles." *Kansas Academy of Science, Transactions* 49 (3): 289–332, figs. 1–7.

Liggett, G. A., S. C. Bennett, K. Shimada, and J. Huenergarde. 1997. "A Late Cretaceous (Cenomanian) Fauna in Russell County, KS." *Kansas Academy of Science, Transactions* (Abstracts) 16: 26.

Marsh, O. C. 1871a. "Scientific Expedition to the Rocky Mountains." *American Journal of Science,* Series 3, 1 (6): 142–143.

———. 1871b. "Notice of Some New Fossil Reptiles from the Cretaceous and Tertiary Formations." *American Journal of Science,* Series 3, 1 (6): 447–459.

———. 1871c. "Note on a New and Gigantic Species of Pterodactyle." *American Journal of Science,* Series 3, 1 (6): 472.

———. 1872. "Discovery of Additional Remains of Pterosauria, with Descriptions of Two New Species." *American Journal of Science,* Series 3, 3 (16): 241–248.

———. 1876a. "Notice of a New Sub-Order of Pterosauria." *American Journal of Science,* Series 3, 11 (65): 507–509.

———. 1876b. "Principal Characters of American Pterodactyls." *American Journal of Science,* Series 3, 12 (72): 479–480.

———. 1881. "Note on American Pterodactyls." *American Journal of Science,* Series 3, 21 (124): 342–343.

———. 1882. "The Wings of Pterodactyles." *American Journal of Science,* Series 3, 23 (136): 251–256, pl. III.

———. 1884. "Principal Characters of American Cretaceous Pterodactyls. Part I: The Skull of *Pteranodon.*" *American Journal of Science,* Series 3, 27 (161): 422–426, pl. 15.

Miller, H. W. 1971a. "The Taxonomy of the *Pteranodon* Species from Kansas." *Kansas Academy of Science, Transactions* 74 (1): 1–19.

———. 1971b. "A Skull of *Pteranodon (Longicepia) longiceps* Marsh Associated with Wing and Body Parts." *Kansas Academy of Science, Transactions* 74 (10): 20–33.

Naish, D., and D. M. Martill. 2003. "Pterosaurs: A Successful Invasion of Prehistoric Skies." *Biologist* 50 (5): 213–216.

Padian, K. 1983. "A Functional Analysis of Flying and Walking in Pterosaurs." *Paleobiology* 9 (3): 218–239.

Russell, D. A. 1988. *A Check List of North American Marine Cretaceous Vertebrates Including Fresh Water Fishes.* Occasional Paper of the Tyrrell Museum of Palaeontology, no. 4. Drumheller, Alta.: Tyrrell Museum of Palaeontology. 57 pp.

Schultze, H.-P., L. Hunt, J. Chorn, and A. M. Neuner. 1985. *Type and Figured Specimens of Fossil Vertebrates in the Collection of the University of Kansas Museum of Natural History.* Part II: *Fossil Amphibians and Reptiles.* Miscellaneous Publications of the University of Kansas Museum of Natural History 77. Lawrence: University of Kansas Museum of Natural History. 66 pp.

Seeley, H. G. 1871. "Additional Evidence of the Structure of the Head in Ornithosaurs from the Cambridge Upper Greensand; Being a Supplement to 'The Ornithosauria.'" *The Annals and Magazine of Natural History,* Series 4, 7: 20–36, pls. 2–3.

———. 1901. *Dragons of the Air: An Account Extinct Flying Reptiles.* London. xiii + 239 pp., 80 figs.

Shor, E. N. 1971. *Fossils and Flies: The Life of a Compleat Scientist— Samuel Wendell Williston, 1851–1918.* Norman: University of Oklahoma Press. 285 pp.

Stein, R. S. 1975. "Dynamic Analysis of *Pteranodon ingens:* A Reptilian Adaptation to Flight." *Journal of Paleontology* 49 (3): 534–548.

Sternberg, C. H. 1909. *The Life of a Fossil Hunter.* New York: Henry Holt and Company. 286 pp. (Reprinted 1990 by Indiana University Press.)

Sternberg, G. F., and M. V. Walker. 1958. "Observation of Articulated Limb Bones of a Recently Discovered *Pteranodon* in the Niobrara Cretaceous of Kansas." *Kansas Academy of Science, Transactions* 61 (1): 81–85.

Stewart, J. D. 1990. "Niobrara Formation Vertebrate Stratigraphy." In S. C. Bennett, ed., *Niobrara Chalk Excursion Guidebook,* pp. 19–30. Lawrence: University of Kansas Museum of Natural History and Kansas Geological Survey.

Wang, X., and Z. Zhou. 2004. "Pterosaur Embryo from the Early Cretaceous." *Nature* 429: 621.

Webb, W. E. 1872. *Buffalo Land: An Authentic Account of the Discoveries, Adventures, and Mishaps of a Scientific and Sporting Party in the Wild West.* Philadelphia: Hubbard Bros. 503 pp.

Wellnhofer, P. 1991. *The Illustrated Encyclopedia of Pterosaurs.* New York: Crescent Books. 192 pp.

Williston, S. W. 1891. "The Skull and Hind Extremity of Pteranodon." *American Naturalist* 25 (300): 1124–1126.

———. 1892. "Kansas Pterodactyls: Part I." *Kansas University Quarterly* 1: 1–13, pl. I.

———. 1893. "Kansas Pterodactyls: Part II." *Kansas University Quarterly* 2: 79–81. 1 fig.

———. 1894. "On Various Vertebrate Remains from the Lowermost Cretaceous of Kansas." *Kansas University Quarterly* 3 (1): 1–4, pl. I.

———. 1895. "Note on the Mandible of *Ornithostoma*." *Kansas University Quarterly* 4: 61.

———. 1896. "On the Skull of *Ornithostoma*." *Kansas University Quarterly* 4 (4): 195–197, with pl. I.

———. 1897. "Restoration of *Ornithostoma (Pteranodon)*." *Kansas University Quarterly* 6: 35–51, with pl. II.

———. 1898. "Crocodiles." *The University Geological Survey of Kansas.* 4 (4): 75– 78, pl. 9.

———. 1902a. "On the Skeleton of *Nyctodactylus*, with Restoration." *American Journal of Anatomy* 1: 297–305.

———. 1902b. "On the Skull of *Nyctodactylus*, an Upper Cretaceous Pterodactyl." *Journal of Geology* 10: 520–531. 2 pls.

———. 1902c. "Winged Reptiles." *Popular Science Monthly* 60: 314–322. 2 figs.

———. 1903. "On the Osteology of *Nyctosaurus (Nyctodactylus)*, with Notes on American Pterosaurs." *Field Museum Publications* (Geological Series) 2 (3): 125–163, pls. XL–XLIV. 2 figs.

———. 1904. "The Fingers of Pterodactyls." *Geology Magazine,* Series 5, 1 (2): 59–60.

———. 1910. "A Mounted Skeleton of *Platecarpus*." *Journal of Geology* 18 (6): 537–541.

———. 1911. "The Wing-Finger of Pterodactyls, with Restoration of *Nyctosaurus*." *Journal of Geology* 19: 696–705.

———. 1912. "A Review of G. B. Eaton's 'Osteology of *Pteranodon*.' " *Journal of Geology* 20: 288.

Wiman, C. J. 1920. "Some Reptiles from the Niobrara Group in Kansas." *Bulletin of the Geological Institute of Uppsala* 18: 9–18, pls. II–IV. 9 figs.

11. FEATHERS AND TEETH

Chinsamy, A., L. D. Martin, and P. Dodson. 1998. "Bone Microstructure of the Diving *Hesperornis* and the Volant *Ichthyornis* from the Niobrara Chalk of Western Kansas." *Cretaceous Research* 19: 225–235.

Clarke, J. A. 2004. "Morphology, Phylogenetic Taxonomy, and Systematics of *Ichthyornis* and *Apatornis* (Avialae: Ornithurae)." *Bulletin of the American Museum of Natural History* 286: 1–179.

Cope, E. D. 1872. "Note of Some Cretaceous Vertebrata in the State Agricultural College of Kansas." *Proceedings of the American Philosophical Society* 12 (87): 168–170.

Everhart, M. J. 2002. "New Data on Cranial Measurements and Body

Length of the Mosasaur, *Tylosaurus nepaeolicus* (Squamata; Mosasauridae), from the Niobrara Formation of Western Kansas." *Kansas Academy of Science, Transactions* 105 (1–2): 33–43.

Fox, R. C. 1984. "*Ichthyornis* (Aves) from the Early Turonian (Late Cretaceous) of Alberta." *Canadian Journal of Earth Sciences* 21: 258–260.

Hanks, D. H., and K. Shimada. 2002. "Vertebrate Fossils, Including Non-Avian Dinosaur Remains and the First Shark-Bitten Bird Bone from a Late Cretaceous (Turonian) Marine Deposit of Northeastern South Dakota." *Journal of Vertebrate Paleontology* 22 (Supplement to Number 3): 62A.

Lane, H. H. 1946. "A Survey of the Fossil Vertebrates of Kansas. Part IV: Birds." *Kansas Academy of Science, Transactions* 49 (4): 390–400.

Marsh, O. C. 1872a. "Discovery of a Remarkable Fossil Bird." *American Journal of Science,* Series 3, 3 (13): 56–57.

———. 1872b. "Notice of a New and Remarkable Fossil Bird." *American Journal of Science,* Series 3, 4 (22): 344.

———. 1872c. "Notice of a New Reptile from the Cretaceous." *American Journal of Science,* Series 3, 4 (23): 406.

———. 1873a. "On a New Sub-class of Fossil Birds (Odontornithes)." *American Journal of Science,* Series 3, 5 (25): 161–162.

———. 1873b. "Fossil Birds from the Cretaceous of North America." *American Journal of Science,* Series 3, 5 (27): 229–231.

———. 1875. "On the Odontornithes, or Birds with Teeth." *American Journal of Science* 10 (59): 403–408, pls. 9–10.

———. 1880. "Synopsis of American Cretaceous Birds." In *Odontornithes: A Monograph on the Extinct Toothed Birds of North America,* pp. 191–199. U.S. Geological Exploration of the 40th Parallel, Clarence King, Geologist-in-charge, vol. 7. xv + 201 pp., 34 pls.

———. 1883. *Birds with Teeth.* United States Geological Survey, 3rd Annual Report of the Secretary of the Interior, 3: 43–88. Washington, D.C.: Government Printing Office.

Martin, L. D. 1984. "A New Hesperornithid and the Relationships of the Mesozoic Birds." *Kansas Academy of Science, Transactions* 87: 141–150.

Martin, L. D., and O. Bonner. 1977. "An Immature Specimen of *Baptornis advenus* from the Cretaceous of Kansas." *The Auk* 94 (4): 787–789.

Martin, L. D., and J. D. Stewart. 1977. "Teeth in *Ichthyornis* (Class: Aves)." *Science* 185 (4284): 1331–1332.

Martin, L. D., and J. Tate, Jr. 1966. "A Bird with Teeth." *Museum Notes,* University of Nebraska State Museum, 29: 1–2.

Martin, L. D., and J. Tate, Jr. 1976. "The Skeleton of *Baptornis advenus* (Aves: Hesperornithiformes)." *Smithsonian Contributions to Paleobiology* 27: 35–66.

Martin, J. E., and P. R. Bjork. 1987. "Gastric Residues Associated with a Mosasaur from the Late Cretaceous (Campanian) Pierre Shale in South Dakota." *Dakoterra* 3: 68–72.

Mudge, B. F. 1866a. "Discovery of Fossil Footmarks in the Liassic(?) Formation in Kansas." *American Journal of Science,* Series 2, 41 (122): 174–176.

———. 1866b. *First Annual Report on the Geology of Kansas.* Lawrence: State Printer. 57 pp.

———. 1877. "Annual Report of the Committee on Geology, for the Year Ending November 1, 1876." *Kansas Academy of Science, Transactions,* Ninth Annual Meeting, pp. 4–5.

Nicholls, E. L. 1988. "Marine Vertebrates of the Pembina Member of the Pierre Shale (Campanian, Upper Cretaceous) of Manitoba and Their Significance to the Biogeography." Ph.D. dissertation, University of Calgary. 317 pp.

Peterson, J. M. 1987. "Science in Kansas: The Early Years, 1804–1875." *Kansas History Magazine* 10 (3): 201–240.

Russell, D. A. 1967. "Cretaceous Vertebrates from the Anderson River N.W.T." *Canadian Journal of Earth Sciences* 4: 21–38.

———. 1993. "Vertebrates in the Western Interior Sea." In W. G. E. Caldwell and E. G. Kauffman, eds., *Evolution of the Western Interior Basin,,* pp. 665–680. Geological Association of Canada, Special Paper 39. St. John's, Nfld.: Geological Association of Canada.

Snow, F. H. 1887. "On the Discovery of a Fossil Bird Track in the Dakota Sandstone." *Kansas Academy of Science*, Transactions 10: 3–6.

Sternberg, C. H. 1909. *The Life of a Fossil Hunter.* New York: Henry Holt and Company. 286 pp. (Republished 1990 by Indiana University Press.)

Tokaryk, T. T., S. L. Cumbaa, and J. E. Storer. 1997. "Early Late Cretaceous Birds from Saskatchewan, Canada: The Oldest Diverse Avifauna Known from North America." *Journal of Vertebrate Paleontology* 17: 172–176.

Walker, M. V. 1967. "Revival of Interest in the Toothed Birds of Kansas." *Kansas Academy of Science, Transactions* 70 (1): 60–66.

Williston, S. W. 1898a. "Addenda to Part I." *University Geological Survey of Kansas* 4: 28–32.

———. 1898b. "Birds." *University Geological Survey of Kansas* 4: 43–49, pls. 5–8.

———. 1898c. "Bird Tracks from the Dakota Cretaceous." *University Geological Survey of Kansas* 4: 50–53, fig. 2.

12. DINOSAURS?

Carpenter, K., D. Dilkes, and D. B. Weishampel. 1995. "The Dinosaurs of the Niobrara Chalk Formation (Upper Cretaceous, Kansas)." *Journal of Vertebrate Paleontology* 15 (2): 275–297.

Hamm, S. A., and M. J. Everhart. 2001. "Notes on the Occurrence of Nodosaurs (Ankylosauridae) in the Smoky Hill Chalk (Upper Cretaceous) of Western Kansas." *Journal of Vertebrate Paleontology* 21 (Supplement to Number 3): 58A.

Hattin, D. E. 1982. *Stratigraphy and Depositional Environment of the Smoky Hill Chalk Member, Niobrara Chalk (Upper Cretaceous) of the Type Area, Western Kansas.* Kansas Geological Survey Bulletin, no. 225. Lawrence: Kansas Geological Survey, University of Kansas. 108 pp.

Marsh, O. C. 1872. "Notice of a New Species of *Hadrosaurus.*" *American Journal of Science,* Series 3, 3 (16): 301.

———. 1890. "Additional Characters of the Ceratopsidae, with Notes on New Cretaceous Dinosaurs." *American Journal of Science* 3 (39): 418–425.

Mehl, M. G. 1931. "Aquatic Dinosaur from the Niobrara of Western Kansas." *Bulletin of the Geologic Society of America* 42: 326–327.

———. 1936. "*Hierosaurus coleii:* A New Aquatic Dinosaur from the Niobrara Cretaceous of Kansas." *Denison University Bulletin, Journal of the Scientific Laboratory* 31: 1–20. 3 pls.

Sternberg, C. H. 1909. "An Armored Dinosaur from the Kansas Chalk." *Kansas Academy of Science, Transactions* 22: 257–258.

Walters, R. F. 1986. "Memorial: Virgil Bedford Cole (1897–1984)." *AAPG Bulletin* 70 (2): 208–209.

Wieland, G. R. 1909. "An Armored Saurian." *American Journal of Science* 27: 250–252.

———. 1911. "Notes on the Armored Dinosauria." *American Journal of Science* 31: 112–124.

13. THE BIG PICTURE

Bennett, S. C. 2000. "Inferring Stratigraphic Position of Fossil Vertebrates from the Niobrara Chalk of Western Kansas." *Current Research in Earth Sciences*, Kansas Geological Survey Bulletin 244 (1): 1–26.

———. 1992. "Sexual Dimorphism of *Pteranodon* and Other Pterosaurs, with Comments on Cranial Crests." *Journal of Vertebrate Paleontology* 12 (4): 422–434.

Carpenter, K. 1990. "Upward continuity of the Niobrara fauna with the Pierre Shale fauna." In S. C. Bennett, ed., *Niobrara Chalk Excursion Guidebook*, pp. 73–81. Lawrence: University of Kansas Museum of Natural History and Kansas Geological Survey.

———. 2003. Vertebrate Biostratigraphy of the Smoky Hill Chalk (Niobrara Formation) and the Sharon Springs Member (Pierre Shale). In P. J. Harries, ed., *Approaches in High-Resolution Stratigraphic Paleontology*, pp. 421–437. Kluwer Academic Publishers, The Netherlands.

Clarke, J. A. 2004. Morphology, phylogenetic taxonomy, and systematics of *Ichthyornis* and *Apatornis* (Avialae: Ornithurae). *Bulletin of the American Museum of Natural History* 286: 1–179.

Davidson, J. P. 2003. "Edward Drinker Cope, Professor Paleozoic and Buffalo Land." *Kansas Academy of Science, Transactions* 106 (3–4): 177–191.

Everhart, M. J. 2000. "Gastroliths Associated with Plesiosaur Remains in the Sharon Springs Member of the Pierre Shale (Late Cretaceous), Western Kansas." *Kansas Academy of Science, Transactions* 103 (1–2): 58–69.

———. 2001. "Revisions to the Biostratigraphy of the Mosasauridae (Squamata) in the Smoky Hill Chalk Member of the Niobrara Chalk (Late Cretaceous) of Kansas." *Kansas Academy of Science, Transactions* 104 (1–2): 56–75.

———. 2002. "Remains of Immature Mosasaurs (Squamata; Mosasauridae) from the Niobrara Chalk (Late Cretaceous) Argue Against Nearshore Nurseries." *Journal of Vertebrate Paleontology* 22 (Supplement to Number 3): 52A.

———. 2003. "First Records of Plesiosaur Remains in the Lower Smoky Hill Chalk Member (Upper Coniacian) of the Niobrara Formation in Western Kansas." *Kansas Academy of Science, Transactions* 106 (3–4): 139–148.

———. 2004a. "Late Cretaceous Interaction between Predators and Prey: Evidence of Feeding by Two Species of Shark on a Mosasaur." *PalArch Vertebrate Paleontology Series* 1 (1): 1–7.

———. 2004b. "Plesiosaurs as the Food of Mosasaurs: New Data on the Stomach Contents of a *Tylosaurus proriger* (Squamata; Mosasauridae) from the Niobrara Formation of Western Kansas." *The Mosasaur* 7: 41–46.

Hattin, D. E. 1982. *Stratigraphy and Depositional Environment of the Smoky Hill Chalk Member, Niobrara Chalk (Upper Cretaceous) of the Type Area, Western Kansas.* Kansas Geological Survey Bulletin, no. 225. Lawrence: Kansas Geological Survey, University of Kansas. 108 pp.

Liggett, G. A., S. C. Bennett, K. Shimada, and J. Huenergarde. 1997. "A Late Cretaceous(Cenomanian) Fauna in Russell County, KS." *Kansas Academy of Science, Transactions* (Abstracts) 16: 26.

Logan, W. N. 1897. "The Upper Cretaceous of Kansas: With an Introduction by Erasmus Haworth." *University Geological Survey of Kansas* 2: 194–234.

Martin, L. D. 1984. "A New Hesperornithid and the Relationships of the Mesozoic Birds." *Kansas Academy of Science, Transactions* 87: 141–150.

Martin, L. D., and J. D. Stewart. 1977. "The Oldest (Turonian) Mosasaurs from Kansas." *Journal of Paleontology* 51 (5): 973–975.

Moodie, R. L. 1912. "The Stomach Stones of Reptiles." *Science* (New Series) 35 (897): 377–378.

Russell, D. A. 1988. *A Check List of North American Marine Cretaceous Vertebrates Including Fresh Water Fishes.* Occasional Paper of the Tyrrell Museum of Palaeontology, no. 4. Drumheller, Alta.: Tyrrell Museum of Palaeontology. 57 pp.

Shimada, K. 1996. "Selachians from the Fort Hays Limestone Member of the Niobrara Chalk (Upper Cretaceous), Ellis County, Kansas." *Kansas Academy of Science, Transactions* 99 (1–2): 1–15.

———. 1997. "Paleoecological Relationships of the Late Cretaceous Lamniform Shark *Cretoxyrhina mantelli* (Agassiz)." *Journal of Paleontology* 71 (5): 926–933.

Shimada, K and M. J. Everhart. 2003. "*Ptychodus mammillaris* (Elasmobranchii) and *Enchodus* cf. *E. schumardi* (Teleostei) from the Fort Hays Limestone Member of the Niobrara Chalk (Upper Cretaceous) in Ellis County, Kansas." *Kansas Academy of Science, Transactions* 106 (3–4): 171–176.

Stewart, J. D. 1990a. "Niobrara Formation Vertebrate Stratigraphy." In S. C. Bennett, ed., *Niobrara Chalk Excursion Guidebook,* pp. 19–30. Lawrence: University of Kansas Museum of Natural History and Kansas Geological Survey.

———. 1990b. "Niobrara Formation Symbiotic Fish in Inoceramid Bivalves." In S. C. Bennett, ed., *Niobrara Chalk Excursion Guidebook,* pp. 31–41. Lawrence: University of Kansas Museum of Natural History and Kansas Geological Survey.

———. 1990c. "Preliminary Account of Holecostome-Inoceramid Commensalism in the Upper Cretaceous of Kansas." In A. J. Boucot, *Evolutionary Paleobiology of Behavior and Coevolution,* pp. 51–58. Amsterdam: Elsevier.

Webb, W. E. 1872. *Buffalo Land: An Authentic Account of the Discoveries, Adventures, and Mishaps of a Scientific and Sporting Party in the Wild West.* Philadelphia: Hubbard Bros. 503 pp.

Williston, S. W. 1897. "The Kansas Niobrara Cretaceous." *University Geological Survey of Kansas* 2: 235–246.

———. 1890a. Structure of the Plesiosaurian Skull. *Science* 16 (405): 262.

———. 1890b. A New Plesiosaur from the Niobrara Cretaceous of

Kansas. *Kansas Academy of Science, Transactions* 12: 174–178. 2
figs.

———. 1897. "The Kansas Niobrara Cretaceous." *University Geological
Survey of Kansas* 2: 235–246.

———. 1898a. "Birds." *University Geological Survey of Kansas* 4: 43–49,
pls. 5–8.

———. 1898b. "Mosasaurs." *University Geological Survey of Kansas* 4
(5): 81–221.

Index

Page numbers in italics refer to illustrations.

MICHAEL J. EVERHART is Adjunct Curator of Paleontology at the Sternberg Museum of Natural History, Fort Hays State University, and an expert on the Late Cretaceous of western Kansas. He is the author of many papers on the fossils of the Smoky Hill Chalk, including mosasaurs and plesiosaurs, as well as the creator of the award-winning Oceans of Kansas Paleontology website: www.oceansofkansas.com.